A SHORT ACCOUNT OF THE HISTORY OF MATHEMATICS

W.W. Rouse Ball

Fellow of Trinity College, Cambridge

MAVEN BOOKS

Chennai   Trichy   Tirunelveli   New Delhi

MAVEN BOOKS

An Imprint of **MJP Publishers**

ISBN 978-93-88694-67-4 **Maven Books**

All rights reserved No. 44, Nallathambi Street,
Printed and bound in India Triplicane, Chennai 600 005

MJP 771 © Publishers, 2020

Publisher : C. Janarthanan

Publisher's Note

The legacy of a country is in its varied cultural heritage, historical literature, developments in the field of economy and science. The top nations in the world are competing in the field of science, economy and literature. This vast legacy has to be conserved and documented so that it can be bestowed to the future generation. The knowledge of this legacy is slowly getting perished in the present generation due to lack of documentation.

Keeping this in mind, the concern with retrospective acquiring of rare books has been accented recently by the burgeoning reprint industry. Maven Books is gratified to retrieve the rare collections with a view to bring back those books that were landmarks in their time.

In this effort, a series of rare books would be republished under the banner, "Maven Books". The books in the reprint series have been carefully selected for their contemporary usefulness as well as their historical importance within the intellectual. We reconstruct the book with slight enhancements made for better presentation, without affecting the contents of the original edition.

Most of the works selected for republishing covers a huge range of subjects, from history to anthropology. We believe this reprint edition will be a service to the numerous researchers and practitioners active in this fascinating field. We allow readers to experience the wonder of peering into a scholarly work of the highest order and seminal significance.

Maven Books

PREFACE.

THE subject-matter of this book is a historical summary of the development of mathematics, illustrated by the lives and discoveries of those to whom the progress of the science is mainly due. It may serve as an introduction to more elaborate works on the subject, but primarily it is intended to give a short and popular account of those leading facts in the history of mathematics which many who are unwilling, or have not the time, to study it systematically may yet desire to know.

The first edition was substantially a transcript of some lectures which I delivered in the year 1888 with the object of giving a sketch of the history, previous to the nineteenth century, that should be intelligible to any one acquainted with the elements of mathematics. In the second edition, issued in 1893, I rearranged parts of it, and introduced a good deal of additional matter.

The scheme of arrangement will be gathered from the table of contents at the end of this preface. Shortly it is as follows. The first chapter contains a brief statement of what is known concerning the mathematics of the Egyptians and Phoenicians; this is introductory to the history of mathematics under Greek influence. The subsequent history is divided into three periods: first, that under Greek influence, chapters II to VII; second, that of the middle ages and renaissance, chapters VIII to XIII; and lastly that of modern times, chapters XIV to XIX.

In discussing the mathematics of these periods I have confined myself to giving the leading events in the history, and frequently have passed in silence over men or works whose influence was comparatively unimportant. Doubtless an exaggerated view of the discoveries of those mathematicians who are mentioned may be caused by the non-allusion to minor writers who preceded and prepared the way for them, but in all historical sketches this is to some extent inevitable, and I have done my best to guard against it by interpolating remarks on the progress of the science at different times. Perhaps also I should here state that generally I have not referred to the results obtained by practical astronomers and physicists unless there was some mathematical interest in them. In quoting results I have commonly made use of modern notation; the reader must therefore recollect that, while the matter is the same as that of any writer to whom allusion is made, his proof is sometimes translated into a more convenient and familiar language.

The greater part of my account is a compilation from existing histories or memoirs, as indeed must be necessarily the case where the works discussed are so numerous and cover so much ground. When authorities disagree I have generally stated only that view which seems to me to be the most probable; but if the question be one of importance, I believe that I have always indicated that there is a difference of opinion about it.

I think that it is undesirable to overload a popular account with a mass of detailed references or the authority for every particular fact mentioned. For the history previous to 1758, I need only refer, once for all, to the closely printed pages of M. Cantor's monumental *Vorlesungen über die Geschichte der Mathematik* (hereafter alluded to as Cantor), which may be regarded as the standard treatise on the subject, but usually I have given references to the other leading authorities on which I have relied or with which I am acquainted. My account for the period subsequent to 1758 is generally based on the memoirs or monographs referred to in the footnotes, but the main facts to 1799 have been also enumerated in a supplementary volume issued by Prof. Cantor last year. I hope that my footnotes will supply the means of studying in detail the history of mathematics at any specified period should the reader desire to do so.

My thanks are due to various friends and correspondents who have called my attention to points in the previous editions. I shall be grateful for notices of additions or corrections which may occur to any of my readers.

W. W. ROUSE BALL.

TRINITY COLLEGE, CAMBRIDGE.

NOTE.

THE fourth edition was stereotyped in 1908, but no material changes have been made since the issue of the second edition in 1893, other duties having, for a few years, rendered it impossible for me to find time for any extensive revision. Such revision and incorporation of recent researches on the subject have now to be postponed till the cost of printing has fallen, though advantage has been taken of reprints to make trivial corrections and additions.

W. W. R. B.

TRINITY COLLEGE, CAMBRIDGE.

TABLE OF CONTENTS.

𝔉irst 𝔓eriod.   𝔐athematics under 𝔊reek 𝔍nfluence.

This period begins with the teaching of Thales, circ. 600 B.C., and ends with the capture of Alexandria by the Mohammedans in or about 641 A.D. The characteristic feature of this period is the development of geometry.

CHAPTER II. THE IONIAN AND PYTHAGOREAN SCHOOLS. CIRC. 600 B.C.–400 B.C.

CHAPTER III. THE SCHOOLS OF ATHENS AND CYZICUS. CIRC. 420–300 B.C.

CHAPTER IV. THE FIRST ALEXANDRIAN SCHOOL. CIRC. 300–30 B.C.

CHAPTER V. THE SECOND ALEXANDRIAN SCHOOL.
30 B.C.–641 A.D.

CHAPTER VI. THE BYZANTINE SCHOOL. 641–1453.

CHAPTER VII. SYSTEMS OF NUMERATION AND PRIMITIVE ARITHMETIC.

Second Period. Mathematics of the Middle Ages and of the Renaissance.

This period begins about the sixth century, and may be said to end with the invention of analytical geometry and of the infinitesimal calculus. The characteristic feature of this period is the creation or development of modern arithmetic, algebra, and trigonometry.

CHAPTER VIII. THE RISE OF LEARNING IN WESTERN EUROPE. CIRC. 600–1200.

CHAPTER IX. THE MATHEMATICS OF THE ARABS.

CHAPTER X. INTRODUCTION OF ARABIAN WORKS INTO EUROPE. CIRC. 1150–1450.

CHAPTER XI. THE DEVELOPMENT OF ARITHMETIC. CIRC. 1300–1637.

CHAPTER XII. THE MATHEMATICS OF THE RENAISSANCE.
CIRC. 1450–1637.

Chapter XIII. The Close of the Renaissance.
Circ. 1586–1637.

Third period. Modern Mathematics.

This period begins with the invention of analytical geometry and the infinitesimal calculus. The mathematics is far more complex than that produced in either of the preceding periods: but it may be generally described as characterized by the development of analysis, and its application to the phenomena of nature.

Chapter XIV. The History of Modern Mathematics.

Chapter XV. History of Mathematics from Descartes to
Huygens. Circ. 1635–1675.

CHAPTER XVI. THE LIFE AND WORKS OF NEWTON.

Chapter XVII. Leibnitz and the Mathematicians of the First Half of the Eighteenth Century.

CHAPTER XVIII. LAGRANGE, LAPLACE, AND THEIR CONTEMPORARIES. CIRC. 1740–1830.

CHAPTER XIX. MATHEMATICS OF THE NINETEENTH CENTURY.

CHAPTER I.

EGYPTIAN AND PHOENICIAN MATHEMATICS.

THE history of mathematics cannot with certainty be traced back to any school or period before that of the Ionian Greeks. The subsequent history may be divided into three periods, the distinctions between which are tolerably well marked. The first period is that of the history of mathematics under Greek influence, this is discussed in chapters II to VII; the second is that of the mathematics of the middle ages and the renaissance, this is discussed in chapters VIII to XIII; the third is that of modern mathematics, and this is discussed in chapters XIV to XIX.

Although the history of mathematics commences with that of the Ionian schools, there is no doubt that those Greeks who first paid attention to the subject were largely indebted to the previous investigations of the Egyptians and Phoenicians. Our knowledge of the mathematical attainments of those races is imperfect and partly conjectural, but, such as it is, it is here briefly summarised. The definite history begins with the next chapter.

On the subject of prehistoric mathematics, we may observe in the first place that, though all early races which have left records behind them knew something of numeration and mechanics, and though the majority were also acquainted with the elements of land-surveying, yet the rules which they possessed were in general founded only on the results of observation and experiment, and were neither deduced from nor did they form part of any science. The fact then that various nations in the vicinity of Greece had reached a high state of civilisation does not justify us in assuming that they had studied mathematics.

The only races with whom the Greeks of Asia Minor (amongst whom our history begins) were likely to have come into frequent contact were those inhabiting the eastern littoral of the Mediterranean; and Greek tradition uniformly assigned the special development of geometry to the Egyptians, and that of the science of numbers either to the Egyptians or to the Phoenicians. I discuss these subjects separately.

First, as to the science of *numbers*. So far as the acquirements of the Phoenicians on this subject are concerned it is impossible to speak with certainty. The magnitude of the commercial transactions of Tyre and Sidon necessitated a considerable development of arithmetic, to which it is probable the name of science might be properly applied. A Babylonian table of the numerical value of the squares of a series of consecutive integers has been found, and this would seem to indicate that properties of numbers were studied. According to Strabo the Tyrians paid particular attention to the sciences of numbers, navigation, and astronomy; they had, we know, considerable commerce with their neighbours and kinsmen the Chaldaeans; and Böckh says that they regularly supplied the weights and measures used in Babylon. Now the Chaldaeans had certainly paid some attention to arithmetic and geometry, as is shown by their astronomical calculations; and, whatever was the extent of their attainments in arithmetic, it is almost certain that the Phoenicians were equally proficient, while it is likely that the knowledge of the latter, such as it was, was communicated to the Greeks. On the whole it seems probable that the early Greeks were largely indebted to the Phoenicians for their knowledge of practical arithmetic or the art of calculation, and perhaps also learnt from them a few properties of numbers. It may be worthy of note that Pythagoras was a Phoenician; and according to Herodotus, but this is more doubtful, Thales was also of that race.

I may mention that the almost universal use of the abacus or swan-pan rendered it easy for the ancients to add and subtract without any knowledge of theoretical arithmetic. These instruments will be described later in chapter VII; it will be sufficient here to say that they afford a concrete way of representing a number in the decimal scale, and enable the results of addition and subtraction to be obtained by a merely mechanical process. This, coupled with a means of representing the result in writing, was all that was required for practical purposes.

We are able to speak with more certainty on the arithmetic of the Egyptians. About forty years ago a hieratic papyrus,[1] forming part

---

[1] See *Ein mathematisches Handbuch der alten Aegypter*, by A. Eisenlohr, second edition, Leipzig, 1891; see also Cantor, chap. i; and *A Short History of Greek Mathematics*, by J. Gow, Cambridge, 1884, arts. 12–14. Besides these authorities the

of the Rhind collection in the British Museum, was deciphered, which has thrown considerable light on their mathematical attainments. The manuscript was written by a scribe named Ahmes at a date, according to Egyptologists, considerably more than a thousand years before Christ, and it is believed to be itself a copy, with emendations, of a treatise more than a thousand years older. The work is called "directions for knowing all dark things," and consists of a collection of problems in arithmetic and geometry; the answers are given, but in general not the processes by which they are obtained. It appears to be a summary of rules and questions familiar to the priests.

The first part deals with the reduction of fractions of the form $2/(2n + 1)$ to a sum of fractions each of whose numerators is unity: for example, Ahmes states that $\frac{2}{29}$ is the sum of $\frac{1}{24}$, $\frac{1}{58}$, $\frac{1}{174}$, and $\frac{1}{232}$; and $\frac{2}{97}$ is the sum of $\frac{1}{56}$, $\frac{1}{679}$, and $\frac{1}{776}$. In all the examples n is less than 50. Probably he had no rule for forming the component fractions, and the answers given represent the accumulated experiences of previous writers: in one solitary case, however, he has indicated his method, for, after having asserted that $\frac{2}{3}$ is the sum of $\frac{1}{2}$ and $\frac{1}{6}$, he adds that therefore two-thirds of one-fifth is equal to the sum of a half of a fifth and a sixth of a fifth, that is, to $\frac{1}{10} + \frac{1}{30}$.

That so much attention was paid to fractions is explained by the fact that in early times their treatment was found difficult. The Egyptians and Greeks simplified the problem by reducing a fraction to the sum of several fractions, in each of which the numerator was unity, the sole exception to this rule being the fraction $\frac{2}{3}$. This remained the Greek practice until the sixth century of our era. The Romans, on the other hand, generally kept the denominator constant and equal to twelve, expressing the fraction (approximately) as so many twelfths. The Babylonians did the same in astronomy, except that they used sixty as the constant denominator; and from them through the Greeks the modern division of a degree into sixty equal parts is derived. Thus in one way or the other the difficulty of having to consider changes in both numerator and denominator was evaded. To-day when using decimals we often keep a fixed denominator, thus reverting to the Roman practice.

After considering fractions Ahmes proceeds to some examples of the fundamental processes of arithmetic. In multiplication he seems to have

papyrus has been discussed in memoirs by L. Rodet, A. Favaro, V. Bobynin, and E. Weyr.

relied on repeated additions. Thus in one numerical example, where he requires to multiply a certain number, say a, by 13, he first multiplies by 2 and gets $2a$, then he doubles the results and gets $4a$, then he again doubles the result and gets $8a$, and lastly he adds together a, $4a$, and $8a$. Probably division was also performed by repeated subtractions, but, as he rarely explains the process by which he arrived at a result, this is not certain. After these examples Ahmes goes on to the solution of some simple numerical equations. For example, he says "heap, its seventh, its whole, it makes nineteen," by which he means that the object is to find a number such that the sum of it and one-seventh of it shall be together equal to 19; and he gives as the answer $16 + \frac{1}{2} + \frac{1}{8}$, which is correct.

The arithmetical part of the papyrus indicates that he had some idea of algebraic symbols. The unknown quantity is always represented by the symbol which means a heap; addition is sometimes represented by a pair of legs walking forwards, subtraction by a pair of legs walking backwards or by a flight of arrows; and equality by the sign $\leftrightharpoons$.

The latter part of the book contains various geometrical problems to which I allude later. He concludes the work with some arithmetico-algebraical questions, two of which deal with arithmetical progressions and seem to indicate that he knew how to sum such series.

Second, as to the science of *geometry*. Geometry is supposed to have had its origin in land-surveying; but while it is difficult to say when the study of numbers and calculation—some knowledge of which is essential in any civilised state—became a science, it is comparatively easy to distinguish between the abstract reasonings of geometry and the practical rules of the land-surveyor. Some methods of land-surveying must have been practised from very early times, but the universal tradition of antiquity asserted that the origin of geometry was to be sought in Egypt. That it was not indigenous to Greece, and that it arose from the necessity of surveying, is rendered the more probable by the derivation of the word from $\gamma\tilde{\eta}$, the earth, and $\mu\epsilon\tau\rho\acute{\epsilon}\omega$, I measure. Now the Greek geometricians, as far as we can judge by their extant works, always dealt with the science as an abstract one: they sought for theorems which should be absolutely true, and, at any rate in historical times, would have argued that to measure quantities in terms of a unit which might have been incommensurable with some of the magnitudes considered would have made their results mere approximations to the truth. The name does not therefore refer to their practice. It is not, however, unlikely that it indicates the use which was made of geometry among the Egyptians from whom the Greeks learned it. This also agrees with the Greek traditions, which in themselves appear probable; for Herodotus states that the periodical inundations of the Nile (which swept away the landmarks in the valley of the river, and by altering its course increased or decreased the taxable value of the adjoining lands) rendered a tolerably accurate system of surveying indispensable, and thus led to a systematic study of the subject by the priests.

We have no reason to think that any special attention was paid to geometry by the Phoenicians, or other neighbours of the Egyptians. A small piece of evidence which tends to show that the Jews had not paid much attention to it is to be found in the mistake made in their sacred books,[1] where it is stated that the circumference of a circle is three times its diameter: the Babylonians[2] also reckoned that π was equal to 3.

Assuming, then, that a knowledge of geometry was first derived by the Greeks from Egypt, we must next discuss the range and nature of Egyptian geometry.[3] That some geometrical results were known at a date anterior to Ahmes's work seems clear if we admit, as we have reason to do, that, centuries before it was written, the following method of obtaining a right angle was used in laying out the ground-plan of certain buildings. The Egyptians were very particular about the exact orientation of their temples; and they had therefore to obtain with accuracy a north and south line, as also an east and west line. By observing the points on the horizon where a star rose and set, and taking a plane midway between them, they could obtain a north and south line. To get an east and west line, which had to be drawn at right angles to this, certain professional "rope-fasteners" were employed. These men used a rope $ABCD$ divided by knots or marks at B and C, so that the lengths AB, BC, CD were in the ratio $3 : 4 : 5$. The length BC was placed along the north and south line, and pegs P and Q inserted at the knots B and C. The piece BA (keeping it stretched all the time) was then rotated round the peg P, and similarly the piece CD was rotated round the peg Q, until the ends A and D coincided; the point thus indicated was marked by a peg R. The result was to form a triangle PQR whose sides RP, PQ, QR were in the ratio $3 : 4 : 5$. The angle of

---

[1] I. Kings, chap. vii, verse 23, and II. Chronicles, chap. iv, verse 2.

[2] See J. Oppert, *Journal Asiatique*, August 1872, and October 1874.

[3] See Eisenlohr; Cantor, chap. ii; Gow, arts. 75, 76; and *Die Geometrie der alten Aegypter*, by E. Weyr, Vienna, 1884.

the triangle at P would then be a right angle, and the line PR would give an east and west line. A similar method is constantly used at the present time by practical engineers for measuring a right angle. The property employed can be deduced as a particular case of Euc. I, 48; and there is reason to think that the Egyptians were acquainted with the results of this proposition and of Euc. I, 47, for triangles whose sides are in the ratio mentioned above. They must also, there is little doubt, have known that the latter proposition was true for an isosceles right-angled triangle, as this is obvious if a floor be paved with tiles of that shape. But though these are interesting facts in the history of the Egyptian arts we must not press them too far as showing that geometry was then studied as a science. Our real knowledge of the nature of Egyptian geometry depends mainly on the Rhind papyrus.

Ahmes commences that part of his papyrus which deals with geometry by giving some numerical instances of the contents of barns. Unluckily we do not know what was the usual shape of an Egyptian barn, but where it is defined by three linear measurements, say a, b, and c, the answer is always given as if he had formed the expression $a \times b \times (c + \frac{1}{2}c)$. He next proceeds to find the areas of certain rectilineal figures; if the text be correctly interpreted, some of these results are wrong. He then goes on to find the area of a circular field of diameter 12—no unit of length being mentioned—and gives the result as $(d - \frac{1}{9}d)^2$, where d is the diameter of the circle: this is equivalent to taking 3.1604 as the value of π, the actual value being very approximately 3.1416. Lastly, Ahmes gives some problems on pyramids. These long proved incapable of interpretation, but Cantor and Eisenlohr have shown that Ahmes was attempting to find, by means of data obtained from the measurement of the external dimensions of a building, the ratio of certain other dimensions which could not be directly measured: his process is equivalent to determining the trigonometrical ratios of certain angles. The data and the results given agree closely with the dimensions of some of the existing pyramids. Perhaps all Ahmes's geometrical results were intended only as approximations correct enough for practical purposes.

It is noticeable that all the specimens of Egyptian geometry which we possess deal only with particular numerical problems and not with general theorems; and even if a result be stated as universally true, it was probably proved to be so only by a wide induction. We shall see later that Greek geometry was from its commencement deductive. There are reasons for thinking that Egyptian geometry and arithmetic made little or no progress subsequent to the date of Ahmes's work; and though for nearly two hundred years after the time of Thales Egypt was recognised by the Greeks as an important school of mathematics, it would seem that, almost from the foundation of the Ionian school, the Greeks outstripped their former teachers.

It may be added that Ahmes's book gives us much that idea of Egyptian mathematics which we should have gathered from statements about it by various Greek and Latin authors, who lived centuries later. Previous to its translation it was commonly thought that these statements exaggerated the acquirements of the Egyptians, and its discovery must increase the weight to be attached to the testimony of these authorities.

We know nothing of the applied mathematics (if there were any) of the Egyptians or Phoenicians. The astronomical attainments of the Egyptians and Chaldaeans were no doubt considerable, though they were chiefly the results of observation: the Phoenicians are said to have confined themselves to studying what was required for navigation. Astronomy, however, lies outside the range of this book.

I do not like to conclude the chapter without a brief mention of the Chinese, since at one time it was asserted that they were familiar with the sciences of arithmetic, geometry, mechanics, optics, navigation, and astronomy nearly three thousand years ago, and a few writers were inclined to suspect (for no evidence was forthcoming) that some knowledge of this learning had filtered across Asia to the West. It is true that at a very early period the Chinese were acquainted with several geometrical or rather architectural implements, such as the rule, square, compasses, and level; with a few mechanical machines, such as the wheel and axle; that they knew of the characteristic property of the magnetic needle; and were aware that astronomical events occurred in cycles. But the careful investigations of L. A. Sédillot[1] have shown that the Chinese made no serious attempt to classify or extend the few rules of arithmetic or geometry with which they were acquainted, or to explain the causes of the phenomena which they observed.

The idea that the Chinese had made considerable progress in theoretical mathematics seems to have been due to a misapprehension of the Jesuit missionaries who went to China in the sixteenth century.

---

[1]See Boncompagni's *Bulletino di bibliografia e di storia delle scienze matematiche e fisiche* for May, 1868, vol. i, pp. 161–166. On Chinese mathematics, mostly of a later date, see Cantor, chap. xxxi.

In the first place, they failed to distinguish between the original science of the Chinese and the views which they found prevalent on their arrival—the latter being founded on the work and teaching of Arab or Hindoo missionaries who had come to China in the course of the thirteenth century or later, and while there introduced a knowledge of spherical trigonometry. In the second place, finding that one of the most important government departments was known as the Board of Mathematics, they supposed that its function was to promote and superintend mathematical studies in the empire. Its duties were really confined to the annual preparation of an almanack, the dates and predictions in which regulated many affairs both in public and domestic life. All extant specimens of these almanacks are defective and, in many respects, inaccurate.

The only geometrical theorem with which we can be certain that the ancient Chinese were acquainted is that in certain cases (namely, when the ratio of the sides is $3 : 4 : 5$, or $1 : 1 : \sqrt{2}$) the area of the square described on the hypotenuse of a right-angled triangle is equal to the sum of the areas of the squares described on the sides. It is barely possible that a few geometrical theorems which can be demonstrated in the quasi-experimental way of superposition were also known to them. Their arithmetic was decimal in notation, but their knowledge seems to have been confined to the art of calculation by means of the swan-pan, and the power of expressing the results in writing. Our acquaintance with the early attainments of the Chinese, slight though it is, is more complete than in the case of most of their contemporaries. It is thus specially instructive, and serves to illustrate the fact that a nation may possess considerable skill in the applied arts while they are ignorant of the sciences on which those arts are founded.

From the foregoing summary it will be seen that our knowledge of the mathematical attainments of those who preceded the Greeks is very limited; but we may reasonably infer that from one source or another the early Greeks learned the use of the abacus for practical calculations, symbols for recording the results, and as much mathematics as is contained or implied in the Rhind papyrus. It is probable that this sums up their indebtedness to other races. In the next six chapters I shall trace the development of mathematics under Greek influence.

FIRST PERIOD.

Mathematics under Greek Influence.

This period begins with the teaching of Thales, circ. 600 B.C., *and ends with the capture of Alexandria by the Mohammedans in or about* 641 A.D. *The characteristic feature of this period is the development of Geometry.*

It will be remembered that I commenced the last chapter by saying that the history of mathematics might be divided into three periods, namely, that of mathematics under Greek influence, that of the mathematics of the middle ages and of the renaissance, and lastly that of modern mathematics. The next four chapters (chapters II, III, IV and V) deal with the history of mathematics under Greek influence: to these it will be convenient to add one (chapter VI) on the Byzantine school, since through it the results of Greek mathematics were transmitted to western Europe; and another (chapter VII) on the systems of numeration which were ultimately displaced by the system introduced by the Arabs. I should add that many of the dates mentioned in these chapters are not known with certainty, and must be regarded as only approximately correct.

There appeared in December 1921, just before this reprint was struck off, Sir T. L. Heath's work in 2 volumes on the History of Greek Mathematics. This may now be taken as the standard authority for this period.

CHAPTER II.

THE IONIAN AND PYTHAGOREAN SCHOOLS.[1]
CIRC. 600 B.C.–400 B.C.

WITH the foundation of the Ionian and Pythagorean schools we emerge from the region of antiquarian research and conjecture into the light of history. The materials at our disposal for estimating the knowledge of the philosophers of these schools previous to about the year 430 B.C. are, however, very scanty Not only have all but fragments of the different mathematical treatises then written been lost, but we possess no copy of the history of mathematics written about 325 B.C. by Eudemus (who was a pupil of Aristotle). Luckily Proclus, who about 450 A.D. wrote a commentary on the earlier part of Euclid's *Elements*, was familiar with Eudemus's work, and freely utilised it in his historical references. We have also a fragment of the *General View of Mathematics* written by Geminus about 50 B.C., in which the methods of proof used by the early Greek geometricians are compared with those current at a later date. In addition to these general statements we have biographies of a few of the leading mathematicians, and some scattered notes in various writers in which allusions are made to the lives and works of others. The original authorities are criticised and discussed at length in the works mentioned in the footnote to the heading of the chapter.

[1] The history of these schools has been discussed by G. Loria in his *Le Scienze Esatte nell' Antica Grecia*, Modena, 1893–1900; by Cantor, chaps. v–viii; by G. J. Allman in his *Greek Geometry from Thales to Euclid*, Dublin, 1889; by J. Gow, in his *Greek Mathematics*, Cambridge, 1884; by C. A. Bretschneider in his *Die Geometrie und die Geometer vor Eukleides*, Leipzig, 1870; and partially by H. Hankel in his posthumous *Geschichte der Mathematik,* Leipzig, 1874.

The Ionian School.

Thales.[1]   The founder of the earliest Greek school of mathematics
and philosophy was *Thales*, one of the seven sages of Greece, who was
born about 640 B.C. at Miletus, and died in the same town about
550 B.C. The materials for an account of his life consist of little more
than a few anecdotes which have been handed down by tradition.

During the early part of his life Thales was engaged partly in commerce and partly in public affairs; and to judge by two stories that have
been preserved, he was then as distinguished for shrewdness in business
and readiness in resource as he was subsequently celebrated in science.
It is said that once when transporting some salt which was loaded on
mules, one of the animals slipping in a stream got its load wet and so
caused some of the salt to be dissolved, and finding its burden thus
lightened it rolled over at the next ford to which it came; to break it
of this trick Thales loaded it with rags and sponges which, by absorb-
ing the water, made the load heavier and soon effectually cured it of
its troublesome habit. At another time, according to Aristotle, when
there was a prospect of an unusually abundant crop of olives Thales
got possession of all the olive-presses of the district; and, having thus
"cornered" them, he was able to make his own terms for lending them
out, or buying the olives, and thus realized a large sum. These tales
may be apocryphal, but it is certain that he must have had considerable reputation as a man of affairs and as a good engineer, since he was
employed to construct an embankment so as to divert the river Halys
in such a way as to permit of the construction of a ford.

Probably it was as a merchant that Thales first went to Egypt, but
during his leisure there he studied astronomy and geometry. He was
middle-aged when he returned to Miletus; he seems then to have abandoned business and public life, and to have devoted himself to the study
of philosophy and science—subjects which in the Ionian, Pythagorean,
and perhaps also the Athenian schools, were closely connected: his
views on philosophy do not here concern us. He continued to live at
Miletus till his death circ. 550 B.C.

We cannot form any exact idea as to how Thales presented his
geometrical teaching. We infer, however, from Proclus that it consisted
of a number of isolated propositions which were not arranged in a logical
sequence, but that the proofs were deductive, so that the theorems were

---

[1]See Loria, book I, chap. ii; Cantor, chap. v; Allman, chap. i.

not a mere statement of an induction from a large number of special instances, as probably was the case with the Egyptian geometricians. The deductive character which he thus gave to the science is his chief claim to distinction.

The following comprise the chief propositions that can now with reasonable probability be attributed to him; they are concerned with the geometry of angles and straight lines.

(i) The angles at the base of an isosceles triangle are equal (Euc. I, 5). Proclus seems to imply that this was proved by taking another exactly equal isosceles triangle, turning it over, and then superposing it on the first—a sort of experimental demonstration.

(ii) If two straight lines cut one another, the vertically opposite angles are equal (Euc. I, 15). Thales may have regarded this as obvious, for Proclus adds that Euclid was the first to give a strict proof of it.

(iii) A triangle is determined if its base and base angles be given (cf. Euc. I, 26). Apparently this was applied to find the distance of a ship at sea—the base being a tower, and the base angles being obtained by observation.

(iv) The sides of equiangular triangles are proportionals (Euc. VI, 4, or perhaps rather Euc. VI, 2). This is said to have been used by Thales when in Egypt to find the height of a pyramid. In a dialogue given by Plutarch, the speaker, addressing Thales, says, "Placing your stick at the end of the shadow of the pyramid, you made by the sun's rays two triangles, and so proved that the [height of the] pyramid was to the [length of the] stick as the shadow of the pyramid to the shadow of the stick." It would seem that the theorem was unknown to the Egyptians, and we are told that the king Amasis, who was present, was astonished at this application of abstract science.

(v) A circle is bisected by any diameter. This may have been enunciated by Thales, but it must have been recognised as an obvious fact from the earliest times.

(vi) The angle subtended by a diameter of a circle at any point in the circumference is a right angle (Euc. III, 31). This appears to have been regarded as the most remarkable of the geometrical achievements of Thales, and it is stated that on inscribing a right-angled triangle in a circle he sacrificed an ox to the immortal gods. It has been conjectured that he may have come to this conclusion by noting that the diagonals of a rectangle are equal and bisect one another, and that therefore a rectangle can be inscribed in a circle. If so, and if he went on to apply proposition (i), he would have discovered that the sum of the angles of a right-angled triangle is equal to two right angles, a fact with which it is believed that he was acquainted. It has been remarked that the shape of the tiles used in paving floors may have suggested these results.

On the whole it seems unlikely that he knew how to draw a perpendicular from a point to a line; but if he possessed this knowledge, it is possible he was also aware, as suggested by some modern commentators, that the sum of the angles of any triangle is equal to two right angles. As far as equilateral and right-angled triangles are concerned, we know from Eudemus that the first geometers proved the general property separately for three species of triangles, and it is not unlikely that they proved it thus. The area about a point can be filled by the angles of six equilateral triangles or tiles, hence the proposition is true for an equilateral triangle. Again, any two equal right-angled triangles can be placed in juxtaposition so as to form a rectangle, the sum of whose angles is four right angles; hence the proposition is true for a right-angled triangle. Lastly, any triangle can be split into the sum of two right-angled triangles by drawing a perpendicular from the biggest angle on the opposite side, and therefore again the proposition is true. The first of these proofs is evidently included in the last, but there is nothing improbable in the suggestion that the early Greek geometers continued to teach the first proposition in the form above given.

Thales wrote on astronomy, and among his contemporaries was more famous as an astronomer than as a geometrician. A story runs that one night, when walking out, he was looking so intently at the stars that he tumbled into a ditch, on which an old woman exclaimed, "How can you tell what is going on in the sky when you can't see what is lying at your own feet?"—an anecdote which was often quoted to illustrate the unpractical character of philosophers.

Without going into astronomical details, it may be mentioned that he taught that a year contained about 365 days, and not (as is said to have been previously reckoned) twelve months of thirty days each. It is said that his predecessors occasionally intercalated a month to keep the seasons in their customary places, and if so they must have realized that the year contains, on the average, more than 360 days. There is some reason to think that he believed the earth to be a disc-like body floating on water. He predicted a solar eclipse which took place at or about the time he foretold; the actual date was either May 28, 585 B.C., or September 30, 609 B.C. But though this prophecy and its fulfilment gave extraordinary prestige to his teaching, and secured him the name of one of the seven sages of Greece, it is most likely that he only made use of one of the Egyptian or Chaldaean registers which stated that solar eclipses recur at intervals of about 18 years 11 days.

Among the pupils of Thales were **Anaximander**, **Anaximenes**, **Mamercus**, and **Mandryatus**. Of the three mentioned last we know next to nothing. *Anaximander* was born in 611 B.C., and died in 545 B.C., and succeeded Thales as head of the school at Miletus. According to Suidas he wrote a treatise on geometry in which, tradition says, he paid particular attention to the properties of spheres, and dwelt at length on the philosophical ideas involved in the conception of infinity in space and time. He constructed terrestrial and celestial globes.

Anaximander is alleged to have introduced the use of the *style* or *gnomon* into Greece. This, in principle, consisted only of a stick stuck upright in a horizontal piece of ground. It was originally used as a sun-dial, in which case it was placed at the centre of three concentric circles, so that every two hours the end of its shadow passed from one circle to another. Such sun-dials have been found at Pompeii and Tusculum. It is said that he employed these styles to determine his meridian (presumably by marking the lines of shadow cast by the style at sunrise and sunset on the same day, and taking the plane bisecting the angle so formed); and thence, by observing the time of year when the noon-altitude of the sun was greatest and least, he got the solstices; thence, by taking half the sum of the noon-altitudes of the sun at the two solstices, he found the inclination of the equator to the horizon (which determined the altitude of the place), and, by taking half their difference, he found the inclination of the ecliptic to the equator. There seems good reason to think that he did actually determine the latitude of Sparta, but it is more doubtful whether he really made the rest of these astronomical deductions.

We need not here concern ourselves further with the successors of Thales. The school he established continued to flourish till about 400 B.C., but, as time went on, its members occupied themselves more and more with philosophy and less with mathematics. We know very little of the mathematicians comprised in it, but they would seem to have devoted most of their attention to astronomy. They exercised but slight influence on the further advance of Greek mathematics, which was made almost entirely under the influence of the Pythagoreans, who not only immensely developed the science of geometry, but created a science of numbers. If Thales was the first to direct general attention to geometry, it was Pythagoras, says Proclus, quoting from Eudemus, who "changed the study of geometry into the form of a liberal education, for he examined its principles to the bottom and investigated its theorems in an ... intellectual manner"; and it is accordingly to Pythagoras that we must now direct attention.

The Pythagorean School.

Pythagoras.[1]   *Pythagoras* was born at Samos about 569 B.C., perhaps of Tyrian parents, and died in 500 B.C. He was thus a contemporary of Thales. The details of his life are somewhat doubtful, but the following account is, I think, substantially correct. He studied first under Pherecydes of Syros, and then under Anaximander; by the latter he was recommended to go to Thebes, and there or at Memphis he spent some years. After leaving Egypt he travelled in Asia Minor, and then settled at Samos, where he gave lectures but without much success. About 529 B.C. he migrated to Sicily with his mother, and with a single disciple who seems to have been the sole fruit of his labours at Samos. Thence he went to Tarentum, but very shortly moved to Croton, a Dorian colony in the south of Italy. Here the schools that he opened were crowded with enthusiastic audiences; citizens of all ranks, especially those of the upper classes, attended, and even the women broke a law which forbade their going to public meetings and flocked to hear him. Amongst his most attentive auditors was Theano, the young and beautiful daughter of his host Milo, whom, in spite of the disparity of their ages, he married. She wrote a biography of her husband, but unfortunately it is lost.

Pythagoras divided those who attended his lectures into two classes, whom we may term probationers and Pythagoreans. The majority were probationers, but it was only to the Pythagoreans that his chief discoveries were revealed. The latter formed a brotherhood with all things in common, holding the same philosophical and political beliefs, engaged in the same pursuits, and bound by oath not to reveal the teaching or secrets of the school; their food was simple; their discipline

[1]See Loria, book I, chap. iii; Cantor, chaps. vi, vii; Allman, chap. ii; Hankel, pp. 92–111; Hoefer, *Histoire des mathématiques*, Paris, third edition, 1886, pp. 87–130; and various papers by S. P. Tannery. For an account of Pythagoras's life, embodying the Pythagorean traditions, see the biography by Iamblichus, of which there are two or three English translations. Those who are interested in esoteric literature may like to see a modern attempt to reproduce the Pythagorean teaching in *Pythagoras*, by E. Schuré, Eng. trans., London, 1906.

severe; and their mode of life arranged to encourage self-command, temperance, purity, and obedience. This strict discipline and secret organisation gave the society a temporary supremacy in the state which brought on it the hatred of various classes; and, finally, instigated by his political opponents, the mob murdered Pythagoras and many of his followers.

Though the political influence of the Pythagoreans was thus destroyed, they seem to have re-established themselves at once as a philosophical and mathematical society, with Tarentum as their headquarters, and they continued to flourish for more than a hundred years.

Pythagoras himself did not publish any books; the assumption of his school was that all their knowledge was held in common and veiled from the outside world, and, further, that the glory of any fresh discovery must be referred back to their founder. Thus Hippasus (circ. 470 B.C.) is said to have been drowned for violating his oath by publicly boasting that he had added the dodecahedron to the number of regular solids enumerated by Pythagoras. Gradually, as the society became more scattered, this custom was abandoned, and treatises containing the substance of their teaching and doctrines were written. The first book of the kind was composed, about 370 B.C., by Philolaus, and we are told that Plato secured a copy of it. We may say that during the early part of the fifth century before Christ the Pythagoreans were considerably in advance of their contemporaries, but by the end of that time their more prominent discoveries and doctrines had become known to the outside world, and the centre of intellectual activity was transferred to Athens.

Though it is impossible to separate precisely the discoveries of Pythagoras himself from those of his school of a later date, we know from Proclus that it was Pythagoras who gave geometry that rigorous character of deduction which it still bears, and made it the foundation of a liberal education; and there is reason to believe that he was the first to arrange the leading propositions of the subject in a logical order. It was also, according to Aristoxenus, the glory of his school that they raised arithmetic above the needs of merchants. It was their boast that they sought knowledge and not wealth, or in the language of one of their maxims, "a figure and a step forwards, not a figure to gain three oboli."

Pythagoras was primarily a moral reformer and philosopher, but his system of morality and philosophy was built on a mathematical foundation. His mathematical researches were, however, designed to lead up to a system of philosophy whose exposition was the main object of his teaching. The Pythagoreans began by dividing the mathematical subjects with which they dealt into four divisions: numbers absolute or arithmetic, numbers applied or music, magnitudes at rest or geometry, and magnitudes in motion or astronomy. This "quadrivium" was long considered as constituting the necessary and sufficient course of study for a liberal education. Even in the case of geometry and arithmetic (which are founded on inferences unconsciously made and common to all men) the Pythagorean presentation was involved with philosophy; and there is no doubt that their teaching of the sciences of astronomy, mechanics, and music (which can rest safely only on the results of conscious observation and experiment) was intermingled with metaphysics even more closely. It will be convenient to begin by describing their treatment of geometry and arithmetic.

First, as to their geometry. Pythagoras probably knew and taught the substance of what is contained in the first two books of Euclid about parallels, triangles, and parallelograms, and was acquainted with a few other isolated theorems including some elementary propositions on irrational magnitudes; but it is suspected that many of his proofs were not rigorous, and in particular that the converse of a theorem was sometimes assumed without a proof. It is hardly necessary to say that we are unable to reproduce the whole body of Pythagorean teaching on this subject, but we gather from the notes of Proclus on Euclid, and from a few stray remarks in other writers, that it included the following propositions, most of which are on the geometry of areas.

(i) It commenced with a number of definitions, which probably were rather statements connecting mathematical ideas with philosophy than explanations of the terms used. One has been preserved in the definition of a point as unity having position.

(ii) The sum of the angles of a triangle was shown to be equal to two right angles (Euc. I, 32); and in the proof, which has been preserved, the results of the propositions Euc. I, 13 and the first part of Euc. I, 29 are quoted. The demonstration is substantially the same as that in Euclid, and it is most likely that the proofs there given of the two propositions last mentioned are also due to Pythagoras himself.

(iii) Pythagoras certainly proved the properties of right-angled triangles which are given in Euc. I, 47 and I, 48. We know that the proofs of these propositions which are found in Euclid were of Euclid's own invention; and a good deal of curiosity has been excited to discover what was the demonstration which was originally offered by Pythagoras of the first of these theorems. It has been conjectured that not improbably it may have been one of the two following.[1]

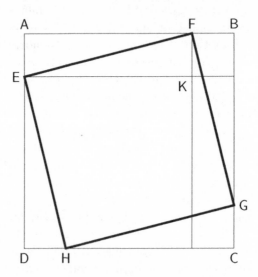

(α) Any square $ABCD$ can be split up, as in Euc. II, 4, into two squares BK and DK and two equal rectangles AK and CK: that is, it is equal to the square on FK, the square on EK, and four times the triangle AEF. But, if points be taken, G on BC, H on CD, and E on DA, so that BG, CH, and DE are each equal to AF, it can be easily shown that $EFGH$ is a square, and that the triangles AEF, BFG, CGH, and DHE are equal: thus the square $ABCD$ is also equal to the square on EF and four times the triangle AEF. Hence the square on EF is equal to the sum of the squares on FK and EK.

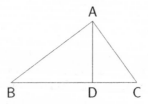

(β) Let ABC be a right-angled triangle, A being the right angle. Draw AD perpendicular to BC. The triangles ABC and DBA are similar,

$$\therefore BC : AB = AB : BD.$$

Similarly $$BC : AC = AC : DC.$$

Hence $$AB^2 + AC^2 = BC(BD + DC) = BC^2.$$

This proof requires a knowledge of the results of Euc. II, 2, VI, 4, and VI, 17, with all of which Pythagoras was acquainted.

(iv) Pythagoras is credited by some writers with the discovery of the theorems Euc. I, 44, and I, 45, and with giving a solution of the problem Euc. II, 14. It is said that on the discovery of the necessary construction for the problem last mentioned he sacrificed an ox, but as his school had all things in common the liberality was less striking than it seems at first. The Pythagoreans of a later date were aware of the extension given in Euc. VI, 25, and Allman thinks that Pythagoras himself was acquainted with it, but this must be regarded as doubtful. It will be noticed that Euc. II, 14 provides a geometrical solution of the equation $x^2 = ab$.

(v) Pythagoras showed that the plane about a point could be completely filled by equilateral triangles, by squares, or by regular hexagons —results that must have been familiar wherever tiles of these shapes were in common use.

(vi) The Pythagoreans were said to have attempted the quadrature of the circle: they stated that the circle was the most perfect of all plane figures.

(vii) They knew that there were five regular solids inscribable in a sphere, which was itself, they said, the most perfect of all solids.

(viii) From their phraseology in the science of numbers and from other occasional remarks, it would seem that they were acquainted with the methods used in the second and fifth books of Euclid, and knew something of irrational magnitudes. In particular, there is reason to believe that Pythagoras proved that the side and the diagonal of a square were incommensurable, and that it was this discovery which led the early Greeks to banish the conceptions of number and measurement from their geometry. A proof of this proposition which may be that due to Pythagoras is given below.[1]

[1]See below, page 49.

Next, as to their theory of numbers.[1] In this Pythagoras was chiefly concerned with four different groups of problems which dealt respectively with polygonal numbers, with ratio and proportion, with the factors of numbers, and with numbers in series; but many of his arithmetical inquiries, and in particular the questions on polygonal numbers and proportion, were treated by geometrical methods.

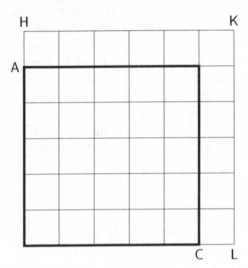

Pythagoras commenced his theory of arithmetic by dividing all numbers into even or odd: the odd numbers being termed *gnomons*. An odd number, such as $2n + 1$, was regarded as the difference of two square numbers $(n + 1)^2$ and n^2; and the sum of the gnomons from 1 to $2n + 1$ was stated to be a *square number*, viz. $(n + 1)^2$, its square root was termed a *side*. Products of two numbers were called *plane*, and if a product had no exact square root it was termed an *oblong*. A product of three numbers was called a *solid number*, and, if the three numbers were equal, a *cube*. All this has obvious reference to geometry, and the opinion is confirmed by Aristotle's remark that when a gnomon is put round a square the figure remains a square though it is increased in dimensions. Thus, in the figure given above in which n is taken equal to 5, the gnomon AKC (containing 11 small squares) when put round the square AC (containing 5^2 small squares) makes a square HL (containing 6^2 small squares). It is possible that several

[1]See the appendix *Sur l'arithmétique pythagorienne* to S. P. Tannery's *La science hellène*, Paris, 1887.

of the numerical theorems due to Greek writers were discovered and proved by an analogous method: the abacus can be used for many of these demonstrations.

The numbers $(2n^2 + 2n + 1)$, $(2n^2 + 2n)$, and $(2n + 1)$ possessed special importance as representing the hypotenuse and two sides of a right-angled triangle: Cantor thinks that Pythagoras knew this fact before discovering the geometrical proposition Euc. I, 47. A more general expression for such numbers is $(m^2 + n^2)$, $2mn$, and $(m^2 - n^2)$, or multiples of them: it will be noticed that the result obtained by Pythagoras can be deduced from these expressions by assuming $m = n + 1$; at a later time Archytas and Plato gave rules which are equivalent to taking $n = 1$; Diophantus knew the general expressions.

After this preliminary discussion the Pythagoreans proceeded to the four special problems already alluded to. Pythagoras was himself acquainted with triangular numbers; polygonal numbers of a higher order were discussed by later members of the school. A triangular number represents the sum of a number of counters laid in rows on a plane; the bottom row containing n, and each succeeding row one less: it is therefore equal to the sum of the series

$$n + (n - 1) + (n - 2) + \ldots + 2 + 1,$$

that is, to $\frac{1}{2}n(n + 1)$. Thus the triangular number corresponding to 4 is 10. This is the explanation of the language of Pythagoras in the well-known passage in Lucian where the merchant asks Pythagoras what he can teach him. Pythagoras replies "I will teach you how to count." *Merchant*, "I know that already." *Pythagoras*, "How do you count?" *Merchant*, "One, two, three, four—" *Pythagoras*, "Stop! what you take to be four is ten, a perfect triangle and our symbol." The Pythagoreans are, on somewhat doubtful authority, said to have classified numbers by comparing them with the sum of their integral subdivisors or factors, calling a number excessive, perfect, or defective, according as the sum of these subdivisors was greater than, equal to, or less than the number: the classification at first being restricted to even numbers. The third group of problems which they considered dealt with numbers which formed a proportion; presumably these were discussed with the aid of geometry as is done in the fifth book of Euclid. Lastly, the Pythagoreans were concerned with series of numbers in arithmetical, geometrical, harmonical, and musical progressions. The three progressions first mentioned are well known; four integers are said to be in musical progression when they are in the ratio $a : 2ab/(a + b) : \frac{1}{2}(a + b) : b$, for example, 6, 8, 9, and 12 are in musical progression.

Of the Pythagorean treatment of the applied subjects of the quadrivium, and the philosophical theories founded on them, we know very little. It would seem that Pythagoras was much impressed by certain numerical relations which occur in nature. It has been suggested that he was acquainted with some of the simpler facts of crystallography. It is thought that he was aware that the notes sounded by a vibrating string depend on the length of the string, and in particular that lengths which gave a note, its fifth and its octave were in the ratio 2 : 3 : 4, forming terms in a musical progression. It would seem, too, that he believed that the distances of the astrological planets from the earth were also in musical progression, and that the heavenly bodies in their motion through space gave out harmonious sounds: hence the phrase the harmony of the spheres. These and similar conclusions seem to have suggested to him that the explanation of the order and harmony of the universe was to be found in the science of numbers, and that numbers are to some extent the cause of form as well as essential to its accurate measurement. He accordingly proceeded to attribute particular properties to particular numbers and geometrical figures. For example, he taught that the cause of colour was to be sought in properties of the number five, that the explanation of fire was to be discovered in the nature of the pyramid, and so on. I should not have alluded to this were it not that the Pythagorean tradition strengthened, or perhaps was chiefly responsible for the tendency of Greek writers to found the study of nature on philosophical conjectures and not on experimental observation—a tendency to which the defects of Hellenic science must be largely attributed.

After the death of Pythagoras his teaching seems to have been carried on by **Epicharmus** and **Hippasus**, and subsequently by **Philolaus** (specially distinguished as an astronomer), **Archippus**, and **Lysis**. About a century after the murder of Pythagoras we find Archytas recognised as the head of the school.

Archytas.[1]   *Archytas*, circ. 400 B.C., was one of the most influential citizens of Tarentum, and was made governor of the city no less

---

[1] See Allman, chap. iv. A catalogue of the works of Archytas is given by Fabricius in his *Bibliotheca Graeca*, vol. i, p. 833: most of the fragments on philosophy were published by Thomas Gale in his *Opuscula Mythologica*, Cambridge, 1670; and by Thomas Taylor as an Appendix to his translation of Iamblichus's *Life of Pythagoras*, London, 1818. See also the references given by Cantor, vol. i, p. 203.

than seven times. His influence among his contemporaries was very great, and he used it with Dionysius on one occasion to save the life of Plato. He was noted for the attention he paid to the comfort and education of his slaves and of children in the city. He was drowned in a shipwreck near Tarentum, and his body washed on shore—a fit punishment, in the eyes of the more rigid Pythagoreans, for his having departed from the lines of study laid down by their founder. Several of the leaders of the Athenian school were among his pupils and friends, and it is believed that much of their work was due to his inspiration.

The Pythagoreans at first made no attempt to apply their knowledge to mechanics, but Archytas is said to have treated it with the aid of geometry. He is alleged to have invented and worked out the theory of the pulley, and is credited with the construction of a flying bird and some other ingenious mechanical toys. He introduced various mechanical devices for constructing curves and solving problems. These were objected to by Plato, who thought that they destroyed the value of geometry as an intellectual exercise, and later Greek geometricians confined themselves to the use of two species of instruments, namely, rulers and compasses. Archytas was also interested in astronomy; he taught that the earth was a sphere rotating round its axis in twenty-four hours, and round which the heavenly bodies moved.

Archytas was one of the first to give a solution of the problem to duplicate a cube, that is, to find the side of a cube whose volume is double that of a given cube. This was one of the most famous problems of antiquity.[1] The construction given by Archytas is equivalent to the following. On the diameter OA of the base of a right circular cylinder describe a semicircle whose plane is perpendicular to the base of the cylinder. Let the plane containing this semicircle rotate round the generator through O, then the surface traced out by the semicircle will cut the cylinder in a tortuous curve. This curve will be cut by a right cone whose axis is OA and semivertical angle is (say) 60° in a point P, such that the projection of OP on the base of the cylinder will be to the radius of the cylinder in the ratio of the side of the required cube to that of the given cube. The proof given by Archytas is of course geometrical;[2] it will be enough here to remark that in the course of it he shews himself acquainted with the results of the propositions Euc. III, 18, Euc. III, 35, and Euc. XI, 19. To shew analytically that the construction is correct,

---

[1] See below, pp. 30, 34, 34.
[2] It is printed by Allman, pp. 111–113.

take OA as the axis of x, and the generator through O as axis of z, then, with the usual notation in polar co-ordinates, and if a be the radius of the cylinder, we have for the equation of the surface described by the semicircle, $r = 2a \sin \theta$; for that of the cylinder, $r \sin \theta = 2a \cos \phi$; and for that of the cone, $\sin \theta \cos \phi = \frac{1}{2}$. These three surfaces cut in a point such that $\sin^3 \theta = \frac{1}{2}$, and, therefore, if ρ be the projection of OP on the base of the cylinder, then $\rho^3 = (r \sin \theta)^3 = 2a^3$. Hence the volume of the cube whose side is ρ is twice that of a cube whose side is a. I mention the problem and give the construction used by Archytas to illustrate how considerable was the knowledge of the Pythagorean school at the time.

Theodorus.   Another Pythagorean of about the same date as Archytas was *Theodorus of Cyrene*, who is said to have proved geometrically that the numbers represented by $\sqrt{3}$, $\sqrt{5}$, $\sqrt{6}$, $\sqrt{7}$, $\sqrt{8}$, $\sqrt{10}$, $\sqrt{11}$, $\sqrt{12}$, $\sqrt{13}$, $\sqrt{14}$, $\sqrt{15}$, and $\sqrt{17}$ are incommensurable with unity. Theaetetus was one of his pupils.

Perhaps **Timaeus** of Locri and **Bryso** of Heraclea should be mentioned as other distinguished Pythagoreans of this time. It is believed that Bryso attempted to find the area of a circle by inscribing and circumscribing squares, and finally obtained polygons between whose areas the area of the circle lay; but it is said that at some point he assumed that the area of the circle was the arithmetic mean between an inscribed and a circumscribed polygon.

Other Greek Mathematical Schools in the Fifth Century B.C.

It would be a mistake to suppose that Miletus and Tarentum were the only places where, in the fifth century, Greeks were engaged in laying a scientific foundation for the study of mathematics. These towns represented the centres of chief activity, but there were few cities or colonies of any importance where lectures on philosophy and geometry were not given. Among these smaller schools I may mention those at Chios, Elea, and Thrace.

The best known philosopher of the *School of Chios* was **Oenopides**, who was born about 500 B.C., and died about 430 B.C. He devoted himself chiefly to astronomy, but he had studied geometry in Egypt, and is credited with the solution of two problems, namely, to draw a straight line from a given external point perpendicular to a given straight line (Euc. I, 12), and at a given point to construct an angle equal to a given angle (Euc. I, 23).

Another important centre was at Elea in Italy. This was founded in Sicily by **Xenophanes**. He was followed by **Parmenides, Zeno,** and **Melissus**. The members of the *Eleatic School* were famous for the difficulties they raised in connection with questions that required the use of infinite series, such, for example, as the well-known paradox of Achilles and the tortoise, enunciated by *Zeno*, one of their most prominent members. Zeno was born in 495 B.C., and was executed at Elea in 435 B.C. in consequence of some conspiracy against the state; he was a pupil of Parmenides, with whom he visited Athens, circ. 455–450 B.C.

Zeno argued that if Achilles ran ten times as fast as a tortoise, yet if the tortoise had (say) 1000 yards start it could never be overtaken: for, when Achilles had gone the 1000 yards, the tortoise would still be 100 yards in front of him; by the time he had covered these 100 yards, it would still be 10 yards in front of him; and so on for ever: thus Achilles would get nearer and nearer to the tortoise, but never overtake it. The fallacy is usually explained by the argument that the time required to overtake the tortoise, can be divided into an infinite number of parts, as stated in the question, but these get smaller and smaller in geometrical progression, and the sum of them all is a finite time: after the lapse of that time Achilles would be in front of the tortoise. Probably Zeno would have replied that this argument rests on the assumption that space is infinitely divisible, which is the question under discussion: he himself asserted that magnitudes are not infinitely divisible.

These paradoxes made the Greeks look with suspicion on the use of infinitesimals, and ultimately led to the invention of the method of exhaustions.

The *Atomistic School*, having its headquarters in Thrace, was another important centre. This was founded by **Leucippus**, who was a pupil of Zeno. He was succeeded by **Democritus** and **Epicurus**. Its most famous mathematician was *Democritus*, born at Abdera in 460 B.C., and said to have died in 370 B.C., who, besides philosophical works, wrote on plane and solid geometry, incommensurable lines, perspective, and numbers. These works are all lost. From the Archimedean MS., discovered by Heiberg in 1906, it would seem that Democritus enunciated, but without a proof, the proposition that the volume of a pyramid is equal to one-third that of a prism of an equal base and of equal height.

But though several distinguished individual philosophers may be mentioned who, during the fifth century, lectured at different cities, they mostly seem to have drawn their inspiration from Tarentum, and towards the end of the century to have looked to Athens as the intellectual capital of the Greek world; and it is to the Athenian schools that we owe the next great advance in mathematics.

CHAPTER III.

THE SCHOOLS OF ATHENS AND CYZICUS.[1]
CIRC. 420 B.C.–300 B.C.

IT was towards the close of the fifth century before Christ that Athens first became the chief centre of mathematical studies. Several causes conspired to bring this about. During that century she had become, partly by commerce, partly by appropriating for her own purposes the contributions of her allies, the most wealthy city in Greece; and the genius of her statesmen had made her the centre on which the politics of the peninsula turned. Moreover, whatever states disputed her claim to political supremacy her intellectual pre-eminence was admitted by all. There was no school of thought which had not at some time in that century been represented at Athens by one or more of its leading thinkers; and the ideas of the new science, which was being so eagerly studied in Asia Minor and Graecia Magna, had been brought before the Athenians on various occasions.

Anaxagoras. Amongst the most important of the philosophers who resided at Athens and prepared the way for the Athenian school I may mention *Anaxagoras of Clazomenae*, who was almost the last philosopher of the Ionian school. He was born in 500 B.C., and died in 428 B.C. He seems to have settled at Athens about 440 B.C., and there taught the results of the Ionian philosophy. Like all members of that school he was much interested in astronomy. He asserted that

---

[1] The history of these schools is discussed at length in G. Loria's *Le Scienze Esatte nell' Antica Grecia*, Modena, 1893–1900; in G. J. Allman's *Greek Geometry from Thales to Euclid*, Dublin, 1889; and in J. Gow's *Greek Mathematics*, Cambridge, 1884; it is also treated by Cantor, chaps. ix, x, and xi; by Hankel, pp. 111–156; and by C. A. Bretschneider in his *Die Geometrie und die Geometer vor Eukleides*, Leipzig, 1870; a critical account of the original authorities is given by S. P. Tannery in his *Géométrie Grecque*, Paris, 1887, and other papers.

the sun was larger than the Peloponnesus: this opinion, together with some attempts he had made to explain various physical phenomena which had been previously supposed to be due to the direct action of the gods, led to a prosecution for impiety, and he was convicted. While in prison he is said to have written a treatise on the quadrature of the circle.

The Sophists.   The sophists can hardly be considered as belonging to the Athenian school, any more than Anaxagoras can; but like him they immediately preceded and prepared the way for it, so that it is desirable to devote a few words to them. One condition for success in public life at Athens was the power of speaking well, and as the wealth and power of the city increased a considerable number of "sophists" settled there who undertook amongst other things to teach the art of oratory. Many of them also directed the general education of their pupils, of which geometry usually formed a part. We are told that two of those who are usually termed sophists made a special study of geometry—these were Hippias of Elis and Antipho, and one made a special study of astronomy—this was Meton, after whom the metonic cycle is named.

Hippias.   The first of these geometricians, *Hippias of Elis* (circ. 420 B.C.), is described as an expert arithmetician, but he is best known to us through his invention of a curve called the quadratrix, by means of which an angle can be trisected, or indeed divided in any given ratio. If the radius of a circle rotate uniformly round the centre O from the position OA through a right angle to OB, and in the same time a straight line drawn perpendicular to OB move uniformly parallel to itself from the position OA to BC, the locus of their intersection will be the quadratrix.

Let OR and MQ be the position of these lines at any time; and let them cut in P, a point on the curve. Then

$$\text{angle } AOP : \text{angle } AOB = OM : OB.$$

Similarly, if OR' be another position of the radius,

$$\text{angle } AOP' : \text{angle } AOB = OM' : OB$$
$$\therefore \quad \text{angle } AOP : \text{angle } AOP' = OM : OM';$$
$$\therefore \quad \text{angle } AOP' : \text{angle } P'OP = OM' : MM.$$

Hence, if the angle AOP be given, and it be required to divide it in any given ratio, it is sufficient to divide OM in that ratio at M' and draw the line $M'P'$; then OP' will divide AOP in the required ratio.

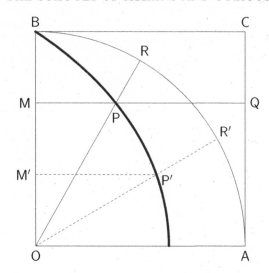

If OA be taken as the initial line, $OP = r$, the angle $AOP = \theta$, and $OA = a$, we have $\theta : \frac{1}{2}\pi = r \sin\theta : a$, and the equation of the curve is $\pi r = 2a\theta \cosec\theta$.

Hippias devised an instrument to construct the curve mechanically; but constructions which involved the use of any mathematical instruments except a ruler and a pair of compasses were objected to by Plato, and rejected by most geometricians of a subsequent date.

Antipho.   The second sophist whom I mentioned was *Antipho* (circ. 420 B.C.). He is one of the very few writers among the ancients who attempted to find the area of a circle by considering it as the limit of an inscribed regular polygon with an infinite number of sides. He began by inscribing an equilateral triangle (or, according to some accounts, a square); on each side he inscribed in the smaller segment an isosceles triangle, and so on *ad infinitum.* This method of attacking the quadrature problem is similar to that described above as used by Bryso of Heraclea.

No doubt there were other cities in Greece besides Athens where similar and equally meritorious work was being done, though the record of it has now been lost; I have mentioned here the investigations of these three writers, chiefly because they were the immediate predecessors of those who created the Athenian school.

The Schools of Athens and Cyzicus.   The history of the Athenian school begins with the teaching of Hippocrates about 420 B.C.; the school was established on a permanent basis by the labours of Plato and Eudoxus; and, together with the neighbouring school of Cyzicus, continued to extend on the lines laid down by these three geometricians until the foundation (about 300 B.C.) of the university at Alexandria drew thither most of the talent of Greece.

Eudoxus, who was amongst the most distinguished of the Athenian mathematicians, is also reckoned as the founder of the school at Cyzicus. The connection between this school and that of Athens was very close, and it is now impossible to disentangle their histories. It is said that Hippocrates, Plato, and Theaetetus belonged to the Athenian school; while Eudoxus, Menaechmus, and Aristaeus belonged to that of Cyzicus. There was always a constant intercourse between the two schools, the earliest members of both had been under the influence either of Archytas or of his pupil Theodorus of Cyrene, and there was no difference in their treatment of the subject, so that they may be conveniently treated together.

Before discussing the work of the geometricians of these schools in detail I may note that they were especially interested in three problems:[1] namely (i), the duplication of a cube, that is, the determination of the side of a cube whose volume is double that of a given cube; (ii) the trisection of an angle; and (iii) the squaring of a circle, that is, the determination of a square whose area is equal to that of a given circle.

Now the first two of these problems (considered analytically) require the solution of a cubic equation; and, since a construction by means of circles (whose equations are of the form $x^2 + y^2 + ax + by + c = 0$) and straight lines (whose equations are of the form $x + \beta y + \gamma = 0$) cannot be equivalent to the solution of a cubic equation, the problems are insoluble if in our constructions we restrict ourselves to the use of circles and straight lines, that is, to Euclidean geometry. If the use of the conic sections be permitted, both of these questions can be solved in many ways. The third problem is equivalent to finding a rectangle whose sides are equal respectively to the radius and to the semiperimeter of the circle. These lines have been long known to be incommensurable, but it is only recently that it has been shewn by Lindemann that their ratio cannot be the root of a rational algebraical equation. Hence this problem also is insoluble by Euclidean geometry. The Athenians and Cyzicians were thus destined to fail in all three

---

[1]On these problems, solutions of them, and the authorities for their history, see my *Mathematical Recreations and Problems*, London, ninth edition, 1920, chap. xiv.

problems, but the attempts to solve them led to the discovery of many new theorems and processes.

Besides attacking these problems the later Platonic school collected all the geometrical theorems then known and arranged them systematically. These collections comprised the bulk of the propositions in Euclid's *Elements*, books I–IX, XI, and XII, together with some of the more elementary theorems in conic sections.

Hippocrates.   *Hippocrates of Chios* (who must be carefully distinguished from his contemporary, Hippocrates of Cos, the celebrated physician) was one of the greatest of the Greek geometricians. He was born about 470 B.C. at Chios, and began life as a merchant. The accounts differ as to whether he was swindled by the Athenian custom-house officials who were stationed at the Chersonese, or whether one of his vessels was captured by an Athenian pirate near Byzantium; but at any rate somewhere about 430 B.C. he came to Athens to try to recover his property in the law courts. A foreigner was not likely to succeed in such a case, and the Athenians seem only to have laughed at him for his simplicity, first in allowing himself to be cheated, and then in hoping to recover his money. While prosecuting his cause he attended the lectures of various philosophers, and finally (in all probability to earn a livelihood) opened a school of geometry himself. He seems to have been well acquainted with the Pythagorean philosophy, though there is no sufficient authority for the statement that he was ever initiated as a Pythagorean.

He wrote the first elementary text-book of geometry, a text-book on which probably Euclid's *Elements* was founded; and therefore he may be said to have sketched out the lines on which geometry is still taught in English schools. It is supposed that the use of letters in diagrams to describe a figure was made by him or introduced about this time, as he employs expressions such as "the point on which the letter A stands" and "the line on which AB is marked." Cantor, however, thinks that the Pythagoreans had previously been accustomed to represent the five vertices of the pentagram-star by the letters v γ ι θ α; and though this was a single instance, perhaps they may have used the method generally. The Indian geometers never employed letters to aid them in the description of their figures. Hippocrates also denoted the square on a line by the word δύναμις, and thus gave the technical meaning to the word *power* which it still retains in algebra: there is reason to think that this use of the word was derived from the Pythagoreans, who are said to have enunciated the result of the proposition Euc. I, 47, in the form that "the total power of the sides of a right-angled triangle is the same as that of the hypotenuse."

In this text-book Hippocrates introduced the method of "reducing" one theorem to another, which being proved, the thing proposed necessarily follows; of this method the *reductio ad absurdum* is an illustration. No doubt the principle had been used occasionally before, but he drew attention to it as a legitimate mode of proof which was capable of numerous applications. He elaborated the geometry of the circle: proving, among other propositions, that similar segments of a circle contain equal angles; that the angle subtended by the chord of a circle is greater than, equal to, or less than a right angle as the segment of the circle containing it is less than, equal to, or greater than a semicircle (Euc. III, 31); and probably several other of the propositions in the third book of Euclid. It is most likely that he also established the propositions that [similar] circles are to one another as the squares of their diameters (Euc. XII, 2), and that similar segments are as the squares of their chords. The proof given in Euclid of the first of these theorems is believed to be due to Hippocrates.

The most celebrated discoveries of Hippocrates were, however, in connection with the quadrature of the circle and the duplication of the cube, and owing to his influence these problems played a prominent part in the history of the Athenian school.

The following propositions will sufficiently illustrate the method by which he attacked the quadrature problem.

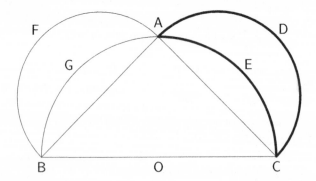

(α) He commenced by finding the area of a lune contained between a semicircle and a quadrantal arc standing on the same chord. This he did as follows. Let *ABC* be an isosceles right-angled triangle inscribed in the semicircle *ABOC*, whose centre is *O*. On *AB* and *AC* as diameters describe semicircles as in the figure. Then, since by Euc. I, 47,

$$\text{sq. on } BC = \text{sq. on } AC + \text{sq. on } AB,$$

therefore, by Euc. XII, 2,

$$\text{area } \tfrac{1}{2} \odot \text{ on } BC = \text{area } \tfrac{1}{2} \odot \text{ on } AC + \text{area } \tfrac{1}{2} \odot \text{ on } AB.$$

Take away the common parts

$$\therefore \text{ area } \triangle ABC = \text{sum of areas of lunes } AECD \text{ and } AFBG.$$

Hence the area of the lune $AECD$ is equal to half that of the triangle ABC.

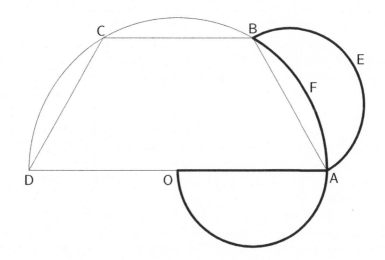

(β) He next inscribed half a regular hexagon $ABCD$ in a semicircle whose centre was O, and on OA, AB, BC, and CD as diameters described semicircles of which those on OA and AB are drawn in the figure. Then AD is double any of the lines OA, AB, BC, and CD,

$$\therefore \text{sq. on } AD = \text{sum of sqs. on } OA, AB, BC, \text{ and } CD,$$
$$\therefore \text{area } \tfrac{1}{2} \odot ABCD = \text{sum of areas of } \tfrac{1}{2} \odot s \text{ on } OA, AB, BC, \text{ and } CD.$$

Take away the common parts

$$\therefore \text{area trapezium } ABCD = 3 \text{ lune } AEBF + \tfrac{1}{2} \odot \text{ on } OA.$$

If therefore the area of this latter lune be known, so is that of the semicircle described on OA as diameter. According to Simplicius, Hippocrates assumed that the area of this lune was the same as the area of the lune found in proposition (α); if this be so, he was of course mistaken, as in this case he is dealing with a lune contained between a semicircle and a sextantal arc standing on the same chord; but it seems more probable that Simplicius misunderstood Hippocrates.

Hippocrates also enunciated various other theorems connected with lunes (which have been collected by Bretschneider and by Allman) of which the theorem last given is a typical example. I believe that they are the earliest instances in which areas bounded by curves were determined by geometry.

The other problem to which Hippocrates turned his attention was the duplication of a cube, that is, the determination of the side of a cube whose volume is double that of a given cube.

This problem was known in ancient times as the Delian problem, in consequence of a legend that the Delians had consulted Plato on the subject. In one form of the story, which is related by Philoponus, it is asserted that the Athenians in 430 B.C., when suffering from the plague of eruptive typhoid fever, consulted the oracle at Delos as to how they could stop it. Apollo replied that they must double the size of his altar which was in the form of a cube. To the unlearned suppliants nothing seemed more easy, and a new altar was constructed either having each of its edges double that of the old one (from which it followed that the volume was increased eightfold) or by placing a similar cubic altar next to the old one. Whereupon, according to the legend, the indignant god made the pestilence worse than before, and informed a fresh deputation that it was useless to trifle with him, as his new altar must be a cube and have a volume exactly double that of his old one. Suspecting a mystery the Athenians applied to Plato, who referred them to the geometricians, and especially to Euclid, who had made a special study of the problem. The introduction of the names of Plato and Euclid is an obvious anachronism. Eratosthenes gives a somewhat similar account of its origin, but with king Minos as the propounder of the problem.

Hippocrates reduced the problem of duplicating the cube to that of finding two means between one straight line (a), and another twice as long ($2a$). If these means be x and y, we have $a : x = x : y = y : 2a$, from which it follows that $x^3 = 2a^3$. It is in this form that the problem is usually presented now. Hippocrates did not succeed in finding a construction for these means.

Plato.   The next philosopher of the Athenian school who re-quires mention here was *Plato*. He was born at Athens in 429 B.C., and was, as is well known, a pupil for eight years of Socrates; much of the teaching of the latter is inferred from Plato's dialogues. After the execution of his master in 399 B.C. Plato left Athens, and being possessed of considerable wealth he spent some years in travelling; it was during this time that he studied mathematics. He visited Egypt with Eudoxus, and Strabo says that in his time the apartments they occupied at Heliopolis were still shewn. Thence Plato went to Cyrene, where he studied under Theodorus. Next he moved to Italy, where he became intimate with Archytas the then head of the Pythagorean school, Eurytas of Metapontum, and Timaeus of Locri. He returned to Athens about the year 380 B.C., and formed a school of students in a suburban gymnasium called the "Academy." He died in 348 B.C.

Plato, like Pythagoras, was primarily a philosopher, and perhaps his philosophy should be regarded as founded on the Pythagorean rather than on the Socratic teaching. At any rate it, like that of the Pythagore-ans, was coloured with the idea that the secret of the universe is to be found in number and in form; hence, as Eudemus says, "he exhibited on every occasion the remarkable connection between mathematics and philosophy." All the authorities agree that, unlike many later philoso-phers, he made a study of geometry or some exact science an indis-pensable preliminary to that of philosophy. The inscription over the entrance to his school ran "Let none ignorant of geometry enter my door," and on one occasion an applicant who knew no geometry is said to have been refused admission as a student.

Plato's position as one of the masters of the Athenian mathematical school rests not so much on his individual discoveries and writings as on the extraordinary influence he exerted on his contemporaries and successors. Thus the objection that he expressed to the use in the con-struction of curves of any instruments other than rulers and compasses was at once accepted as a canon which must be observed in such prob-lems. It is probably due to Plato that subsequent geometricians began the subject with a carefully compiled series of definitions, postulates, and axioms. He also systematized the methods which could be used in attacking mathematical questions, and in particular directed atten-tion to the value of analysis. The analytical method of proof begins by assuming that the theorem or problem is solved, and thence de-ducing some result: if the result be false, the theorem is not true or the problem is incapable of solution: if the result be true, and if the steps be reversible, we get (by reversing them) a synthetic proof; but if the steps be not reversible, no conclusion can be drawn. Numerous illustrations of the method will be found in any modern text-book on geometry. If the classification of the methods of legitimate induction given by Mill in his work on logic had been universally accepted and every new discovery in science had been justified by a reference to the rules there laid down, he would, I imagine, have occupied a position in reference to modern science somewhat analogous to that which Plato occupied in regard to the mathematics of his time.

The following is the only extant theorem traditionally attributed to Plato. If CAB and DAB be two right-angled triangles, having one side, AB, common, their other sides, AD and BC, parallel, and their hypotenuses, AC and BD, at right angles, then, if these hypotenuses cut in P, we have $PC : PB = PB : PA = PA : PD$. This theorem was used in duplicating the cube, for, if such triangles can be constructed having $PD = 2PC$, the problem will be solved. It is easy to make an instrument by which the triangles can be constructed.

Eudoxus.[1]   Of *Eudoxus*, the third great mathematician of the Athenian school and the founder of that at Cyzicus, we know very little. He was born in Cnidus in 408 B.C. Like Plato, he went to Tarentum and studied under Archytas the then head of the Pythagoreans. Subsequently he travelled with Plato to Egypt, and then settled at Cyzicus, where he founded the school of that name. Finally he and his pupils moved to Athens. There he seems to have taken some part in public affairs, and to have practised medicine; but the hostility of Plato and his own unpopularity as a foreigner made his position uncomfortable, and he returned to Cyzicus or Cnidus shortly before his death. He died while on a journey to Egypt in 355 B.C.

His mathematical work seems to have been of a high order of excellence. He discovered most of what we now know as the fifth book of Euclid, and proved it in much the same form as that in which it is there given.

He discovered some theorems on what was called "the golden section." The problem to cut a line AB in the golden section, that is, to divide it, say at H, in extreme and mean ratio

A ●————————————————● H ————————● B

---

[1] The works of Eudoxus were discussed in considerable detail by H. Künssberg of Dinkelsbühl in 1888 and 1890; see also the authorities mentioned above in the footnote on p. 27.

(that is, so that $AB : AH = AH : HB$) is solved in Euc. II, 11, and probably was known to the Pythagoreans at an early date. If we denote AB by l, AH by a, and HB by b, the theorems that Eudoxus proved are equivalent to the following algebraical identities. (i) $(a + \frac{1}{2}l)^2 = 5(\frac{1}{2}l)^2$. (ii) Conversely, if (i) be true, and AH be taken equal to a, then AB will be divided at H in a golden section. (iii) $(b + \frac{1}{2}a)^2 = 5(\frac{1}{2}a^2)$. (iv) $l^2 + b^2 = 3a^2$. (v) $l + a : l = l : a$, which gives another golden section. These propositions were subsequently put by Euclid as the first five propositions of his thirteenth book, but they might have been equally well placed towards the end of the second book. All of them are obvious algebraically, since $l = a + b$ and $a^2 = bl$.

Eudoxus further established the "method of exhaustions"; which depends on the proposition that "if from the greater of two unequal magnitudes there be taken more than its half, and from the remainder more than its half, and so on, there will at length remain a magnitude less than the least of the proposed magnitudes." This proposition was placed by Euclid as the first proposition of the tenth book of his *Elements*, but in most modern school editions it is printed at the beginning of the twelfth book. By the aid of this theorem the ancient geometers were able to avoid the use of infinitesimals: the method is rigorous, but awkward of application. A good illustration of its use is to be found in the demonstration of Euc. XII, 2, namely, that the square of the radius of one circle is to the square of the radius of another circle as the area of the first circle is to an area which is neither less nor greater than the area of the second circle, and which therefore must be exactly equal to it: the proof given by Euclid is (as was usual) completed by a *reductio ad absurdum*. Eudoxus applied the principle to shew that the volume of a pyramid or a cone is one-third that of the prism or the cylinder on the same base and of the same altitude (Euc. XII, 7 and 10). It is believed that he proved that the volumes of two spheres were to one another as the cubes of their radii; some writers attribute the proposition Euc. XII, 2 to him, and not to Hippocrates.

Eudoxus also considered certain curves other than the circle. There is no authority for the statement made in some old books that these were conic sections, and recent investigations have shewn that the assertion (which I repeated in the earlier editions of this book) that they were plane sections of the anchor-ring is also improbable. It seems most likely that they were tortuous curves; whatever they were, he applied them in explaining the apparent motions of the planets as seen from the earth.

Eudoxus constructed an orrery, and wrote a treatise on practical astronomy, in which he supposed a number of moving spheres to which the sun, moon, and stars were attached, and which by their rotation produced the effects observed. In all he required twenty-seven spheres. As observations became more accurate, subsequent astronomers who accepted the theory had continually to introduce fresh spheres to make the theory agree with the facts. The work of Aratus on astronomy, which was written about 300 B.C. and is still extant, is founded on that of Eudoxus.

Plato and Eudoxus were contemporaries. Among Plato's pupils were the mathematicians **Leodamas**, **Neocleides**, **Amyclas**, and to their school also belonged **Leon**, **Theudius** (both of whom wrote text-books on plane geometry), **Cyzicenus**, **Thasus**, **Hermotimus**, **Philippus**, and **Theaetetus**. Among the pupils of Eudoxus are reckoned **Menaechmus**, his brother **Dinostratus** (who applied the quadratrix to the duplication and trisection problems), and **Aristaeus**.

Menaechmus. Of the above-mentioned mathematicians *Menaechmus* requires special mention. He was born about 375 B.C., and died about 325 B.C. Probably he succeeded Eudoxus as head of the school at Cyzicus, where he acquired great reputation as a teacher of geometry, and was for that reason appointed one of the tutors of Alexander the Great. In answer to his pupil's request to make his proofs shorter, Menaechmus made the well-known reply that though in the country there are private and even royal roads, yet in geometry there is only one road for all.

Menaechmus was the first to discuss the conic sections, which were long called the Menaechmian triads. He divided them into three classes, and investigated their properties, not by taking different plane sections of a fixed cone, but by keeping his plane fixed and cutting it by different cones. He shewed that the section of a right cone by a plane perpendicular to a generator is an ellipse, if the cone be acute-angled; a parabola, if it be right-angled; and a hyperbola, if it be obtuse-angled; and he gave a mechanical construction for curves of each class. It seems almost certain that he was acquainted with the fundamental properties of these curves; but some writers think that he failed to connect them with the sections of the cone which he had discovered, and there is no doubt that he regarded the latter not as plane loci but as curves drawn on the surface of a cone.

He also shewed how these curves could be used in either of the two following ways to give a solution of the problem to duplicate a cube. In the first of these, he pointed out that two parabolas having a common vertex, axes at right angles, and such that the latus rectum of the one is double that of the other will intersect in another point whose abscissa (or ordinate) will give a solution; for (using analysis) if the equations of the parabolas be $y^2 = 2ax$ and $x^2 = ay$, they intersect in a point whose abscissa is given by $x^3 = 2a^3$. It is probable that this method was suggested by the form in which Hippocrates had cast the problem; namely, to find x and y so that $a : x = x : y = y : 2a$, whence we have $x^2 = ay$ and $y^2 = 2ax$.

The second solution given by Menaechmus was as follows. Describe a parabola of latus rectum l. Next describe a rectangular hyperbola, the length of whose real axis is $4l$, and having for its asymptotes the tangent at the vertex of the parabola and the axis of the parabola. Then the ordinate and the abscissa of the point of intersection of these curves are the mean proportionals between l and $2l$. This is at once obvious by analysis. The curves are $x^2 = ly$ and $xy = 2l^2$. These cut in a point determined by $x^3 = 2l^3$ and $y^3 = 4l^3$. Hence $l : x = x : y = y : 2l$.

Aristaeus and **Theaetetus**.   Of the other members of these schools, *Aristaeus* and *Theaetetus*, whose works are entirely lost, were mathematicians of repute. We know that Aristaeus wrote on the five regular solids and on conic sections, and that Theaetetus developed the theory of incommensurable magnitudes. The only theorem we can now definitely ascribe to the latter is that given by Euclid in the ninth proposition of the tenth book of the *Elements*, namely, that the squares on two commensurable right lines have one to the other a ratio which a square number has to a square number (and conversely); but the squares on two incommensurable right lines have one to the other a ratio which cannot be expressed as that of a square number to a square number (and conversely). This theorem includes the results given by Theodorus.[1]

The contemporaries or successors of these mathematicians wrote some fresh text-books on the elements of geometry and the conic sections, introduced problems concerned with finding loci, and systematized the knowledge already acquired, but they originated no new methods of research.

Aristotle.   An account of the Athenian school would be incomplete if there were no mention of *Aristotle*, who was born at Stagira in Macedonia in 384 B.C. and died at Chalcis in Euboea in 322 B.C. Aris-

---

[1] See above, p. 24.

totle, however, deeply interested though he was in natural philosophy, was chiefly concerned with mathematics and mathematical physics as supplying illustrations of correct reasoning. A small book containing a few questions on mechanics which is sometimes attributed to him is of doubtful authority; but, though in all probability it is not his work, it is interesting, partly as shewing that the principles of mechanics were beginning to excite attention, and partly as containing the earliest known employment of letters to indicate magnitudes.

The most instructive parts of the book are the dynamical proof of the parallelogram of forces for the direction of the resultant, and the statement, in effect, that if α be a force, β the mass to which it is applied, γ the distance through which it is moved, and δ the time of the motion, then α will move $\frac{1}{2}\beta$ through 2γ in the time δ, or through γ in the time $\frac{1}{2}\delta$: but the author goes on to say that it does not follow that $\frac{1}{2}\alpha$ will move β through $\frac{1}{2}\gamma$ in the time δ, because $\frac{1}{2}\alpha$ may not be able to move β at all; for 100 men may drag a ship 100 yards, but it does not follow that one man can drag it one yard. The first part of this statement is correct and is equivalent to the statement that an impulse is proportional to the momentum produced, but the second part is wrong.

The author also states the fact that what is gained in power is lost in speed, and therefore that two weights which keep a [weightless] lever in equilibrium are inversely proportional to the arms of the lever; this, he says, is the explanation why it is easier to extract teeth with a pair of pincers than with the fingers. Among other questions raised, but not answered, are why a projectile should ever stop, and why carriages with large wheels are easier to move than those with small.

I ought to add that the book contains some gross blunders, and as a whole is not as able or suggestive as might be inferred from the above extracts. In fact, here as elsewhere, the Greeks did not sufficiently realise that the fundamental facts on which the mathematical treatment of mechanics must be based can be established only by carefully devised observations and experiments.

CHAPTER IV.

THE FIRST ALEXANDRIAN SCHOOL.[1]
CIRC. 300 B.C.–30 B.C.

THE earliest attempt to found a university, as we understand the word, was made at Alexandria. Richly endowed, supplied with lecture rooms, libraries, museums, laboratories, and gardens, it became at once the intellectual metropolis of the Greek race, and remained so for a thousand years. It was particularly fortunate in producing within the first century of its existence three of the greatest mathematicians of antiquity—Euclid, Archimedes, and Apollonius. They laid down the lines on which mathematics subsequently developed, and treated it as a subject distinct from philosophy: hence the foundation of the Alexandrian Schools is rightly taken as the commencement of a new era. Thenceforward, until the destruction of the city by the Arabs in 641 A.D., the history of mathematics centres more or less round that of Alexandria; for this reason the Alexandrian Schools are commonly taken to include all Greek mathematicians of their time.

The city and university of Alexandria were created under the following circumstances. Alexander the Great had ascended the throne of Macedonia in 336 B.C. at the early age of twenty, and by 332 B.C. he had conquered or subdued Greece, Asia Minor, and Egypt. Following

---

[1] The history of the Alexandrian Schools is discussed by G. Loria in his *Le Scienze Esatte nell' Antica Grecia*, Modena, 1893–1900; by Cantor, chaps. xii–xxiii; and by J. Gow in his *History of Greek Mathematics*, Cambridge, 1884. The subject of Greek algebra is treated by E. H. F. Nesselmann in his *Die Algebra der Griechen*, Berlin, 1842; see also L. Matthiessen, *Grundzüge der antiken und modernen Algebra der litteralen Gleichungen*, Leipzig, 1878. The Greek treatment of the conic sections forms the subject of *Die Lehre von den Kegelschnitten in Altertum*, by H. G. Zeuthen, Copenhagen, 1886. The materials for the history of these schools have been subjected to a searching criticism by S. P. Tannery, and most of his papers are collected in his *Géométrie Grecque*, Paris, 1887.

the plan he adopted whenever a commanding site had been left unoc-
cupied, he founded a new city on the Mediterranean near one mouth
of the Nile; and he himself sketched out the ground-plan, and arranged
for drafts of Greeks, Egyptians, and Jews to be sent to occupy it. The
city was intended to be the most magnificent in the world, and, the
better to secure this, its erection was left in the hands of Dinocrates,
the architect of the temple of Diana at Ephesus.

After Alexander's death in 323 B.C. his empire was divided, and
Egypt fell to the lot of Ptolemy, who chose Alexandria as the capital
of his kingdom. A short period of confusion followed, but as soon as
Ptolemy was settled on the throne, say about 306 B.C., he determined
to attract, so far as he was able, learned men of all sorts to his new city;
and he at once began the erection of the university buildings on a piece
of ground immediately adjoining his palace. The university was ready
to be opened somewhere about 300 B.C., and Ptolemy, who wished to
secure for its staff the most eminent philosophers of the time, naturally
turned to Athens to find them. The great library which was the central
feature of the scheme was placed under Demetrius Phalereus, a distinguished Athenian, and so rapidly did it grow that within forty years
it (together with the Egyptian annexe) possessed about 600,000 rolls.
The mathematical department was placed under Euclid, who was thus
the first, as he was one of the most famous, of the mathematicians of
the Alexandrian school.

It happens that contemporaneously with the foundation of this
school the information on which our history is based becomes more
ample and certain. Many of the works of the Alexandrian mathematicians are still extant; and we have besides an invaluable treatise by
Pappus, described below, in which their best-known treatises are col-
lated, discussed, and criticized. It curiously turns out that just as we
begin to be able to speak with confidence on the subject-matter which
was taught, we find that our information as to the personality of the
teachers becomes vague; and we know very little of the lives of the
mathematicians mentioned in this and the next chapter, even the dates
at which they lived being frequently in doubt.

The third century before Christ.

Euclid.[1]—This century produced three of the greatest mathematicians of antiquity, namely Euclid, Archimedes, and Apollonius. The earliest of these was *Euclid*. Of his life we know next to nothing, save that he was of Greek descent, and was born about 330 B.C.; he died about 275 B.C. It would appear that he was well acquainted with the Platonic geometry, but he does not seem to have read Aristotle's works; and these facts are supposed to strengthen the tradition that he was educated at Athens. Whatever may have been his previous training and career, he proved a most successful teacher when settled at Alexandria. He impressed his own individuality on the teaching of the new university to such an extent that to his successors and almost to his contemporaries the name Euclid meant (as it does to us) the book or books he wrote, and not the man himself. Some of the medieval writers went so far as to deny his existence, and with the ingenuity of philologists they explained that the term was only a corruption of ὑκλι a key, and δις geometry. The former word was presumably derived from κλείς. I can only explain the meaning assigned to δις by the conjecture that as the Pythagoreans said that the number two symbolized a line, possibly a schoolman may have thought that it could be taken as indicative of geometry.

From the meagre notices of Euclid which have come down to us we find that the saying that there is no royal road in geometry was attributed to Euclid as well as to Menaechmus; but it is an epigrammatic remark which has had many imitators. According to tradition, Euclid was noticeable for his gentleness and modesty. Of his teaching, an anecdote has been preserved. Stobaeus, who is a somewhat doubtful authority, tells us that, when a lad who had just begun geometry asked, "What do I gain by learning all this stuff?" Euclid insisted that knowledge was worth acquiring for its own sake, but made his slave give the boy some coppers, "since," said he, "he must make a profit out of what he learns."

---

[1]Besides Loria, book ii, chap. i; Cantor, chaps. xii, xiii; and Gow, pp. 72–86, 195–221; see the articles *Eucleides* by A. De Morgan in Smith's *Dictionary of Greek and Roman Biography*, London, 1849; the article on *Irrational Quantity* by A. De Morgan in the *Penny Cyclopaedia*, London, 1839; *Litterargeschichtliche Studien über Euklid*, by J. L. Heiberg, Leipzig, 1882; and above all *Euclid's Elements*, translated with an introduction and commentary by T. L. Heath, 3 volumes, Cambridge, 1908. The latest complete edition of all Euclid's works is that by J. L. Heiberg and H. Menge, Leipzig, 1883–96.

Euclid was the author of several works, but his reputation rests mainly on his *Elements*. This treatise contains a systematic exposition of the leading propositions of elementary metrical geometry (exclusive of conic sections) and of the theory of numbers. It was at once adopted by the Greeks as the standard text-book on the elements of pure mathematics, and it is probable that it was written for that purpose and not as a philosophical attempt to shew that the results of geometry and arithmetic are necessary truths.

The modern text[1] is founded on an edition or commentary prepared by Theon, the father of Hypatia (circ. 380 A.D.). There is at the Vatican a copy (circ. 1000 A.D.) of an older text, and we have besides quotations from the work and references to it by numerous writers of various dates. From these sources we gather that the definitions, axioms, and postulates were rearranged and slightly altered by subsequent editors, but that the propositions themselves are substantially as Euclid wrote them.

As to the matter of the work. The geometrical part is to a large extent a compilation from the works of previous writers. Thus the substance of books I and II (except perhaps the treatment of parallels) is probably due to Pythagoras; that of book III to Hippocrates; that of book V to Eudoxus; and the bulk of books IV, VI, XI, and XII to the later Pythagorean or Athenian schools. But this material was rearranged, obvious deductions were omitted (for instance, the proposition that the perpendiculars from the angular points of a triangle on the opposite sides meet in a point was cut out), and in some cases new proofs substituted. Book X, which deals with irrational magnitudes, may be founded on the lost book of Theaetetus; but probably much of it is original, for Proclus says that while Euclid arranged the propositions of Eudoxus he completed many of those of Theaetetus. The whole was presented as a complete and consistent body of theorems.

The form in which the propositions are presented, consisting of enunciation, statement, construction, proof, and conclusion, is due to Euclid: so also is the synthetical character of the work, each proof being written out as a logically correct train of reasoning but without any clue to the method by which it was obtained.

---

[1]Most of the modern text-books in English are founded on Simson's edition, issued in 1758. *Robert Simson*, who was born in 1687 and died in 1768, was professor of mathematics at the University of Glasgow, and left several valuable works on ancient geometry.

The defects of Euclid's *Elements* as a text-book of geometry have been often stated; the most prominent are these. (i) The definitions and axioms contain many assumptions which are not obvious, and in particular the postulate or axiom about parallel lines is not self-evident.[1] (ii) No explanation is given as to the reason why the proofs take the form in which they are presented, that is, the synthetical proof is given but not the analysis by which it was obtained. (iii) There is no attempt made to generalize the results arrived at; for instance, the idea of an angle is never extended so as to cover the case where it is equal to or greater than two right angles: the second half of the thirty-third proposition in the sixth book, as now printed, appears to be an exception, but it is due to Theon and not to Euclid. (iv) The principle of superposition as a method of proof might be used more frequently with advantage. (v) The classification is imperfect. And (vi) the work is unnecessarily long and verbose. Some of those objections do not apply to certain of the recent school editions of the *Elements*.

On the other hand, the propositions in Euclid are arranged so as to form a chain of geometrical reasoning, proceeding from certain almost obvious assumptions by easy steps to results of considerable complexity. The demonstrations are rigorous, often elegant, and not too difficult for a beginner. Lastly, nearly all the elementary metrical (as opposed to the graphical) properties of space are investigated, while the fact that for two thousand years it was the usual text-book on the subject raises a strong presumption that it is not unsuitable for the purpose.

On the Continent rather more than a century ago, Euclid was generally superseded by other text-books. In England determined efforts have lately been made with the same purpose, and numerous other works on elementary geometry have been produced in the last decade. The change is too recent to enable us to say definitely what its effect may be. But as far as I can judge, boys who have learnt their geometry on the new system know more facts, but have missed the mental and logical training which was inseparable from a judicious study of Euclid's treatise.

I do not think that all the objections above stated can fairly be urged against Euclid himself. He published a collection of problems, generally known as the $\Delta\epsilon\delta o\mu\acute{\epsilon}\nu a$ or *Data*. This contains 95 illustrations of the kind of deductions which frequently have to be made in analysis;

---

[1] We know, from the researches of Lobatschewsky and Riemann, that it is incapable of proof.

such as that, if one of the data of the problem under consideration be that one angle of some triangle in the figure is constant, then it is legitimate to conclude that the ratio of the area of the rectangle under the sides containing the angle to the area of the triangle is known [prop. 66]. Pappus says that the work was written for those "who wish to acquire the power of solving problems." It is in fact a gradual series of exercises in geometrical analysis. In short the *Elements* gave the principal results, and were intended to serve as a training in the science of reasoning, while the *Data* were intended to develop originality.

Euclid also wrote a work called Περὶ Διαιρέσεων or *De Divisionibus*, known to us only through an Arabic translation which may be itself imperfect.[1] This is a collection of 36 problems on the division of areas into parts which bear to one another a given ratio. It is not unlikely that this was only one of several such collections of examples—possibly including the *Fallacies* and the *Porisms*—but even by itself it shews that the value of exercises and riders was fully recognized by Euclid.

I may here add a suggestion made by De Morgan, whose comments on Euclid's writings were notably ingenious and informing. From internal evidence he thought it likely that the *Elements* were written towards the close of Euclid's life, and that their present form represents the first draft of the proposed work, which, with the exception of the tenth book, Euclid did not live to revise. This opinion is generally discredited, and there is no extrinsic evidence to support it.

The geometrical parts of the *Elements* are so well known that I need do no more than allude to them. Euclid admitted only those constructions which could be made by the use of a ruler and compasses.[2] He also excluded practical work and hypothetical constructions. The first four books and book VI deal with plane geometry; the theory of proportion (of any magnitudes) is discussed in book V; and books XI and XII treat of solid geometry. On the hypothesis that the *Elements* are the first draft of Euclid's proposed work, it is possible that book XIII

---

[1] R. C. Archibald, *Euclid's Book on Divisions*, Cambridge, 1915.

[2] The ruler must be of unlimited length and not graduated; the compasses also must be capable of being opened as wide as is desired. *Lorenzo Mascheroni* (who was born at Castagneta on May 14, 1750, and died at Paris on July 30, 1800) set himself the task to obtain by means of constructions made only with a pair of compasses as many Euclidean results as possible. Mascheroni's treatise on the geometry of the compass, which was published at Pavia in 1795, is a curious *tour de force*: he was professor first at Bergamo and afterwards at Pavia, and left numerous minor works. Similar limitations have been proposed by other writers.

is a sort of appendix containing some additional propositions which would have been put ultimately in one or other of the earlier books. Thus, as mentioned above, the first five propositions which deal with a line cut in golden section might be added to the second book. The next seven propositions are concerned with the relations between certain incommensurable lines in plane figures (such as the radius of a circle and the sides of an inscribed regular triangle, pentagon, hexagon, and decagon) which are treated by the methods of the tenth book and as an illustration of them. Constructions of the five regular solids are discussed in the last six propositions, and it seems probable that Euclid and his contemporaries attached great importance to this group of problems. Bretschneider inclined to think that the thirteenth book is a summary of part of the lost work of Aristaeus: but the illustrations of the methods of the tenth book are due most probably to Theaetetus.

Books VII, VIII, IX, and X of the *Elements* are given up to the theory of numbers. The mere art of calculation or λογιστική was taught to boys when quite young, it was stigmatized by Plato as childish, and never received much attention from Greek mathematicians; nor was it regarded as forming part of a course of mathematics. We do not know how it was taught, but the abacus certainly played a prominent part in it. The scientific treatment of numbers was called ἀριθμητική, which I have here generally translated as the science of numbers. It had special reference to ratio, proportion, and the theory of numbers. It is with this alone that most of the extant Greek works deal.

In discussing Euclid's arrangement of the subject, we must therefore bear in mind that those who attended his lectures were already familiar with the art of calculation. The system of numeration adopted by the Greeks is described later,[1] but it was so clumsy that it rendered the scientific treatment of numbers much more difficult than that of geometry; hence Euclid commenced his mathematical course with plane geometry. At the same time it must be observed that the results of the second book, though geometrical in form, are capable of expression in algebraical language, and the fact that numbers could be represented by lines was probably insisted on at an early stage, and illustrated by concrete examples. This graphical method of using lines to represent numbers possesses the obvious advantage of leading to proofs which are true for all numbers, rational or irrational. It will be noticed that among other propositions in the second book we get geometrical proofs

[1] See below, chapter vii.

of the distributive and commutative laws, of rules for multiplication, and finally geometrical solutions of the equations $a(a-x) = x^2$, that is $x^2 + ax - a^2 = 0$ (Euc. II, 11), and $x^2 - ab = 0$ (Euc. II, 14): the solution of the first of these equations is given in the form $\sqrt{a^2 + (\frac{1}{2}a)^2} - \frac{1}{2}a$. The solutions of the equations $ax^2 - bx + c = 0$ and $ax^2 + bx - c = 0$ are given later in Euc. VI, 28 and VI, 29; the cases when $a = 1$ can be deduced from the identities proved in Euc. II, 5 and 6, but it is doubtful if Euclid recognized this.

The results of the fifth book, in which the theory of proportion is considered, apply to any magnitudes, and therefore are true of numbers as well as of geometrical magnitudes. In the opinion of many writers this is the most satisfactory way of treating the theory of proportion on a scientific basis; and it was used by Euclid as the foundation on which he built the theory of numbers. The theory of proportion given in this book is believed to be due to Eudoxus. The treatment of the same subject in the seventh book is less elegant, and is supposed to be a reproduction of the Pythagorean teaching. This double discussion of proportion is, as far as it goes, in favour of the conjecture that Euclid did not live to revise the work.

In books VII, VIII, and IX Euclid discusses the theory of rational numbers. He commences the seventh book with some definitions founded on the Pythagorean notation. In propositions 1 to 3 he shews that if, in the usual process for finding the greatest common measure of two numbers, the last divisor be unity, the numbers must be prime; and he thence deduces the rule for finding their G.C.M. Propositions 4 to 22 include the theory of fractions, which he bases on the theory of proportion; among other results he shews that $ab = ba$ [prop. 16]. In propositions 23 to 34 he treats of prime numbers, giving many of the theorems in modern text-books on algebra. In propositions 35 to 41 he discusses the least common multiple of numbers, and some miscellaneous problems.

The eighth book is chiefly devoted to numbers in continued proportion, that is, in a geometrical progression; and the cases where one or more is a product, square, or cube are specially considered.

In the ninth book Euclid continues the discussion of geometrical progressions, and in proposition 35 he enunciates the rule for the summation of a series of n terms, though the proof is given only for the case where n is equal to 4. He also develops the theory of primes, shews that the number of primes is infinite [prop. 20], and discusses the properties of odd and even numbers. He concludes by shewing that a number of the form $2^{n-1}(2^n - 1)$, where $2^n - 1$ is a prime, is a "perfect" number [prop. 36].

In the tenth book Euclid deals with certain irrational magnitudes; and, since the Greeks possessed no symbolism for surds, he was forced to adopt a geometrical representation. Propositions 1 to 21 deal generally with incommensurable magnitudes. The rest of the book, namely, propositions 22 to 117, is devoted to the discussion of every possible variety of lines which can be represented by $\sqrt{(\sqrt{a} \pm \sqrt{b})}$, where a and b denote commensurable lines. There are twenty-five species of such lines, and that Euclid could detect and classify them all is in the opinion of so competent an authority as Nesselmann the most striking illustration of his genius. No further advance in the theory of incommensurable magnitudes was made until the subject was taken up by Leonardo and Cardan after an interval of more than a thousand years.

In the last proposition of the tenth book [prop. 117] the side and diagonal of a square are proved to be incommensurable. The proof is so short and easy that I may quote it. If possible let the side be to the diagonal in a commensurable ratio, namely, that of two integers, a and b. Suppose this ratio reduced to its lowest terms so that a and b have no common divisor other than unity, that is, they are prime to one another. Then (by Euc. I, 47) $b^2 = 2a^2$; therefore b^2 is an even number; therefore b is an even number; hence, since a is prime to b, a must be an odd number. Again, since it has been shewn that b is an even number, b may be represented by $2n$; therefore $(2n)^2 = 2a^2$; therefore $a^2 = 2n^2$; therefore a^2 is an even number; therefore a is an even number. Thus the same number a must be both odd and even, which is absurd; therefore the side and diagonal are incommensurable. Hankel believes that this proof was due to Pythagoras, and this is not unlikely. This proposition is also proved in another way in Euc. X, 9, and for this and other reasons it is now usually believed to be an interpolation by some commentator on the *Elements*.

In addition to the *Elements* and the two collections of riders above mentioned (which are extant) Euclid wrote the following books on geometry: (i) an elementary treatise on *conic sections* in four books; (ii) a book on *surface loci*, probably confined to curves on the cone and cylinder; (iii) a collection of *geometrical fallacies*, which were to be used as exercises in the detection of errors; and (iv) a treatise on *porisms* arranged in three books. All of these are lost, but the work on porisms was discussed at such length by Pappus, that some writers have thought it possible to restore it. In particular, Chasles in 1860 published what he considered to be substantially a reproduction of it. In this will be found the conceptions of cross ratios and projection, and those ideas of modern geometry which were used so extensively by Chasles and other writers of the nineteenth century. It should be realized, however, that the statements of the classical writers concerning this book are either very brief or have come to us only in a mutilated form, and De Morgan frankly says that he found them unintelligible, an opinion in which most of those who read them will, I think, concur.

Euclid published a book on optics, treated geometrically, which contains 61 propositions founded on 12 assumptions. It commences with the assumption that objects are seen by rays emitted from the eye in straight lines, "for if light proceeded from the object we should not, as we often do, fail to perceive a needle on the floor." A work called *Catoptrica* is also attributed to him by some of the older writers; the text is corrupt and the authorship doubtful; it consists of 31 propositions dealing with reflexions in plane, convex, and concave mirrors. The geometry of both books is Euclidean in form.

Euclid has been credited with an ingenious demonstration[1] of the principle of the lever, but its authenticity is doubtful. He also wrote the *Phaenomena*, a treatise on geometrical astronomy. It contains references to the work of Autolycus[2] and to some book on spherical geometry by an unknown writer. Pappus asserts that Euclid also composed a book on the elements of music: this may refer to the *Sectio Canonis*, which is by Euclid, and deals with musical intervals.

To these works I may add the following little problem, which occurs in the Palatine Anthology and is attributed by tradition to Euclid. "A mule and a donkey were going to market laden with wheat. The mule said, 'If you gave me one measure I should carry twice as much as you, but if I gave you one we should bear equal burdens.' Tell me, learned geometrician, what were their burdens." It is impossible to say whether the question is due to Euclid, but there is nothing improbable in the suggestion.

It will be noticed that Euclid dealt only with magnitudes, and did

---

[1] It is given (from the Arabic) by F. Woepcke in the *Journal Asiatique*, series 4, vol. xviii, October 1851, pp. 225–232.

[2] *Autolycus* lived at Pitane in Aeolis and flourished about 330 B.C. His two works on astronomy, containing 43 propositions, are said to be the oldest extant Greek mathematical treatises. They exist in manuscript at Oxford. They were edited, with a Latin translation, by F. Hultsch, Leipzig, 1885.

not concern himself with their numerical measures, but it would seem from the works of Aristarchus and Archimedes that this was not the case with all the Greek mathematicians of that time. As one of the works of the former is extant it will serve as another illustration of Greek mathematics of this period.

Aristarchus. *Aristarchus* of Samos, born in 310 B.C. and died in 250 B.C., was an astronomer rather than a mathematician. He asserted, at any rate as a working hypothesis, that the sun was the centre of the universe, and that the earth revolved round the sun. This view, in spite of the simple explanation it afforded of various phenomena, was generally rejected by his contemporaries. But his propositions[1] on the measurement of the sizes and distances of the sun and moon were accurate in principle, and his results were accepted by Archimedes in his Ψαμμίτης, mentioned below, as approximately correct. There are 19 theorems, of which I select the seventh as a typical illustration, because it shews the way in which the Greeks evaded the difficulty of finding the numerical value of surds.

Aristarchus observed the angular distance between the moon when dichotomized and the sun, and found it to be twenty-nine thirtieths of a right angle. It is actually about 89°21′, but of course his instruments were of the roughest description. He then proceeded to shew that the distance of the sun is greater than eighteen and less than twenty times the distance of the moon in the following manner.

Let S be the sun, E the earth, and M the moon. Then when the moon is dichotomized, that is, when the bright part which we see is exactly a half-circle, the angle between MS and ME is a right angle. With E as centre, and radii ES and EM describe circles, as in the figure below. Draw EA perpendicular to ES. Draw EF bisecting the angle AES, and EG bisecting the angle AEF, as in the figure. Let EM (produced) cut AF in H. The angle AEM is by hypothesis $\frac{1}{30}$th of a right angle. Hence we have

$$\text{angle } AEG : \text{angle } AEH = \tfrac{1}{4} \text{ rt. } \angle : \tfrac{1}{30} \text{ rt. } \angle = 15 : 2,$$
$$\therefore AG : AH \ [= \tan AEG : \tan AEH] \ > 15 : 2. \qquad (\alpha)$$

---

[1] Περὶ μεγέθων καὶ ἀποστημάτων Ἡλίου καὶ Σελήνης, edited by E. Nizze, Stralsund, 1856. Latin translations were issued by F. Commandino in 1572 and by J. Wallis in 1688; and a French translation was published by F. d'Urban in 1810 and 1823.

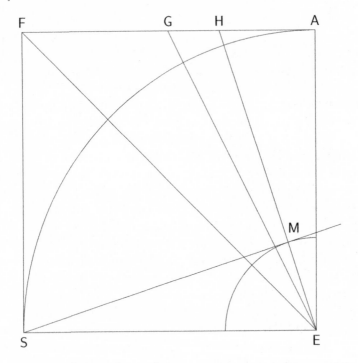

Again $FG^2 : AG^2 = EF^2 : EA^2$ (Euc. VI, 3) $= 2 : 1$ (Euc. I, 47),

$$\therefore FG^2 : AG^2 > 49 : 25,$$
$$\therefore FG : AG > 7 : 5,$$
$$\therefore AF : AG > 12 : 5,$$
$$\therefore AE : AG > 12 : 5. \qquad (\beta)$$

Compounding the ratios (α) and (β), we have

$$AE : AH > 18 : 1.$$

But the triangles EMS and EAH are similar,

$$\therefore ES : EM > 18 : 1.$$

I will leave the second half of the proposition to amuse any reader who may care to prove it: the analysis is straightforward. In a somewhat similar way Aristarchus found the ratio of the radii of the sun, earth, and moon.

We know very little of **Conon** and **Dositheus**, the immediate successors of Euclid at Alexandria, or of their contemporaries **Zeuxippus** and **Nicoteles**, who most likely also lectured there, except that Archimedes, who was a student at Alexandria probably shortly after Euclid's death, had a high opinion of their ability and corresponded with the three first mentioned. Their work and reputation has been completely overshadowed by that of Archimedes.

Archimedes.[1]   *Archimedes*, who probably was related to the royal family at Syracuse, was born there in 287 B.C. and died in 212 B.C. He went to the university of Alexandria and attended the lectures of Conon, but, as soon as he had finished his studies, returned to Sicily where he passed the remainder of his life. He took no part in public affairs, but his mechanical ingenuity was astonishing, and, on any difficulties which could be overcome by material means arising, his advice was generally asked by the government.

Archimedes, like Plato, held that it was undesirable for a philosopher to seek to apply the results of science to any practical use; but in fact he did introduce a large number of new inventions. The stories of the detection of the fraudulent goldsmith and of the use of burning-glasses to destroy the ships of the Roman blockading squadron will recur to most readers. Perhaps it is not as well known that Hiero, who had built a ship so large that he could not launch it off the slips, applied to Archimedes. The difficulty was overcome by means of an apparatus of cogwheels worked by an endless screw, but we are not told exactly how the machine was used. It is said that it was on this occasion, in acknowledging the compliments of Hiero, that Archimedes made the well-known remark that had he but a fixed fulcrum he could move the earth.

Most mathematicians are aware that the Archimedean screw was another of his inventions. It consists of a tube, open at both ends, and bent into the form of a spiral like a corkscrew. If one end be immersed in water, and the axis of the instrument (*i.e.* the axis of the cylinder on the surface of which the tube lies) be inclined to the vertical at a sufficiently big angle, and the instrument turned round it, the water

[1]Besides Loria, book ii, chap. iii, Cantor, chaps. xiv, xv, and Gow, pp. 221–244, see *Quaestiones Archimedeae*, by J. L. Heiberg, Copenhagen, 1879; and Marie, vol. i, pp. 81–134. The best editions of the extant works of Archimedes are those by J. L. Heiberg, in 3 vols., Leipzig, 1880–81, and by Sir Thomas L. Heath, Cambridge, 1897. In 1906 a manuscript, previously unknown, was discovered at Constantinople, containing propositions on hydrostatics and on methods; see *Eine neue Schrift des Archimedes*, by J. L. Heiberg and H. G. Zeuthen, Leipzig, 1907, and the *Method of Archimedes*, by Sir Thomas L. Heath, Cambridge, 1912.

will flow along the tube and out at the other end. In order that it may work, the inclination of the axis of the instrument to the vertical must be greater than the pitch of the screw. It was used in Egypt to drain the fields after an inundation of the Nile, and was also frequently applied to take water out of the hold of a ship.

The story that Archimedes set fire to the Roman ships by means of burning-glasses and concave mirrors is not mentioned till some centuries after his death, and is generally rejected. The mirror of Archimedes is said to have been made in the form of a hexagon surrounded by rings of polygons; and Buffon[1] in 1747 contrived, by the use of a single composite mirror made on this model, to set fire to wood at a distance of 150 feet, and to melt lead at a distance of 140 feet. This was in April and as far north as Paris, so in a Sicilian summer the use of several such mirrors might be a serious annoyance to a blockading fleet, if the ships were sufficiently near. It is perhaps worth mentioning that a similar device is said to have been used in the defence of Constantinople in 514 A.D., and is alluded to by writers who either were present at the siege or obtained their information from those who were engaged in it.

But whatever be the truth as to this story, there is no doubt that Archimedes devised the catapults which kept the Romans, who were then besieging Syracuse, at bay for a considerable time. These were constructed so that the range could be made either short or long at pleasure, and so that they could be discharged through a small loophole without exposing the artillery-men to the fire of the enemy. So effective did they prove that the siege was turned into a blockade, and three years elapsed before the town was taken.

Archimedes was killed during the sack of the city which followed its capture, in spite of the orders, given by the consul Marcellus who was in command of the Romans, that his house and life should be spared. It is said that a soldier entered his study while he was regarding a geometrical diagram drawn in sand on the floor, which was the usual way of drawing figures in classical times. Archimedes told him to get off the diagram, and not spoil it. The soldier, feeling insulted at having orders given to him and ignorant of who the old man was, killed him. According to another and more probable account, the cupidity of the troops was excited by seeing his instruments, constructed of polished brass which they supposed to be made of gold.

---

[1]See *Mémoires de l'académie royale des sciences* for 1747, Paris, 1752, pp. 82–101.

The Romans erected a splendid tomb to Archimedes, on which was engraved (in accordance with a wish he had expressed) the figure of a sphere inscribed in a cylinder, in commemoration of the proof he had given that the volume of a sphere was equal to two-thirds that of the circumscribing right cylinder, and its surface to four times the area of a great circle. Cicero[1] gives a charming account of his efforts (which were successful) to rediscover the tomb in 75 B.C.

It is difficult to explain in a concise form the works or discoveries of Archimedes, partly because he wrote on nearly all the mathematical subjects then known, and partly because his writings are contained in a series of disconnected monographs. Thus, while Euclid aimed at producing systematic treatises which could be understood by all students who had attained a certain level of education, Archimedes wrote a number of brilliant essays addressed chiefly to the most educated mathematicians of the day. The work for which he is perhaps now best known is his treatment of the mechanics of solids and fluids; but he and his contemporaries esteemed his geometrical discoveries of the quadrature of a parabolic area and of a spherical surface, and his rule for finding the volume of a sphere as more remarkable; while at a somewhat later time his numerous mechanical inventions excited most attention.

(i) On *plane geometry* the extant works of Archimedes are three in number, namely, (a) the *Measure of the Circle*, (b) the *Quadrature of the Parabola*, and (c) one on *Spirals*.

(a) The *Measure of the Circle* contains three propositions. In the first proposition Archimedes proves that the area is the same as that of a right-angled triangle whose sides are equal respectively to the radius a and the circumference of the circle, *i.e.* the area is equal to $\frac{1}{2}a(2\pi a)$. In the second proposition he shows that $\pi a^2 : (2a)^2 = 11 : 14$ very nearly; and next, in the third proposition, that π is less than $3\frac{1}{7}$ and greater than $3\frac{10}{71}$. These theorems are of course proved geometrically. To demonstrate the two latter propositions, he inscribes in and circumscribes about a circle regular polygons of ninety-six sides, calculates their perimeters, and then assumes the circumference of the circle to lie between them: this leads to the result $6336/2017\frac{1}{4} < \pi < 14688/4673\frac{1}{2}$, from which he deduces the limits given above. It would seem from the proof that he had some (at present unknown) method of extracting the square roots of numbers approximately. The table which he formed of the numerical values of the chords of a circle is essentially a table of

[1] See his *Tusculanarum Disputationum*, v. 23.

natural sines, and may have suggested the subsequent work on these lines of Hipparchus and Ptolemy.

(*b*) The *Quadrature of the Parabola* contains twenty-four propositions. Archimedes begins this work, which was sent to Dositheus, by establishing some properties of conics [props. 1–5]. He then states correctly the area cut off from a parabola by any chord, and gives a proof which rests on a preliminary mechanical experiment of the ratio of areas which balance when suspended from the arms of a lever [props. 6–17]; and, lastly, he gives a geometrical demonstration of this result [props. 18–24]. The latter is, of course, based on the method of exhaustions, but for brevity I will, in quoting it, use the method of limits.

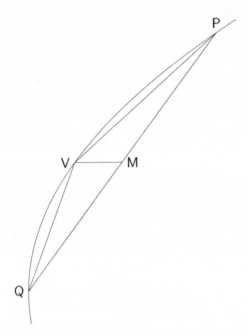

Let the area of the parabola (see figure above) be bounded by the chord PQ. Draw VM the diameter to the chord PQ, then (by a previous proposition), V is more remote from PQ than any other point in the arc PVQ. Let the area of the triangle PVQ be denoted by $\triangle$. In the segments bounded by VP and VQ inscribe triangles in the same way as the triangle PVQ was inscribed in the given segment. Each of these triangles is (by a previous proposition of his) equal to $\frac{1}{8}\triangle$, and their sum is therefore $\frac{1}{4}\triangle$. Similarly in the four segments left inscribe triangles; their sum will be $\frac{1}{16}\triangle$. Proceeding in this way the area of the given segment is shown to be equal to the limit of

$$\triangle + \frac{\triangle}{4} + \frac{\triangle}{16} + \cdots + \frac{\triangle}{4^n} + \cdots,$$

when n is indefinitely large.

The problem is therefore reduced to finding the sum of a geometrical series. This he effects as follows. Let $A, B, C, \ldots, J, K$ be a series of magnitudes such that each is one-fourth of that which precedes it. Take magnitudes $b, c, \ldots, k$ equal respectively to $\frac{1}{3}B, \frac{1}{3}C, \ldots, \frac{1}{3}K$. Then

$$B + b = \tfrac{1}{3}A, \quad C + c = \tfrac{1}{3}B, \quad \ldots, \quad K + k = \tfrac{1}{3}J.$$

Hence $(B + C + \ldots + K) + (b + c + \ldots + k) = \frac{1}{3}(A + B + \ldots + J)$; but, by hypothesis, $(b + c + \ldots + j + k) = \frac{1}{3}(B + C + \ldots + J) + \frac{1}{3}K$;

$$\therefore (B + C + \ldots + K) + \tfrac{1}{3}K = \tfrac{1}{3}A.$$

$$\therefore A + B + C + \ldots + K = \tfrac{4}{3}A - \tfrac{1}{3}K.$$

Hence the sum of these magnitudes exceeds four times the third of the largest of them by one-third of the smallest of them.

Returning now to the problem of the quadrature of the parabola A stands for $\triangle$, and ultimately K is indefinitely small; therefore the area of the parabolic segment is four-thirds that of the triangle PVQ, or two-thirds that of a rectangle whose base is PQ and altitude the distance of V from PQ.

While discussing the question of quadratures it may be added that in the fifth and sixth propositions of his work on conoids and spheroids he determined the area of an ellipse.

(c) The work on *Spirals* contains twenty-eight propositions on the properties of the curve now known as the spiral of Archimedes. It was sent to Dositheus at Alexandria accompanied by a letter, from which it appears that Archimedes had previously sent a note of his results to Conon, who had died before he had been able to prove them. The spiral is defined by saying that the vectorial angle and radius vector both increase uniformly, hence its equation is $r = c\theta$. Archimedes finds most of its properties, and determines the area inclosed between the curve and two radii vectores. This he does (in effect) by saying, in the language of the infinitesimal calculus, that an element of area is $> \frac{1}{2}r^2 d\theta$ and $< \frac{1}{2}(r + dr)^2 d\theta$: to effect the sum of the elementary areas he gives two lemmas in which he sums (geometrically) the series $a^2 + (2a)^2 + (3a)^2 + \ldots + (na)^2$ [prop. 10], and $a + 2a + 3a + \ldots + na$ [prop. 11].

(d) In addition to these he wrote a small treatise on *geometrical methods*, and works on *parallel lines, triangles, the properties of right-angled triangles, data, the heptagon inscribed in a circle*, and *systems of circles touching one another*; possibly he wrote others too. These are all lost, but it is probable that fragments of four of the propositions in the last-mentioned work are preserved in a Latin translation from an Arabic manuscript entitled *Lemmas of Archimedes*.

(ii) On *geometry of three dimensions* the extant works of Archimedes are two in number, namely (a), the *Sphere and Cylinder*, and (b) *Conoids and Spheroids*.

(a) The *Sphere and Cylinder* contains sixty propositions arranged in two books. Archimedes sent this like so many of his works to Dositheus at Alexandria; but he seems to have played a practical joke on his friends there, for he purposely misstated some of his results "to deceive those vain geometricians who say they have found everything, but never give their proofs, and sometimes claim that they have discovered what is impossible." He regarded this work as his masterpiece. It is too long for me to give an analysis of its contents, but I remark in passing that in it he finds expressions for the surface and volume of a pyramid, of a cone, and of a sphere, as well as of the figures produced by the revolution of polygons inscribed in a circle about a diameter of the circle. There are several other propositions on areas and volumes of which perhaps the most striking is the tenth proposition of the second book, namely, that "of all spherical segments whose surfaces are equal the hemisphere has the greatest volume." In the second proposition of the second book he enunciates the remarkable theorem that a line of length a can be divided so that $a - x : b = 4a^2 : 9x^2$ (where b is a given length), only if b be less than $\frac{1}{3}a$; that is to say, the cubic equation $x^3 - ax^2 + \frac{4}{9}a^2b = 0$ can have a real and positive root only if a be greater than $3b$. This proposition was required to complete his solution of the problem to divide a given sphere by a plane so that the volumes of the segments should be in a given ratio. One very simple cubic equation occurs in the *Arithmetic* of Diophantus, but with that exception no such equation appears again in the history of European mathematics for more than a thousand years.

(b) The *Conoids and Spheroids* contains forty propositions on quadrics of revolution (sent to Dositheus in Alexandria) mostly concerned with an investigation of their volumes.

(c) Archimedes also wrote a treatise on certain *semi-regular polyhedrons*, that is, solids contained by regular but dissimilar polygons. This is lost, but references to it are given by Pappus.

(iii) On *arithmetic* Archimedes wrote two papers. One (addressed to Zeuxippus) was on the principles of numeration; this is now lost. The other (addressed to Gelon) was called $\Psi\alpha\mu\mu\dot{\iota}\tau\eta\varsigma$ (*the sand-reckoner*), and in this he meets an objection which had been urged against his first paper.

The object of the first paper had been to suggest a convenient system by which numbers of any magnitude could be represented; and it would seem that some philosophers at Syracuse had doubted whether the system was practicable. Archimedes says people talk of the sand on the Sicilian shore as something beyond the power of calculation, but he can estimate it; and, further, he will illustrate the power of his method by finding a superior limit to the number of grains of sand which would fill the whole universe, *i.e.* a sphere whose centre is the earth, and radius the distance of the sun. He begins by saying that in ordinary Greek nomenclature it was only possible to express numbers from 1 up to 10^8: these are expressed in what he says he may call units of the first order. If 10^8 be termed a unit of the second order, any number from 10^8 to 10^{16} can be expressed as so many units of the second order plus so many units of the first order. If 10^{16} be a unit of the third order any number up to 10^{24} can be then expressed, and so on. Assuming that 10,000 grains of sand occupy a sphere whose radius is not less than $\frac{1}{80}$th of a finger-breadth, and that the diameter of the universe is not greater than 10^{10} stadia, he finds that the number of grains of sand required to fill the solar universe is less than 10^{51}.

Probably this system of numeration was suggested merely as a scientific curiosity. The Greek system of numeration with which we are acquainted had been only recently introduced, most likely at Alexandria, and was sufficient for all the purposes for which the Greeks then required numbers; and Archimedes used that system in all his papers. On the other hand, it has been conjectured that Archimedes and Apollonius had some symbolism based on the decimal system for their own investigations, and it is possible that it was the one here sketched out. The units suggested by Archimedes form a geometrical progression, having 10^8 for the radix. He incidentally adds that it will be convenient to remember that the product of the mth and nth terms of a geometrical progression, whose first term is unity, is equal to the $(m + n)$th term of the series, that is, that $r^m \times r^n = r^{m+n}$.

To these two arithmetical papers I may add the following celebrated problem[1] which he sent to the Alexandrian mathematicians. The sun had a herd of bulls and cows, all of which were either white, grey, dun, or piebald: the number of piebald bulls was less than the number of white bulls by 5/6ths of the number of grey bulls, it was less than the number of grey bulls by 9/20ths of the number of dun bulls, and it was less than the number of dun bulls by 13/42nds of the number of white bulls; the number of white cows was 7/12ths of the number of grey cattle (bulls and cows), the number of grey cows was 9/20ths of the number of dun cattle, the number of dun cows was 11/30ths of the number of piebald cattle, and the number of piebald cows was 13/42nds of the number of white cattle. The problem was to find the composition of the herd. The problem is indeterminate, but the solution in lowest integers is

white bulls, ...	10,366,482;	white cows, ...	7,206,360;
grey bulls, ...	7,460,514;	grey cows, ...	4,893,246;
dun bulls, ...	7,358,060;	dun cows, ...	3,515,820;
piebald bulls, ...	4,149,387;	piebald cows, ...	5,439,213.

In the classical solution, attributed to Archimedes, these numbers are multiplied by 80.

Nesselmann believes, from internal evidence, that the problem has been falsely attributed to Archimedes. It certainly is unlike his extant work, but it was attributed to him among the ancients, and is generally thought to be genuine, though possibly it has come down to us in a modified form. It is in verse, and a later copyist has added the additional conditions that the sum of the white and grey bulls shall be a square number, and the sum of the piebald and dun bulls a triangular number.

It is perhaps worthy of note that in the enunciation the fractions are represented as a sum of fractions whose numerators are unity: thus Archimedes wrote 1/7+1/6 instead of 13/42, in the same way as Ahmes would have done.

(iv) On *mechanics* the extant works of Archimedes are two in number, namely, (a) his *Mechanics*, and (c) his *Hydrostatics*.

---

[1]See a memoir by B. Krumbiegel and A. Amthor, *Zeitschrift für Mathematik, Abhandlungen zur Geschichte der Mathematik*, Leipzig, vol. xxv, 1880, pp. 121–136, 153–171.

(*a*) The *Mechanics* is a work on statics with special reference to the equilibrium of plane laminas and to properties of their centres of gravity; it consists of twenty-five propositions in two books. In the first part of book I, most of the elementary properties of the centre of gravity are proved [props. 1–8]; and in the remainder of book I, [props. 9–15] and in book II the centres of gravity of a variety of plane areas, such as parallelograms, triangles, trapeziums, and parabolic areas are determined.

As an illustration of the influence of Archimedes on the history of mathematics, I may mention that the science of statics rested on his theory of the lever until 1586, when Stevinus published his treatise on statics.

His reasoning is sufficiently illustrated by an outline of his proof for the case of two weights, P and Q, placed at their centres of gravity, A and B, on a weightless bar AB. He wants to shew that the centre of gravity of P and Q is at a point O on the bar such that $P.OA = Q.OB$.

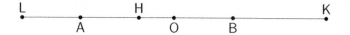

On the line AB (produced if necessary) take points H and K, so that $HB = BK = AO$; and a point L so that $LA = OB$. It follows that LH will be bisected at A, HK at B, and LK at O; also $LH : HK = AH : HB = OB : AO = P : Q$. Hence, by a previous proposition, we may consider that the effect of P is the same as that of a heavy uniform bar LH of weight P, and the effect of Q is the same as that of a similar heavy uniform bar HK of weight Q. Hence the effect of the weights is the same as that of a heavy uniform bar LK. But the centre of gravity of such a bar is at its middle point O.

(*b*) Archimedes also wrote a treatise on *levers* and perhaps, on all the mechanical machines. The book is lost, but we know from Pappus that it contained a discussion of how a given weight could be moved with a given power. It was in this work probably that Archimedes discussed the theory of a certain compound pulley consisting of three or more simple pulleys which he had invented, and which was used in some public works in Syracuse. It is well known[1] that he boasted that, if he had but a fixed fulcrum, he could move the whole earth; and a

---

[1]See above, p. 53.

commentator of later date says that he added he would do it by using a compound pulley.

(c) His work on *floating bodies* contains nineteen propositions in two books, and was the first attempt to apply mathematical reasoning to hydrostatics. The story of the manner in which his attention was directed to the subject is told by Vitruvius. Hiero, the king of Syracuse, had given some gold to a goldsmith to make into a crown. The crown was delivered, made up, and of the proper weight, but it was suspected that the workman had appropriated some of the gold, replacing it by an equal weight of silver. Archimedes was thereupon consulted. Shortly afterwards, when in the public baths, he noticed that his body was pressed upwards by a force which increased the more completely he was immersed in the water. Recognising the value of the observation, he rushed out, just as he was, and ran home through the streets, shouting εὕρηκα, εὕρηκα, "I have found it, I have found it." There (to follow a later account) on making accurate experiments he found that when equal weights of gold and silver were weighed in water they no longer appeared equal: each seemed lighter than before by the weight of the water it displaced, and as the silver was more bulky than the gold its weight was more diminished. Hence, if on a balance he weighed the crown against an equal weight of gold and then immersed the whole in water, the gold would outweigh the crown if any silver had been used in its construction. Tradition says that the goldsmith was found to be fraudulent.

Archimedes began the work by proving that the surface of a fluid at rest is spherical, the centre of the sphere being at the centre of the earth. He then proved that the pressure of the fluid on a body, wholly or partially immersed, is equal to the weight of the fluid displaced; and thence found the position of equilibrium of a floating body, which he illustrated by spherical segments and paraboloids of revolution floating on a fluid. Some of the latter problems involve geometrical reasoning of considerable complexity.

The following is a fair specimen of the questions considered. A solid in the shape of a paraboloid of revolution of height h and latus rectum $4a$ floats in water, with its vertex immersed and its base wholly above the surface. If equilibrium be possible when the axis is not vertical, then the density of the body must be less than $(h - 3a)^2/h^3$ [book II, prop. 4]. When it is recollected that Archimedes was unacquainted with trigonometry or analytical geometry, the fact that he could discover and prove a proposition such as that just quoted will serve as an illustration of his powers of analysis.

It will be noticed that the mechanical investigations of Archimedes were concerned with statics. It may be added that though the Greeks attacked a few problems in dynamics, they did it with but indifferent success: some of their remarks were acute, but they did not sufficiently realise that the fundamental facts on which the theory must be based can be established only by carefully devised observations and experiments. It was not until the time of Galileo and Newton that this was done.

(v) We know, both from occasional references in his works and from remarks by other writers, that Archimedes was largely occupied in *astronomical observations*. He wrote a book, Περὶ Σφειροποιίας, on the construction of a celestial sphere, which is lost; and he constructed a sphere of the stars, and an orrery. These, after the capture of Syracuse, were taken by Marcellus to Rome, and were preserved as curiosities for at least two or three hundred years.

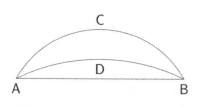

This mere catalogue of his works will show how wonderful were his achievements; but no one who has not actually read some of his writings can form a just appreciation of his extraordinary ability. This will be still further increased if we recollect that the only principles used by Archimedes, in addition to those contained in Euclid's *Elements* and *Conic sections*, are that of all lines like ACB, ADB, ... connecting two points A and B, the straight line is the shortest, and of the curved lines, the inner one ADB is shorter than the outer one ACB; together with two similar statements for space of three dimensions.

In the old and medieval world Archimedes was reckoned as the first of mathematicians, but possibly the best tribute to his fame is the fact that those writers who have spoken most highly of his work and ability are those who have been themselves the most distinguished men of their own generation.

Apollonius.[1]   The third great mathematician of this century

---

[1]In addition to Zeuthen's work and the other authorities mentioned in the footnote on p. 41, see *Litterargeschichtliche Studien über Euklid*, by J. L. Heiberg, Leipzig, 1882.   Editions of the extant works of Apollonius were issued by

was *Apollonius of Perga*, who is chiefly celebrated for having produced a systematic treatise on the conic sections which not only included all that was previously known about them, but immensely extended the knowledge of these curves. This work was accepted at once as the standard text-book on the subject, and completely superseded the previous treatises of Menaechmus, Aristaeus, and Euclid which until that time had been in general use.

We know very little of Apollonius himself. He was born about 260 B.C., and died about 200 B.C. He studied in Alexandria for many years, and probably lectured there; he is represented by Pappus as "vain, jealous of the reputation of others, and ready to seize every opportunity to depreciate them." It is curious that while we know next to nothing of his life, or of that of his contemporary Eratosthenes, yet their nicknames, which were respectively *epsilon* and *beta*, have come down to us. Dr. Gow has ingeniously suggested that the lecture rooms at Alexandria were numbered, and that they always used the rooms numbered 5 and 2 respectively.

Apollonius spent some years at Pergamum in Pamphylia, where a university had been recently established and endowed in imitation of that at Alexandria. There he met Eudemus and Attalus, to whom he subsequently sent each book of his conics as it came out with an explanatory note. He returned to Alexandria, and lived there till his death, which was nearly contemporaneous with that of Archimedes.

In his great work on *conic sections* Apollonius so thoroughly investigated the properties of these curves that he left but little for his successors to add. But his proofs are long and involved, and I think most readers will be content to accept a short analysis of his work, and the assurance that his demonstrations are valid. Dr. Zeuthen believes that many of the properties enunciated were obtained in the first instance by the use of co-ordinate geometry, and that the demonstrations were translated subsequently into geometrical form. If this be so, we must suppose that the classical writers were familiar with some branches of analytical geometry—Dr. Zeuthen says the use of orthogonal and oblique co-ordinates, and of transformations depending on abridged notation—that this knowledge was confined to a limited school, and was finally lost. This is a mere conjecture and is unsupported by any direct evidence, but it has been accepted by some writers

J. L. Heiberg in two volumes, Leipzig, 1890, 1893; and by E. Halley, Oxford, 1706 and 1710: an edition of the conics was published by T. L. Heath, Cambridge, 1896.

as affording an explanation of the extent and arrangement of the work.

The treatise contained about four hundred propositions, and was divided into eight books; we have the Greek text of the first four of these, and we also possess copies of the commentaries by Pappus and Eutocius on the whole work. In the ninth century an Arabic translation was made of the first seven books, which were the only ones then extant; we have two manuscripts of this version. The eighth book is lost.

In the letter to Eudemus which accompanied the first book Apollonius says that he undertook the work at the request of Naucrates, a geometrician who had been staying with him at Alexandria, and, though he had given some of his friends a rough draft of it, he had preferred to revise it carefully before sending it to Pergamum. In the note which accompanied the next book, he asks Eudemus to read it and communicate it to others who can understand it, and in particular to Philonides, a certain geometrician whom the author had met at Ephesus.

The first four books deal with the elements of the subject, and of these the first three are founded on Euclid's previous work (which was itself based on the earlier treatises by Menaechmus and Aristaeus). Heracleides asserts that much of the matter in these books was stolen from an unpublished work of Archimedes, but a critical examination by Heiberg has shown that this is improbable.

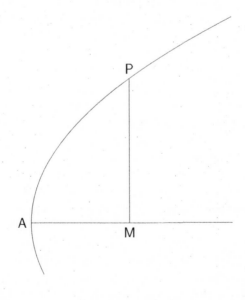

Apollonius begins by defining a cone on a circular base. He then investigates the different plane sections of it, and shows that they are divisible into three kinds of curves which he calls ellipses, parabolas, and hyperbolas. He proves the proposition that, if A, A' be the vertices of a conic, and if P be any point on it, and PM the perpendicular drawn from P on AA', then (in the usual notation) the ratio $MP^2 : AM.MA'$ is constant in an ellipse or hyperbola, and the ratio $MP^2 : AM$ is constant in a parabola. These are the characteristic properties on which almost all the rest of the work is based. He next shows that, if A be the vertex, l the latus rectum, and if AM and MP be the abscissa and ordinate of any point on a conic (see above figure), then MP^2 is less than, equal to, or greater than $l.AM$ according as the conic is an ellipse, parabola, or hyperbola; hence the names which he gave to the curves and by which they are still known.

He had no idea of the directrix, and was not aware that the parabola had a focus, but, with the exception of the propositions which involve these, his first three books contain most of the propositions which are found in modern text-books. In the fourth book he develops the theory of lines cut harmonically, and treats of the points of intersection of systems of conics. In the fifth book he commences with the theory of maxima and minima; applies it to find the centre of curvature at any point of a conic, and the evolute of the curve; and discusses the number of normals which can be drawn from a point to a conic. In the sixth book he treats of similar conics. The seventh and eighth books were given up to a discussion of conjugate diameters; the latter of these was conjecturally restored by E. Halley in his edition of 1710.

The verbose explanations make the book repulsive to most modern readers; but the arrangement and reasoning are unexceptional, and it has been not unfitly described as the crown of Greek geometry. It is the work on which the reputation of Apollonius rests, and it earned for him the name of "the great geometrician."

Besides this immense treatise he wrote numerous shorter works; of course the books were written in Greek, but they are usually referred to by their Latin titles: those about which we now know anything are enumerated below. He was the author of a work on the problem "given two co-planar straight lines Aa and Bb, drawn through fixed points A and B; to draw a line Oab from a given point O outside them cutting them in a and b, so that Aa shall be to Bb in a given ratio." He reduced the question to seventy-seven separate cases and gave an appropriate solution, with the aid of conics, for each case; this was published by

E. Halley (translated from an Arabic copy) in 1706. He also wrote a treatise *De Sectione Spatii* (restored by E. Halley in 1706) on the same problem under the condition that the rectangle $Aa . Bb$ was given. He wrote another entitled *De Sectione Determinata* (restored by R. Simson in 1749), dealing with problems such as to find a point P in a given straight line AB, so that PA^2 shall be to PB in a given ratio. He wrote another *De Tactionibus* (restored by Vieta in 1600) on the construction of a circle which shall touch three given circles. Another work was his *De Inclinationibus* (restored by M. Ghetaldi in 1607) on the problem to draw a line so that the intercept between two given lines, or the circumferences of two given circles, shall be of a given length. He was also the author of a treatise in three books on plane loci, *De Locis Planis* (restored by Fermat in 1637, and by R. Simson in 1746), and of another on the *regular solids*. And, lastly, he wrote a treatise on *unclassed incommensurables*, being a commentary on the tenth book of Euclid. It is believed that in one or more of the lost books he used the method of conical projections.

Besides these geometrical works he wrote on the *methods of arithmetical calculation*. All that we know of this is derived from some remarks of Pappus. Friedlein thinks that it was merely a sort of ready-reckoner. It seems, however, more probable that Apollonius here suggested a system of numeration similar to that proposed by Archimedes, but proceeding by tetrads instead of octads, and described a notation for it. It will be noticed that our modern notation goes by hexads, a million $= 10^6$, a billion $= 10^{12}$, a trillion $= 10^{18}$, etc. It is not impossible that Apollonius also pointed out that a decimal system of notation, involving only nine symbols, would facilitate numerical multiplications.

Apollonius was interested in astronomy, and wrote a book on the *stations and regressions of the planets* of which Ptolemy made some use in writing the *Almagest*. He also wrote a treatise on the use and theory of the screw in statics.

This is a long list, but I should suppose that most of these works were short tracts on special points.

Like so many of his predecessors, he too gave a construction for finding two mean proportionals between two given lines, and thereby duplicating the cube. It was as follows. Let OA and OB be the given lines. Construct a rectangle $OADB$, of which they are adjacent sides. Bisect AB in C. Then, if with C as centre we can describe a circle cutting OA produced in a, and cutting OB produced in b, so that aDb shall be a straight line, the problem is effected. For it is easily shewn

that
$$Oa \cdot Aa + CA^2 = Ca^2.$$

Similarly $$Ob \cdot Bb + CB^2 = Cb^2.$$

Hence $$Oa \cdot Aa = Ob \cdot Bb.$$

That is, $$Oa : Ob = Bb : Aa.$$

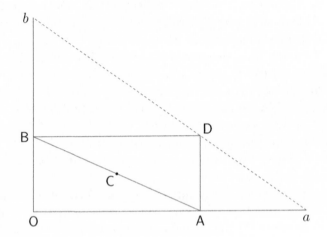

But, by similar triangles,

$$BD : Bb = Oa : Ob = Aa : AD.$$

Therefore $$Oa : Bb = Bb : Aa = Aa : OB,$$

that is, Bb and Oa are the two mean proportionals between OA and OB. It is impossible to construct the circle whose centre is C by Euclidean geometry, but Apollonius gave a mechanical way of describing it. This construction is quoted by several Arabic writers.

In one of the most brilliant passages of his *Aperçu historique* Chasles remarks that, while Archimedes and Apollonius were the most able geometricians of the old world, their works are distinguished by a contrast which runs through the whole subsequent history of geometry. Archimedes, in attacking the problem of the quadrature of curvilinear areas, established the principles of the geometry which rests on measurements; this naturally gave rise to the infinitesimal calculus, and in fact the method of exhaustions as used by Archimedes does not differ in principle from the method of limits as used by Newton. Apollonius, on the other hand, in investigating the properties of conic sections by means of transversals involving the ratio of rectilineal distances and of perspective, laid the foundations of the geometry of form and position.

Eratosthenes.[1]   Among the contemporaries of Archimedes and Apollonius I may mention *Eratosthenes*. Born at Cyrene in 275 B.C., he was educated at Alexandria—perhaps at the same time as Archimedes, of whom he was a personal friend—and Athens, and was at an early age entrusted with the care of the university library at Alexandria, a post which probably he occupied till his death. He was the Admirable Crichton of his age, and distinguished for his athletic, literary, and scientific attainments: he was also something of a poet. He lost his sight by ophthalmia, then as now a curse of the valley of the Nile, and, refusing to live when he was no longer able to read, he committed suicide in 194 B.C.

In science he was chiefly interested in astronomy and geodesy, and he constructed various astronomical instruments which were used for some centuries at the university. He suggested the calendar (now known as Julian), in which every fourth year contains 366 days; and he determined the obliquity of the ecliptic as $23°51'20''$. He measured the length of a degree on the earth's surface, making it to be about 79 miles, which is too long by nearly 10 miles, and thence calculated the circumference of the earth to be 252,000 stadia. If we take the Olympic stadium of $202\frac{1}{4}$ yards, this is equivalent to saying that the radius is about 4600 miles, but there was also an Egyptian stadium, and if he used this he estimated the radius as 3925 miles, which is very near the truth. The principle used in the determination is correct.

Of Eratosthenes's work in mathematics we have two extant illustrations: one in a description of an instrument to duplicate a cube, and the other in a rule he gave for constructing a table of prime numbers. The former is given in many books. The latter, called the "sieve of Eratosthenes," was as follows: write down all the numbers from 1 upwards; then every second number from 2 is a multiple of 2 and may be cancelled; every third number from 3 is a multiple of 3 and may be cancelled; every fifth number from 5 is a multiple of 5 and may be cancelled; and so on. It has been estimated that it would involve working for about 300 hours to thus find the primes in the numbers from 1 to 1,000,000. The labour of determining whether any particular number

---

[1] The works of Eratosthenes exist only in fragments. A collection of these was published by G. Bernhardy at Berlin in 1822: some additional fragments were printed by E. Hillier, Leipzig, 1872.

is a prime may be, however, much shortened by observing that if a number can be expressed as the product of two factors, one must be less and the other greater than the square root of the number, unless the number is the square of a prime, in which case the two factors are equal. Hence every composite number must be divisible by a prime which is not greater than its square root.

The second century before Christ.

The third century before Christ, which opens with the career of Euclid and closes with the death of Apollonius, is the most brilliant era in the history of Greek mathematics. But the great mathematicians of that century were geometricians, and under their influence attention was directed almost solely to that branch of mathematics. With the methods they used, and to which their successors were by tradition confined, it was hardly possible to make any further great advance: to fill up a few details in a work that was completed in its essential parts was all that could be effected. It was not till after the lapse of nearly 1800 years that the genius of Descartes opened the way to any further progress in geometry, and I therefore pass over the numerous writers who followed Apollonius with but slight mention. Indeed it may be said roughly that during the next thousand years Pappus was the sole geometrician of great original ability; and during this long period almost the only other pure mathematicians of exceptional genius were Hipparchus and Ptolemy, who laid the foundations of trigonometry, and Diophantus, who laid those of algebra.

Early in the second century, circ. 180 B.C., we find the names of three mathematicians—Hypsicles, Nicomedes, and Diocles—who in their own day were famous.

Hypsicles.   The first of these was *Hypsicles*, who added a fourteenth book to Euclid's *Elements* in which the regular solids were discussed. In another small work, entitled *Risings*, we find for the first time in Greek mathematics a right angle divided in the Babylonian manner into ninety degrees; possibly Eratosthenes may have previously estimated angles by the number of degrees they contain, but this is only a matter of conjecture.

Nicomedes.   The second was *Nicomedes*, who invented the curve known as the *conchoid* or the shell-shaped curve. If from a fixed point S a line be drawn cutting a given fixed straight line in Q, and if P be taken on SQ so that the length QP is constant (say d), then the locus of

P is the conchoid. Its equation may be put in the form $r = a \sec \theta \pm d$. It is easy with its aid to trisect a given angle or to duplicate a cube; and this no doubt was the cause of its invention.

Diocles.   The third of these mathematicians was *Diocles*, the inventor of the curve known as the *cissoid* or the ivy-shaped curve, which, like the conchoid, was used to give a solution of the duplication problem. He defined it thus: let AOA' and BOB' be two fixed diameters of a circle at right angles to one another. Draw two chords QQ' and RR' parallel to BOB' and equidistant from it. Then the locus of the intersection of AR and QQ' will be the cissoid. Its equation can be expressed in the form $y^2(2a - x) = x^3$. The curve may be used to duplicate the cube. For, if OA and OE be the two lines between which it is required to insert two geometrical means, and if, in the figure constructed as above, $A'E$ cut the cissoid in P, and AP cut OB in D, we have $OD^3 = OA^2 . OE$. Thus OD is one of the means required, and the other mean can be found at once.

Diocles also solved (by the aid of conic sections) a problem which had been proposed by Archimedes, namely, to draw a plane which will divide a sphere into two parts whose volumes shall bear to one another a given ratio.

Perseus. Zenodorus.   About a quarter of a century later, say about 150 B.C., *Perseus* investigated the various plane sections of the anchor-ring, and *Zenodorus* wrote a treatise on isoperimetrical figures. Part of the latter work has been preserved; one proposition which will serve to show the nature of the problems discussed is that "of segments of circles, having equal arcs, the semicircle is the greatest."

Towards the close of this century we find two mathematicians who, by turning their attention to new subjects, gave a fresh stimulus to the study of mathematics. These were Hipparchus and Hero.

Hipparchus.[1]   *Hipparchus* was the most eminent of Greek astronomers—his chief predecessors being Eudoxus, Aristarchus, Archimedes, and Eratosthenes. Hipparchus is said to have been born about 160 B.C. at Nicaea in Bithynia; it is probable that he spent some years at Alexandria, but finally he took up his abode at Rhodes where he made most of his observations. Delambre has obtained an ingenious

---

[1]See C. Manitius, *Hipparchi in Arati et Eudoxi Phaenomena Commentarii*, Leipzig, 1894, and J. B. J. Delambre, *Histoire de l'astronomie ancienne*, Paris, 1817, vol. i, pp. 106–189. S. P. Tannery in his *Recherches sur l'histoire de l'astronomie ancienne*, Paris, 1893, argues that the work of Hipparchus has been overrated, but I have adopted the view of the majority of writers on the subject.

confirmation of the tradition which asserted that Hipparchus lived in the second century before Christ. Hipparchus in one place says that the longitude of a certain star η Canis observed by him was exactly 90°, and it should be noted that he was an extremely careful observer. Now in 1750 it was 116°4'10", and, as the first point of Aries regredes at the rate of 50.2" a year, the observation was made about 120 B.C.

Except for a short commentary on a poem of Aratus dealing with astronomy all his works are lost, but Ptolemy's great treatise, the *Almagest*, described below, was founded on the observations and writings of Hipparchus, and from the notes there given we infer that the chief discoveries of Hipparchus were as follows. He determined the duration of the year to within six minutes of its true value. He calculated the inclination of the ecliptic and equator as 23°51'; it was actually at that time 23°46'. He estimated the annual precession of the equinoxes as 59"; it is 50.2". He stated the lunar parallax as 57', which is nearly correct. He worked out the eccentricity of the solar orbit as 1/24; it is very approximately 1/30. He determined the perigee and mean motion of the sun and of the moon, and he calculated the extent of the shifting of the plane of the moon's motion. Finally he obtained the synodic periods of the five planets then known. I leave the details of his observations and calculations to writers who deal specially with astronomy such as Delambre; but it may be fairly said that this work placed the subject for the first time on a scientific basis.

To account for the lunar motion Hipparchus supposed the moon to move with uniform velocity in a circle, the earth occupying a position near (but not at) the centre of this circle. This is equivalent to saying that the orbit is an epicycle of the first order. The longitude of the moon obtained on this hypothesis is correct to the first order of small quantities for a few revolutions. To make it correct for any length of time Hipparchus further supposed that the apse line moved forward about 3° a month, thus giving a correction for eviction. He explained the motion of the sun in a similar manner. This theory accounted for all the facts which could be determined with the instruments then in use, and in particular enabled him to calculate the details of eclipses with considerable accuracy.

He commenced a series of planetary observations to enable his successors to frame a theory to account for their motions; and with great perspicacity he predicted that to do this it would be necessary to introduce epicycles of a higher order, that is, to introduce three or more circles the centre of each successive one moving uniformly along the circumference of the preceding one.

He also formed a list of 1080 of the fixed stars. It is said that the sudden appearance in the heavens of a new and brilliant star called his attention to the need of such a catalogue; and the appearance of such a star during his lifetime is confirmed by Chinese records.

No further advance in the theory of astronomy was made until the time of Copernicus, though the principles laid down by Hipparchus were extended and worked out in detail by Ptolemy.

Investigations such as these naturally led to *trigonometry*, and Hipparchus must be credited with the invention of that subject. It is known that in plane trigonometry he constructed a table of chords of arcs, which is practically the same as one of natural sines; and that in spherical trigonometry he had some method of solving triangles: but his works are lost, and we can give no details. It is believed, however, that the elegant theorem, printed as Euc. VI, D, and generally known as Ptolemy's Theorem, is due to Hipparchus and was copied from him by Ptolemy. It contains implicitly the addition formulae for $\sin(A \pm B)$ and $\cos(A \pm B)$; and Carnot showed how the whole of elementary plane trigonometry could be deduced from it.

I ought also to add that Hipparchus was the first to indicate the position of a place on the earth by means of its latitude and longitude.

Hero.[1]   The second of these mathematicians was *Hero* of Alexandria, who placed engineering and land-surveying on a scientific basis. He was a pupil of Ctesibus, who invented several ingenious machines, and is alluded to as if he were a mathematician of note. It is not likely that Hero flourished before 80 B.C., but the precise period at which he lived is uncertain.

In pure mathematics Hero's principal and most characteristic work consists of (i) some elementary geometry, with applications to the determination of the areas of fields of given shapes; (ii) propositions on finding the volumes of certain solids, with applications to theatres,

---

[1] See *Recherches sur la vie et les ouvrages d'Héron d'Alexandrie* by T. H. Martin in vol. iv of *Mémoires présentés ... à l'académie d'inscriptions*, Paris, 1854; see also Loria, book iii, chap. v, pp. 107–128, and Cantor, chaps. xviii, xix. On the work entitled *Definitions*, which is attributed to Hero, see S. P. Tannery, chaps. xiii, xiv, and an article by G. Friedlein in Boncompagni's *Bulletino di bibliografia* March 1871, vol. iv, pp. 93–126. Editions of the extant works of Hero were published in Teubner's series, Leipzig, 1899, 1900, 1903. An English translation of the Πνευματικά was published by B. Woodcroft and J. G. Greenwood, London, 1851: drawings of the apparatus are inserted.

baths, banquet-halls, and so on; (iii) a rule to find the height of an inaccessible object; and (iv) tables of weights and measures. He invented a solution of the duplication problem which is practically the same as that which Apollonius had already discovered. Some commentators think that he knew how to solve a quadratic equation even when the coefficients were not numerical; but this is doubtful. He proved the formula that the area of a triangle is equal to $\{s(s-a)(s-b)(s-c)\}^{1/2}$, where s is the semiperimeter, and a, b, c, the lengths of the sides, and gave as an illustration a triangle whose sides were in the ratio 13:14:15. He seems to have been acquainted with the trigonometry of Hipparchus, and the values of $\cot 2\pi/n$ are computed for various values of n, but he nowhere quotes a formula or expressly uses the value of the sine; it is probable that like the later Greeks he regarded trigonometry as forming an introduction to, and being an integral part of, astronomy.

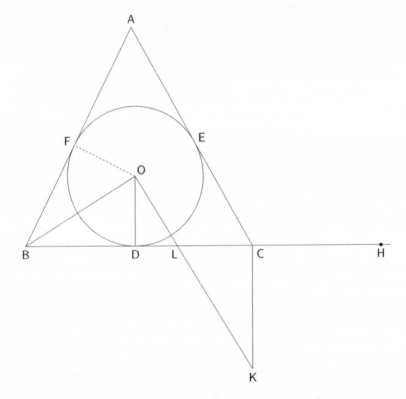

The following is the manner in which he solved[1] the problem to

[1] In his *Dioptra*, Hultsch, part viii, pp. 235–237. It should be stated that some

find the area of a triangle ABC the length of whose sides are a, b, c. Let s be the semiperimeter of the triangle. Let the inscribed circle touch the sides in D, E, F, and let O be its centre. On BC produced take H so that $CH = AF$, therefore $BH = s$. Draw OK at right angles to OB, and CK at right angles to BC; let them meet in K. The area ABC or $\triangle$ is equal to the sum of the areas OBC, OCA, $OAB = \frac{1}{2}ar + \frac{1}{2}br + \frac{1}{2}cr = sr$, that is, is equal to $BH \cdot OD$. He then shews that the angle $OAF =$ angle CBK; hence the triangles OAF and CBK are similar.

$$\therefore BC : CK = AF : OF = CH : OD,$$
$$\therefore BC : CH = CK : OD = CL : LD,$$
$$\therefore BH : CH = CD : LD,$$
$$\therefore BH^2 : CH \cdot BH = CD \cdot BD : LD \cdot BD = CD \cdot BD : OD^2.$$

Hence

$$\triangle = BH \cdot OD = \{CH \cdot BH \cdot CD \cdot BD\}^{\frac{1}{2}} = \{(s-a)s(s-c)(s-b)\}^{\frac{1}{2}}.$$

In applied mathematics Hero discussed the centre of gravity, the five simple machines, and the problem of moving a given weight with a given power; and in one place he suggested a way in which the power of a catapult could be tripled. He also wrote on the theory of hydraulic machines. He described a theodolite and cyclometer, and pointed out various problems in surveying for which they would be useful. But the most interesting of his smaller works are his $\Pi\nu\epsilon\nu\mu\alpha\tau\iota\kappa\acute{a}$ and $A\mathring{\nu}\tau\acute{o}\mu\alpha\tau\alpha$, containing descriptions of about 100 small machines and mechanical toys, many of which are ingenious. In the former there is an account of a small stationary steam-engine which is of the form now known as Avery's patent: it was in common use in Scotland at the beginning of this century, but is not so economical as the form introduced by Watt. There is also an account of a double forcing pump to be used as a fire-engine. It is probable that in the hands of Hero these instruments never got beyond models. It is only recently that general attention has been directed to his discoveries, though Arago had alluded to them in his *éloge* on Watt.

All this is very different from the classical geometry and arithmetic of Euclid, or the mechanics of Archimedes. Hero did nothing to extend a knowledge of abstract mathematics; he learnt all that the text-books

critics think that this is an interpolation, and is not due to Hero.

of the day could teach him, but he was interested in science only on account of its practical applications, and so long as his results were true he cared nothing for the logical accuracy of the process by which he arrived at them. Thus, in finding the area of a triangle, he took the square root of the product of four lines. The classical Greek geometricians permitted the use of the square and the cube of a line because these could be represented geometrically, but a figure of four dimensions is inconceivable, and certainly they would have rejected a proof which involved such a conception.

The first century before Christ.

The successors of Hipparchus and Hero did not avail themselves of the opportunity thus opened of investigating new subjects, but fell back on the well-worn subject of geometry. Amongst the more eminent of these later geometricians were Theodosius and Dionysodorus, both of whom flourished about 50 B.C.

Theodosius.   *Theodosius* was the author of a complete treatise on the geometry of the sphere, and of two works on astronomy.[1]

Dionysodorus.   *Dionysodorus* is known to us only by his solution[2] of the problem to divide a hemisphere by a plane parallel to its base into two parts, whose volumes shall be in a given ratio. Like the solution by Diocles of the similar problem for a sphere above alluded to, it was effected by the aid of conic sections. Pliny says that Dionysodorus determined the length of the radius of the earth approximately as 42,000 stadia, which, if we take the Olympic stadium of $202\frac{1}{4}$ yards, is a little less than 5000 miles; we do not know how it was obtained. This may be compared with the result given by Eratosthenes and mentioned above.

End of the First Alexandrian School.

The administration of Egypt was definitely undertaken by Rome in 30 B.C. The closing years of the dynasty of the Ptolemies and the earlier years of the Roman occupation of the country were marked by much disorder, civil and political. The studies of the university were

---

[1] The work on the sphere was edited by I. Barrow, Cambridge, 1675, and by E. Nizze, Berlin, 1852. The works on astronomy were published by Dasypodius in 1572.

[2] It is reproduced in H. Suter's *Geschichte der mathematischen Wissenschaften*, second edition, Zürich, 1873, p. 101.

naturally interrupted, and it is customary to take this time as the close of the first Alexandrian school.

CHAPTER V.

THE SECOND ALEXANDRIAN SCHOOL.[1]
30 B.C.–641 A.D.

I CONCLUDED the last chapter by stating that the first school of Alexandria may be said to have come to an end at about the same time as the country lost its nominal independence. But, although the schools at Alexandria suffered from the disturbances which affected the whole Roman world in the transition, in fact if not in name, from a republic to an empire, there was no break of continuity; the teaching in the university was never abandoned; and as soon as order was again established, students began once more to flock to Alexandria. This time of confusion was, however, contemporaneous with a change in the prevalent views of philosophy which thenceforward were mostly neo-platonic or neo-pythagorean, and it therefore fitly marks the commencement of a new period. These mystical opinions reacted on the mathematical school, and this may partially account for the paucity of good work.

Though Greek influence was still predominant and the Greek language always used, Alexandria now became the intellectual centre for most of the Mediterranean nations which were subject to Rome. It should be added, however, that the direct connection with it of many of the mathematicians of this time is at least doubtful, but their knowledge was ultimately obtained from the Alexandrian teachers, and they are usually described as of the second Alexandrian school. Such mathematics as were taught at Rome were derived from Greek sources, and we may therefore conveniently consider their extent in connection with this chapter.

[1]For authorities, see footnote above on p. 41. All dates given hereafter are to be taken as *anno domini* unless the contrary is expressly stated.

The first century after Christ.

There is no doubt that throughout the first century after Christ geometry continued to be that subject in science to which most attention was devoted. But by this time it was evident that the geometry of Archimedes and Apollonius was not capable of much further extension; and such geometrical treatises as were produced consisted mostly of commentaries on the writings of the great mathematicians of a preceding age. In this century the only original works of any ability of which we know anything were two by Serenus and one by Menelaus.

Serenus. Menelaus. Those by *Serenus* of Antissa or of Antinoe, circ. 70, are on the *plane sections of the cone and cylinder*,[1] in the course of which he lays down the fundamental proposition of transversals. That by *Menelaus* of Alexandria, circ. 98, is on *spherical trigonometry*, investigated in the Euclidean method.[2] The fundamental theorem on which the subject is based is the relation between the six segments of the sides of a spherical triangle, formed by the arc of a great circle which cuts them [book III, prop. 1]. Menelaus also wrote on the calculation of chords, that is, on plane trigonometry; this is lost.

Nicomachus. Towards the close of this century, circ. 100, a Jew, *Nicomachus*, of Gerasa, published an *Arithmetic*,[3] which (or rather the Latin translation of it) remained for a thousand years a standard authority on the subject. Geometrical demonstrations are here abandoned, and the work is a mere classification of the results then known, with numerical illustrations: the evidence for the truth of the propositions enunciated, for I cannot call them proofs, being in general an induction from numerical instances. The object of the book is the study of the properties of numbers, and particularly of their ratios. Nicomachus commences with the usual distinctions between even, odd, prime, and perfect numbers; he next discusses fractions in a somewhat clumsy manner; he then turns to polygonal and to solid numbers; and finally treats of ratio, proportion, and the progressions. Arithmetic of this kind is usually termed Boethian, and the work of Boethius on it was a recognised text-book in the middle ages.

---

[1] These have been edited by J. L. Heiberg, Leipzig, 1896; and by E. Halley, Oxford, 1710.

[2] This was translated by E. Halley, Oxford, 1758.

[3] The work has been edited by R. Hoche, Leipzig, 1866.

The second century after Christ.

Theon.   Another text-book on arithmetic on much the same lines as that of Nicomachus was produced by *Theon* of Smyrna, circ. 130. It formed the first book of his work[1] on mathematics, written with the view of facilitating the study of Plato's writings.

Thymaridas.   Another mathematician, reckoned by some writers as of about the same date as Theon, was *Thymaridas*, who is worthy of notice from the fact that he is the earliest known writer who explicitly enunciates an algebraical theorem. He states that, if the sum of any number of quantities be given, and also the sum of every pair which contains one of them, then this quantity is equal to one $(n-2)$th part of the difference between the sum of these pairs and the first given sum. Thus, if

$$x_1 + x_2 + \ldots + x_n = S,$$

and if $x_1 + x_2 = s_2, \quad x_1 + x_3 = s_3, \ldots,$ and $x_1 + x_n = s_n,$

then $x_1 = (s_2 + s_3 + \ldots + s_n - S)/(n-2).$

He does not seem to have used a symbol to denote the unknown quantity, but he always represents it by the same word, which is an approximation to symbolism.

Ptolemy.[2]   About the same time as these writers *Ptolemy* of Alexandria, who died in 168, produced his great work on astronomy, which will preserve his name as long as the history of science endures. This treatise is usually known as the *Almagest*: the name is derived from the Arabic title *al midschisti*, which is said to be a corruption of $\mu\epsilon\gamma\iota\sigma\tau\eta$ [$\mu\alpha\theta\eta\mu\alpha\tau\iota\kappa\eta$] $\sigma\upsilon\nu\tau\alpha\xi\iota\varsigma$. The work is founded on the writings of Hipparchus, and, though it did not sensibly advance the theory of the subject, it presents the views of the older writer with a completeness and elegance which will always make it a standard treatise. We gather from it that Ptolemy made observations at Alexandria from the years

---

[1]The Greek text of those parts which are now extant, with a French translation, was issued by J. Dupuis, Paris, 1892.

[2]See the article *Ptolemaeus Claudius*, by A. De Morgan in Smith's *Dictionary of Greek and Roman Biography*, London, 1849; S. P. Tannery, *Recherches sur l'histoire de l'astronomie ancienne*, Paris, 1893; and J. B. J. Delambre, *Histoire de l'astronomie ancienne*, Paris, 1817, vol. ii. An edition of all the works of Ptolemy which are now extant was published at Bâle in 1551. The *Almagest* with various minor works was edited by M. Halma, 12 vols. Paris, 1813–28, and a new edition, in two volumes, by J. L. Heiberg, Leipzig, 1898, 1903, 1907.

125 to 150; he, however, was but an indifferent practical astronomer, and the observations of Hipparchus are generally more accurate than those of his expounder.

The work is divided into thirteen books. In the first book Ptolemy discusses various preliminary matters; treats of trigonometry, plane or spherical; gives a table of chords, that is, of natural sines (which is substantially correct and is probably taken from the lost work of Hipparchus); and explains the obliquity of the ecliptic; in this book he uses degrees, minutes, and seconds as measures of angles. The second book is devoted chiefly to phenomena depending on the spherical form of the earth: he remarks that the explanations would be much simplified if the earth were supposed to rotate on its axis once a day, but states that this hypothesis is inconsistent with known facts. In the third book he explains the motion of the sun round the earth by means of excentrics and epicycles: and in the fourth and fifth books he treats the motion of the moon in a similar way. The sixth book is devoted to the theory of eclipses; and in it he gives $3°8'30''$, that is $3\frac{17}{120}$, as the approximate value of π, which is equivalent to taking it equal to 3.1416. The seventh and eighth books contain a catalogue (probably copied from Hipparchus) of 1028 fixed stars determined by indicating those, three or more, that appear to be in a plane passing through the observer's eye: and in another work Ptolemy added a list of annual sidereal phenomena. The remaining books are given up to the theory of the planets.

This work is a splendid testimony to the ability of its author. It became at once the standard authority on astronomy, and remained so till Copernicus and Kepler shewed that the sun and not the earth must be regarded as the centre of the solar system.

The idea of excentrics and epicycles on which the theories of Hipparchus and Ptolemy are based has been often ridiculed in modern times. No doubt at a later time, when more accurate observations had been made, the necessity of introducing epicycle on epicycle in order to bring the theory into accordance with the facts made it very complicated. But De Morgan has acutely observed that in so far as the ancient astronomers supposed that it was necessary to resolve every celestial motion into a series of uniform circular motions they erred greatly, but that, if the hypothesis be regarded as a convenient way of expressing known facts, it is not only legitimate but convenient. The theory suffices to describe either the angular motion of the heavenly bodies or their change in distance. The ancient astronomers were concerned only with the former question, and it fairly met their needs; for the latter question it is less convenient. In fact it was as good a theory as for their purposes and with their instruments and knowledge it was possible to frame, and corresponds to the expression of a given function as a sum of sines or cosines, a method which is of frequent use in modern analysis.

In spite of the trouble taken by Delambre it is almost impossible to separate the results due to Hipparchus from those due to Ptolemy. But Delambre and De Morgan agree in thinking that the observations quoted, the fundamental ideas, and the explanation of the apparent solar motion are due to Hipparchus; while all the detailed explanations and calculations of the lunar and planetary motions are due to Ptolemy.

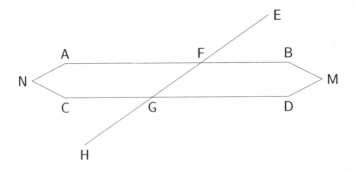

The *Almagest* shews that Ptolemy was a geometrician of the first rank, though it is with the application of geometry to astronomy that he is chiefly concerned. He was also the author of numerous other treatises. Amongst these is one on *pure geometry* in which he proposed to cancel Euclid's postulate on parallel lines, and to prove it in the following manner. Let the straight line $EFGH$ meet the two straight lines AB and CD so as to make the sum of the angles BFG and FGD equal to two right angles. It is required to prove that AB and CD are parallel. If possible let them not be parallel, then they will meet when produced say at M (or N). But the angle AFG is the supplement of BFG, and is therefore equal to FGD: similarly the angle FGC is equal to the angle BFG. Hence the sum of the angles AFG and FGC is equal to two right angles, and the lines BA and DC will therefore if produced meet at N (or M). But two straight lines cannot enclose a space, therefore AB and CD cannot meet when produced, that is, they are parallel. Conversely, if AB and CD be parallel, then AF and CG are not less parallel than FB and GD; and therefore whatever be the sum of the angles AFG and FGC such also must be the sum of the angles FGD and BFG. But the sum of the four angles is equal to four right angles, and therefore the sum of the angles BFG and FGD must be equal to two right angles.

Ptolemy wrote another work to shew that there could not be more than three dimensions in space: he also discussed *orthographic* and *stereographic projections* with special reference to the construction of sun-dials. He wrote on geography, and stated that the length of one degree of latitude is 500 stadia. A book on *sound* is sometimes attributed to him, but on doubtful authority.

The third century after Christ.

Pappus.   Ptolemy had shewn not only that geometry could be applied to astronomy, but had indicated how new methods of analysis like trigonometry might be thence developed. He found however no successors to take up the work he had commenced so brilliantly, and we must look forward 150 years before we find another geometrician of any eminence. That geometrician was *Pappus* who lived and taught at Alexandria about the end of the third century. We know that he had numerous pupils, and it is probable that he temporarily revived an interest in the study of geometry.

Pappus wrote several books, but the only one which has come down to us is his $\Sigma v \nu a \gamma \omega \gamma \acute{\eta}$,[1] a collection of mathematical papers arranged in eight books of which the first and part of the second have been lost. This collection was intended to be a synopsis of Greek mathematics together with comments and additional propositions by the editor. A careful comparison of various extant works with the account given of them in this book shews that it is trustworthy, and we rely largely on it for our knowledge of other works now lost. It is not arranged chronologically, but all the treatises on the same subject are grouped together, and it is most likely that it gives roughly the order in which the classical authors were read at Alexandria. Probably the first book, which is now lost, was on arithmetic. The next four books deal with geometry exclusive of conic sections; the sixth with astronomy including, as subsidiary subjects, optics and trigonometry; the seventh with analysis, conics, and porisms; and the eighth with mechanics.

The last two books contain a good deal of original work by Pappus; at the same time it should be remarked that in two or three cases he

---

[1] It has been published by F. Hultsch, Berlin, 1876–8.

has been detected in appropriating proofs from earlier authors, and it is possible he may have done this in other cases.

Subject to this suspicion we may say that Pappus's best work is in geometry. He discovered the directrix in the conic sections, but he investigated only a few isolated properties: the earliest comprehensive account was given by Newton and Boscovich. As an illustration of his power I may mention that he solved [book VII, prop. 107] the problem to inscribe in a given circle a triangle whose sides produced shall pass through three collinear points. This question was in the eighteenth century generalised by Cramer by supposing the three given points to be anywhere; and was considered a difficult problem.[1] It was sent in 1742 as a challenge to Castillon, and in 1776 he published a solution. Lagrange, Euler, Lhulier, Fuss, and Lexell also gave solutions in 1780. A few years later the problem was set to a Neapolitan lad A. Giordano, who was only 16 but who had shewn marked mathematical ability, and he extended it to the case of a polygon of n sides which pass through n given points, and gave a solution both simple and elegant. Poncelet extended it to conics of any species and subject to other restrictions.

In mechanics Pappus shewed that the centre of mass of a triangular lamina is the same as that of an inscribed triangular lamina whose vertices divide each of the sides of the original triangle in the same ratio. He also discovered the two theorems on the surface and volume of a solid of revolution which are still quoted in text-books under his name: these are that the volume generated by the revolution of a curve about an axis is equal to the product of the area of the curve and the length of the path described by its centre of mass; and the surface is equal to the product of the perimeter of the curve and the length of the path described by its centre of mass.

The problems above mentioned are but samples of many brilliant but isolated theorems which were enunciated by Pappus. His work as a whole and his comments shew that he was a geometrician of power; but it was his misfortune to live at a time when but little interest was taken in geometry, and when the subject, as then treated, had been practically exhausted.

Possibly a small tract[2] on multiplication and division of sexagesimal

---

[1] For references to this problem see a note by H. Brocard in *L'Intermédiaire des mathématiciens*, Paris, 1904, vol. xi, pp. 219–220.

[2] It was edited by C. Henry, Halle, 1879, and is valuable as an illustration of practical Greek arithmetic.

fractions, which would seem to have been written about this time, is due to Pappus.

The fourth century after Christ.

Throughout the second and third centuries, that is, from the time of Nicomachus, interest in geometry had steadily decreased, and more and more attention had been paid to the theory of numbers, though the results were in no way commensurate with the time devoted to the subject. It will be remembered that Euclid used lines as symbols for any magnitudes, and investigated a number of theorems about numbers in a strictly scientific manner, but he confined himself to cases where a geometrical representation was possible. There are indications in the works of Archimedes that he was prepared to carry the subject much further: he introduced numbers into his geometrical discussions and divided lines by lines, but he was fully occupied by other researches and had no time to devote to arithmetic. Hero abandoned the geometrical representation of numbers, but he, Nicomachus, and other later writers on arithmetic did not succeed in creating any other symbolism for numbers in general, and thus when they enunciated a theorem they were content to verify it by a large number of numerical examples. They doubtless knew how to solve a quadratic equation with numerical coefficients—for, as pointed out above, geometrical solutions of the equations $ax^2 - bx + c = 0$ and $ax^2 + bx - c = 0$ are given in Euc. VI, 28 and 29—but probably this represented their highest attainment.

It would seem then that, in spite of the time given to their study, arithmetic and algebra had not made any sensible advance since the time of Archimedes. The problems of this kind which excited most interest in the third century may be illustrated from a collection of questions, printed in the Palatine Anthology, which was made by **Metrodorus** at the beginning of the next century, about 310. Some of them are due to the editor, but some are of an anterior date, and they fairly illustrate the way in which arithmetic was leading up to algebraical methods. The following are typical examples. "Four pipes discharge into a cistern: one fills it in one day; another in two days; the third in three days; the fourth in four days: if all run together how soon will they fill the cistern?" "Demochares has lived a fourth of his life as a boy; a fifth as a youth; a third as a man; and has spent thirteen years in his dotage: how old is he?" "Make a crown of gold, copper, tin, and iron weighing 60 minae: gold and copper shall be two-thirds of it; gold and tin three-fourths of it; and gold and iron three-fifths of it: find the weights of the gold, copper, tin, and iron which are required." The last is a numerical illustration of Thymaridas's theorem quoted above.

It is believed that these problems were solved by *rhetorical algebra*, that is, by a process of algebraical reasoning expressed in words and without the use of any symbols. This, according to Nesselmann, is the first stage in the development of algebra, and we find it used both by Ahmes and by the earliest Arabian, Persian, and Italian algebraists: examples of its use in the solution of a geometrical problem and in the rule for the solution of a quadratic equation are given later.[1] On this view then a rhetorical algebra had been gradually evolved by the Greeks, or was then in process of evolution. Its development was however very imperfect. Hankel, who is no unfriendly critic, says that the results attained as the net outcome of the work of six centuries on the theory of numbers are, whether we look at the form or the substance, unimportant or even childish, and are not in any way the commencement of a science.

In the midst of this decaying interest in geometry and these feeble attempts at algebraic arithmetic, a single algebraist of marked originality suddenly appeared who created what was practically a new science. This was Diophantus who introduced a system of abbreviations for those operations and quantities which constantly recur, though in using them he observed all the rules of grammatical syntax. The resulting science is called by Nesselmann *syncopated algebra*: it is a sort of shorthand. Broadly speaking, it may be said that European algebra did not advance beyond this stage until the close of the sixteenth century.

Modern algebra has progressed one stage further and is entirely *symbolic*; that is, it has a language of its own and a system of notation which has no obvious connection with the things represented, while the operations are performed according to certain rules which are distinct from the laws of grammatical construction.

Diophantus.[2]   All that we know of *Diophantus* is that he lived at Alexandria, and that most likely he was not a Greek. Even the date of his career is uncertain; it cannot reasonably be put before the middle of the third century, and it seems probable that he was alive in the early

---

[1] See below, pp. 168, 174.

[2] A critical edition of the collected works of Diophantus was edited by S. P. Tannery, 2 vols., Leipzig, 1893; see also *Diophantos of Alexandria*, by T. L. Heath, Cambridge, 1885; and Loria, book V, chap. V, pp. 95–158.

years of the fourth century, that is, shortly after the death of Pappus. He was 84 when he died.

In the above sketch of the lines on which algebra has developed I credited Diophantus with the invention of syncopated algebra. This is a point on which opinions differ, and some writers believe that he only systematized the knowledge which was familiar to his contemporaries. In support of this latter opinion it may be stated that Cantor thinks that there are traces of the use of algebraic symbolism in Pappus, and Freidlein mentions a Greek papyrus in which the signs / and $\supset$ are used for addition and subtraction respectively; but no other direct evidence for the non-originality of Diophantus has been produced, and no ancient author gives any sanction to this opinion.

Diophantus wrote a short essay on polygonal numbers; a treatise on algebra which has come down to us in a mutilated condition; and a work on porisms which is lost.

The *Polygonal Numbers* contains ten propositions, and was probably his earliest work. In this he reverts to the classical system by which numbers are represented by lines, a construction is (if necessary) made, and a strictly deductive proof follows: it may be noticed that in it he quotes propositions, such as Euc. II, 3, and II, 8, as referring to numbers and not to magnitudes.

His chief work is his *Arithmetic*. This is really a treatise on algebra; algebraic symbols are used, and the problems are treated analytically. Diophantus tacitly assumes, as is done in nearly all modern algebra, that the steps are reversible. He applies this algebra to find solutions (though frequently only particular ones) of several problems involving numbers. I propose to consider successively the notation, the methods of analysis employed, and the subject-matter of this work.

First, as to the notation. Diophantus always employed a symbol to represent the unknown quantity in his equations, but as he had only one symbol he could not use more than one unknown at a time.[1] The unknown quantity is called $\acute{o}$ $\dot{\alpha}\rho\iota\theta\mu\acute{o}\varsigma$, and is represented by ς' or $\varsigma^{o\prime}$. It is usually printed as ς. In the plural it is denoted by $\varsigma\varsigma$ or $\overline{\varsigma\varsigma}^{oi}$. This symbol may be a corruption of α^{ρ}, or perhaps it may be the final sigma of this word, or possibly it may stand for the word $\sigma\omega\rho\acute{o}\varsigma$ a heap.[2] The square of the unknown is called $\delta\acute{v}\nu\alpha\mu\iota\varsigma$, and denoted by $\delta^{\bar{v}}$: the cube

---

[1] See, however, below, page 90, example (iii), for an instance of how he treated a problem involving two unknown quantities.

[2] See above, page 4.

κύβος, and denoted by κ^{v}; and so on up to the sixth power.

The coefficients of the unknown quantity and its powers are numbers, and a numerical coefficient is written immediately after the quantity it multiplies: thus $\varsigma'\bar{\alpha} = x$, and $\varsigma\varsigma^{o\iota}\overline{\iota\alpha} = \overline{\varsigma\varsigma}\,\overline{\iota\alpha} = 11x$. An absolute term is regarded as a certain number of units or μονάδες which are represented by $\mu^{\hat{o}}$: thus $\mu^{\hat{o}}\bar{\alpha} = 1$, $\mu^{\hat{o}}\overline{\iota\alpha} = 11$.

There is no sign for addition beyond juxtaposition. Subtraction is represented by φ, and this symbol affects all the symbols that follow it. Equality is represented by ι. Thus

$$\kappa^{\hat{v}}\bar{\alpha}\,\overline{\varsigma\varsigma\eta}\,\varphi\,\delta^{\hat{o}}\bar{\epsilon}\,\mu^{\hat{o}}\bar{\alpha}\,\iota\,\varsigma\bar{\alpha}$$

represents $(x^3 + 8x) - (5x^2 + 1) = x.$

Diophantus also introduced a somewhat similar notation for fractions involving the unknown quantity, but into the details of this I need not here enter.

It will be noticed that all these symbols are mere abbreviations for words, and Diophantus reasons out his proofs, writing these abbreviations in the middle of his text. In most manuscripts there is a marginal summary in which the symbols alone are used and which is really symbolic algebra; but probably this is the addition of some scribe of later times.

This introduction of a contraction or a symbol instead of a word to represent an unknown quantity marks a greater advance than anyone not acquainted with the subject would imagine, and those who have never had the aid of some such abbreviated symbolism find it almost impossible to understand complicated algebraical processes. It is likely enough that it might have been introduced earlier, but for the unlucky system of numeration adopted by the Greeks by which they used all the letters of the alphabet to denote particular numbers and thus made it impossible to employ them to represent any number.

Next, as to the knowledge of algebraic methods shewn in the book. Diophantus commences with some definitions which include an explanation of his notation, and in giving the symbol for *minus* he states that a subtraction multiplied by a subtraction gives an addition; by this he means that the product of $-b$ and $-d$ in the expansion of $(a-b)(c-d)$ is $+bd$, but in applying the rule he always takes care that the numbers a, b, c, d are so chosen that a is greater than b and c is greater than d.

The whole of the work itself, or at least as much as is now extant, is devoted to solving problems which lead to equations. It contains rules for solving a simple equation of the first degree and a binomial quadratic. Probably the rule for solving any quadratic equation was given in that part of the work which is now lost, but where the equation is of the form $ax^2 + bx + c = 0$ he seems to have multiplied by a and then "completed the square" in much the same way as is now done: when the roots are negative or irrational the equation is rejected as "impossible," and even when both roots are positive he never gives more than one, always taking the positive value of the square root. Diophantus solves one cubic equation, namely, $x^3 + x = 4x^2 + 4$ [book VI, prob. 19].

The greater part of the work is however given up to indeterminate equations between two or three variables. When the equation is between two variables, then, if it be of the first degree, he assumes a suitable value for one variable and solves the equation for the other. Most of his equations are of the form $y^2 = Ax^2 + Bx + C$. Whenever A or C is equal to zero, he is able to solve the equation completely. When this is not the case, then, if $A = a^2$, he assumes $y = ax + m$; if $C = c^2$, he assumes $y = mx + c$; and lastly, if the equation can be put in the form $y^2 = (ax \pm b)^2 + c^2$, he assumes $y = mx$: where in each case m has some particular numerical value suitable to the problem under consideration. A few particular equations of a higher order occur, but in these he generally alters the problem so as to enable him to reduce the equation to one of the above forms.

The simultaneous indeterminate equations involving three variables, or "double equations" as he calls them, which he considers are of the forms $y^2 = Ax^2 + Bx + C$ and $z^2 = ax^2 + bx + c$. If A and a both vanish, he solves the equations in one of two ways. It will be enough to give one of his methods which is as follows: he subtracts and thus gets an equation of the form $y^2 - z^2 = mx + n$; hence, if $y \pm z = \lambda$, then $y \mp z = (mx + n)/\lambda$; and solving he finds y and z. His treatment of "double equations" of a higher order lacks generality and depends on the particular numerical conditions of the problem.

Lastly, as to the matter of the book. The problems he attacks and the analysis he uses are so various that they cannot be described concisely and I have therefore selected five typical problems to illustrate his methods. What seems to strike his critics most is the ingenuity with which he selects as his unknown some quantity which leads to equations such as he can solve, and the artifices by which he finds numerical solutions of his equations.

I select the following as characteristic examples.

(i) *Find four numbers, the sum of every arrangement three at a time being given; say* 22, 24, 27, *and* 20 [book I, prob. 17].

Let x be the sum of all four numbers; hence the numbers are $x - 22$, $x - 24$, $x - 27$, and $x - 20$.

$$\therefore x = (x - 22) + (x - 24) + (x - 27) + (x - 20).$$
$$\therefore x = 31.$$

$\therefore$ the numbers are 9, 7, 4, and 11.

(ii) *Divide a number, such as* 13 *which is the sum of two squares* 4 *and* 9, *into two other squares* [book II, prob. 10].

He says that since the given squares are 2^2 and 3^2 he will take $(x+2)^2$ and $(mx-3)^2$ as the required squares, and will assume $m = 2$.

$$\therefore (x + 2)^2 + (2x - 3)^2 = 13.$$
$$\therefore x = 8/5.$$

$\therefore$ the required squares are 324/25 and 1/25.

(iii) *Find two squares such that the sum of the product and either is a square* [book II, prob. 29].

Let x^2 and y^2 be the numbers. Then $x^2y^2 + y^2$ and $x^2y^2 + x^2$ are squares. The first will be a square if x^2+1 be a square, which he assumes may be taken equal to $(x - 2)^2$, hence $x = 3/4$. He has now to make $9(y^2 + 1)/16$ a square, to do this he assumes that $9y^2 + 9 = (3y - 4)^2$, hence $y = 7/24$. Therefore the squares required are 9/16 and 49/576.

It will be recollected that Diophantus had only one symbol for an unknown quantity; and in this example he begins by calling the unknowns x^2 and 1, but as soon as he has found x he then replaces the 1 by the symbol for the unknown quantity, and finds it in its turn.

(iv) *To find a* [*rational*] *right-angled triangle such that the line bisecting an acute angle is rational* [book VI, prob. 18].

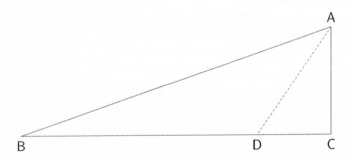

His solution is as follows. Let ABC be the triangle of which C is the right-angle. Let the bisector $AD = 5x$, and let $DC = 3x$, hence $AC = 4x$. Next let BC be a multiple of 3, say 3, $\therefore BD = 3 - 3x$, hence $AB = 4 - 4x$ (by Euc. VI, 3). Hence $(4 - 4x)^2 = 3^2 + (4x)^2$ (Euc. I, 47), $\therefore x = 7/32$. Multiplying by 32 we get for the sides of the triangle 28, 96, and 100; and for the bisector 35.

(v) *A man buys x measures of wine, some at* 8 *drachmae a measure, the rest at* 5. *He pays for them a square number of drachmae, such that, if* 60 *be added to it, the resulting number is* x^2. *Find the number he bought at each price* [book V, prob. 33].

The price paid was $x^2 - 60$, hence $8x > x^2 - 60$ and $5x < x^2 - 60$. From this it follows that x must be greater than 11 and less than 12.

Again $x^2 - 60$ is to be a square; suppose it is equal to $(x - m)^2$ then $x = (m^2 + 60)/2m$, we have therefore

$$11 < \frac{m^2 + 60}{2m} < 12;$$
$$\therefore 19 < m < 21.$$

Diophantus therefore assumes that m is equal to 20, which gives him $x = 11\frac{1}{2}$; and makes the total cost, *i.e.* $x^2 - 60$, equal to $72\frac{1}{4}$ drachmae.

He has next to divide this cost into two parts which shall give the cost of the 8 drachmae measures and the 5 drachmae measures respectively. Let these parts be y and z.

Then
$$\tfrac{1}{5}z + \tfrac{1}{8}(72\tfrac{1}{4} - z) = \tfrac{1}{2}.$$
Therefore
$$z = \frac{5 \times 79}{12}, \text{ and } y = \frac{8 \times 59}{12}.$$

Therefore the number of 5 drachmae measures was 79/12, and of 8 drachmae measures was 59/12.

From the enunciation of this problem it would seem that the wine was of a poor quality, and Tannery ingeniously suggested that the prices mentioned for such a wine are higher than were usual until after the end of the second century. He therefore rejected the view which was formerly held that Diophantus lived in that century, but he did not seem to be aware that De Morgan had previously shewn that this opinion was untenable. Tannery inclined to think that Diophantus lived half a century earlier than I have supposed.

I mentioned that Diophantus wrote a third work entitled *Porisms*. The book is lost, but we have the enunciations of some of the propositions, and though we cannot tell whether they were rigorously proved by Diophantus they confirm our opinion of his ability and sagacity. It has been suggested that some of the theorems which he assumes in his arithmetic were proved in the porisms. Among the more striking of these results are the statements that the difference of the cubes of two numbers can be always expressed as the sum of the cubes of two other numbers; that no number of the form $4n - 1$ can be expressed as the sum of two squares; and that no number of the form $8n - 1$ (or possibly $24n + 7$) can be expressed as the sum of three squares: to these we may perhaps add the proposition that any number can be expressed as a square or as the sum of two or three or four squares.

The writings of Diophantus exercised no perceptible influence on Greek mathematics; but his *Arithmetic*, when translated into Arabic in the tenth century, influenced the Arabian school, and so indirectly affected the progress of European mathematics. An imperfect copy of the original work was discovered in 1462; it was translated into Latin and published by Xylander in 1575; the translation excited general interest, and by that time the European algebraists had, on the whole, advanced beyond the point at which Diophantus had left off.

Iamblichus.   *Iamblichus*, circ. 350, to whom we owe a valuable work on the Pythagorean discoveries and doctrines, seems also to have studied the properties of numbers. He enunciated the theorem that if a number which is equal to the sum of three integers of the form $3n$, $3n - 1$, $3n - 2$ be taken, and if the separate digits of this number be added, and if the separate digits of the result be again added, and so on, then the final result will be 6: for instance, the sum of 54, 53, and 52 is 159, the sum of the separate digits of 159 is 15, the sum of the separate digits of 15 is 6. To any one confined to the usual Greek numerical notation this must have been a difficult result to prove: possibly it was reached empirically.

The names of two commentators will practically conclude the long roll of Alexandrian mathematicians.

Theon.   The first of these is *Theon of Alexandria*, who flourished about 370. He was not a mathematician of special note, but we are indebted to him for an edition of Euclid's *Elements* and a commentary on the *Almagest*; the latter[1] gives a great deal of miscellaneous

---

[1] It was translated with comments by M. Halma and published at Paris in 1821.

information about the numerical methods used by the Greeks.

Hypatia.   The other was *Hypatia* the daughter of Theon. She was more distinguished than her father, and was the last Alexandrian mathematician of any general reputation: she wrote a commentary on the *Conics* of Apollonius and possibly some other works, but none of her writings are now extant. She was murdered at the instigation of the Christians in 415.

The fate of Hypatia may serve to remind us that the Eastern Christians, as soon as they became the dominant party in the state, showed themselves bitterly hostile to all forms of learning. That very singleness of purpose which had at first so materially aided their progress developed into a one-sidedness which refused to see any good outside their own body; and all who did not actively assist them were persecuted. The final establishment of Christianity in the East marks the end of the Greek scientific schools, though nominally they continued to exist for two hundred years more.

The Athenian School (in the fifth century).[1]

The hostility of the Eastern church to Greek science is further illustrated by the fall of the later Athenian school. This school occupies but a small space in our history. Ever since Plato's time a certain number of professional mathematicians had lived at Athens; and about the year 420 this school again acquired considerable reputation, largely in consequence of the numerous students who after the murder of Hypatia migrated there from Alexandria. Its most celebrated members were Proclus, Damascius, and Eutocius.

Proclus.   *Proclus* was born at Constantinople in February 412 and died at Athens on April 17, 485. He wrote a commentary[2] on the first book of Euclid's *Elements*, which contains a great deal of valuable information on the history of Greek mathematics: he is verbose and dull, but luckily he has preserved for us quotations from other and better authorities. Proclus was succeeded as head of the school by **Marinus**, and Marinus by **Isidorus**.

Damascius. Eutocius.   Two pupils of Isidorus, who in their turn subsequently lectured at Athens, may be mentioned in passing. One

---

[1]See *Untersuchungen über die neu aufgefundenen Scholien des Proklus*, by J. H. Knoche, Herford, 1865.

[2]It has been edited by G. Friedlein, Leipzig, 1873.

of these, *Damascius* of Damascus, circ. 490, is commonly said to have added to Euclid's *Elements* a fifteenth book on the inscription of one regular solid in another, but his authorship of this has been questioned by some writers. The other, *Eutocius*, circ. 510, wrote commentaries on the first four books of the *Conics* of Apollonius and on various works of Archimedes.

This later Athenian school was carried on under great difficulties owing to the opposition of the Christians. Proclus, for example, was repeatedly threatened with death because he was "a philosopher." His remark, "after all my body does not matter, it is the spirit that I shall take with me when I die," which he made to some students who had offered to defend him, has been often quoted. The Christians, after several ineffectual attempts, at last got a decree from Justinian in 529 that "heathen learning" should no longer be studied at Athens. That date therefore marks the end of the Athenian school.

The church at Alexandria was less influential, and the city was more remote from the centre of civil power. The schools there were thus suffered to continue, though their existence was of a precarious character. Under these conditions mathematics continued to be read in Egypt for another hundred years, but all interest in the study had gone.

Roman Mathematics[1]

I ought not to conclude this part of the history without any mention of Roman mathematics, for it was through Rome that mathematics first passed into the curriculum of medieval Europe, and in Rome all modern history has its origin. There is, however, very little to say on the subject. The chief study of the place was in fact the art of government, whether by law, by persuasion, or by those material means on which all government ultimately rests. There were, no doubt, professors who could teach the results of Greek science, but there was no demand for a school of mathematics. Italians who wished to learn more than the elements of the science went to Alexandria or to places which drew their inspiration from Alexandria.

The subject as taught in the mathematical schools at Rome seems to have been confined in arithmetic to the art of calculation (no doubt by the aid of the abacus) and perhaps some of the easier parts of the work

---

[1] The subject is discussed by Cantor, chaps. xxv, xxvi, and xxvii; also by Hankel, pp. 294–304.

of Nicomachus, and in geometry to a few practical rules; though some of the arts founded on a knowledge of mathematics (especially that of surveying) were carried to a high pitch of excellence. It would seem also that special attention was paid to the representation of numbers by signs. The manner of indicating numbers up to ten by the use of fingers must have been in practice from quite early times, but about the first century it had been developed by the Romans into a finger-symbolism by which numbers up to 10,000 or perhaps more could be represented: this would seem to have been taught in the Roman schools. It is described by Bede, and therefore would seem to have been known as far west as Britain; Jerome also alludes to it; its use has still survived in the Persian bazaars.

I am not acquainted with any Latin work on the principles of mechanics, but there were numerous books on the practical side of the subject which dealt elaborately with architectural and engineering problems. We may judge what they were like by the *Mathematici Veteres*, which is a collection of various short treatises on catapults, engines of war, &c.: and by the Κεστοί, written by Sextus Julius Africanus about the end of the second century, part of which is included in the *Mathematici Veteres*, which contains, amongst other things, rules for finding the breadth of a river when the opposite bank is occupied by an enemy, how to signal with a semaphore, &c.

In the sixth century Boethius published a geometry containing a few propositions from Euclid and an arithmetic founded on that of Nicomachus; and about the same time Cassiodorus discussed the foundation of a liberal education which, after the preliminary trivium of grammar, logic, and rhetoric, meant the quadrivium of arithmetic, geometry, music, and astronomy. These works were written at Rome in the closing years of the Athenian and Alexandrian schools, and I therefore mention them here, but as their only value lies in the fact that they became recognized text-books in medieval education I postpone their consideration to chapter VIII.

Theoretical mathematics was in fact an exotic study at Rome; not only was the genius of the people essentially practical, but, alike during the building of their empire, while it lasted, and under the Goths, all the conditions were unfavourable to abstract science.

On the other hand, Alexandria was exceptionally well placed to be a centre of science. From the foundation of the city to its capture by the Mohammedans it was disturbed neither by foreign nor by civil war, save only for a few years when the rule of the Ptolemies gave way to that of Rome: it was wealthy, and its rulers took a pride in endowing the university: and lastly, just as in commerce it became the meeting-place of the east and the west, so it had the good fortune to be the dwelling-place alike of Greeks and of various Semitic people; the one race shewed a peculiar aptitude for geometry, the other for sciences which rest on measurement. Here too, however, as time went on the conditions gradually became more unfavourable, the endless discussions on theological dogmas and the increasing insecurity of the empire tending to divert men's thoughts into other channels.

End of the Second Alexandrian School.

The precarious existence and unfruitful history of the last two centuries of the second Alexandrian School need no record. In 632 Mohammed died, and within ten years his successors had subdued Syria, Palestine, Mesopotamia, Persia, and Egypt. The precise date on which Alexandria fell is doubtful, but the most reliable Arab historians give December 10, 641—a date which at any rate is correct within eighteen months.

With the fall of Alexandria the long history of Greek mathematics came to a conclusion. It seems probable that the greater part of the famous university library and museum had been destroyed by the Christians a hundred or two hundred years previously, and what remained was unvalued and neglected. Some two or three years after the first capture of Alexandria a serious revolt occurred in Egypt, which was ultimately put down with great severity. I see no reason to doubt the truth of the account that after the capture of the city the Mohammedans destroyed such university buildings and collections as were still left. It is said that, when the Arab commander ordered the library to be burnt, the Greeks made such energetic protests that he consented to refer the matter to the caliph Omar. The caliph returned the answer, "As to the books you have mentioned, if they contain what is agreeable with the book of God, the book of God is sufficient without them; and, if they contain what is contrary to the book of God, there is no need for them; so give orders for their destruction." The account goes on to say that they were burnt in the public baths of the city, and that it took six months to consume them all.

CHAPTER VI.

THE BYZANTINE SCHOOL.
641–1453.

It will be convenient to consider the Byzantine school in connection with the history of Greek mathematics. After the capture of Alexandria by the Mohammedans the majority of the philosophers, who previously had been teaching there, migrated to Constantinople, which then became the centre of Greek learning in the East and remained so for 800 years. But though the history of the Byzantine school stretches over so many years—a period about as long as that from the Norman Conquest to the present day—it is utterly barren of any scientific interest; and its chief merit is that it preserved for us the works of the different Greek schools. The revelation of these works to the West in the fifteenth century was one of the most important sources of the stream of modern European thought, and the history of the Byzantine school may be summed up by saying that it played the part of a conduit-pipe in conveying to us the results of an earlier and brighter age.

The time was one of constant war, and men's minds during the short intervals of peace were mainly occupied with theological subtleties and pedantic scholarship. I should not have mentioned any of the following writers had they lived in the Alexandrian period, but in default of any others they may be noticed as illustrating the character of the school. I ought also, perhaps, to call the attention of the reader explicitly to the fact that I am here departing from chronological order, and that the mathematicians mentioned in this chapter were contemporaries of those discussed in the chapters devoted to the mathematics of the middle ages. The Byzantine school was so isolated that I deem this the best arrangement of the subject.

Hero.   One of the earliest members of the Byzantine school was *Hero of Constantinople*, circ. 900, sometimes called the younger to distinguish him from Hero of Alexandria. Hero would seem to have written on geodesy and mechanics as applied to engines of war.

During the tenth century two emperors, Leo VI. and Constantine VII., shewed considerable interest in astronomy and mathematics, but the stimulus thus given to the study of these subjects was only temporary.

Psellus. In the eleventh century *Michael Psellus*, born in 1020, wrote a pamphlet[1] on the quadrivium: it is now in the National Library at Paris.

In the fourteenth century we find the names of three monks who paid attention to mathematics.

Planudes. Barlaam. Argyrus. The first of the three was *Maximus Planudes.*[2] He wrote a commentary on the first two books of the *Arithmetic* of Diophantus; a work on Hindoo arithmetic in which he used the Arabic numerals; and another on proportions which is now in the National Library at Paris. The next was a Calabrian monk named *Barlaam*, who was born in 1290 and died in 1348. He was the author of a work, *Logistic*, on the Greek methods of calculation from which we derive a good deal of information as to the way in which the Greeks treated numerical fractions.[3] Barlaam seems to have been a man of great intelligence. He was sent as an ambassador to the Pope at Avignon, and acquitted himself creditably of a difficult mission; while there he taught Greek to Petrarch. He was famous at Constantinople for the ridicule he threw on the preposterous pretensions of the monks at Mount Athos who taught that those who joined them could, by steadily regarding their bodies, see a mystic light which was the essence of God. Barlaam advised them to substitute the light of reason for that of their bodies—a piece of advice which nearly cost him his life. The last of these monks was *Isaac Argyrus*, who died in 1372. He wrote three astronomical tracts, the manuscripts of which are in the libraries at the Vatican, Leyden, and Vienna: one on geodesy, the manuscript of which is at the Escurial: one on geometry, the manuscript of which is in the National Library at Paris: one on the arithmetic of Nicomachus, the

---

[1] It was printed at Bâle in 1536. Psellus also wrote a *Compendium Mathematicum* which was printed at Leyden in 1647.

[2] His arithmetical commentary was published by Xylander, Bâle, 1575: his work on Hindoo arithmetic, edited by C. J. Gerhardt, was published at Halle, 1865.

[3] Barlaam's *Logistic*, edited by Dasypodius, was published at Strassburg, 1572; another edition was issued at Paris in 1600.

manuscript of which is in the National Library at Paris: and one on trigonometry, the manuscript of which is in the Bodleian at Oxford.

Rhabdas.   In the fourteenth or perhaps the fifteenth century *Nicholas Rhabdas* of *Smyrna* wrote two papers[1] on arithmetic which are now in the National Library at Paris. He gave an account of the finger-symbolism[2] which the Romans had introduced into the East and was then current there.

Pachymeres.   Early in the fifteenth century *Pachymeres* wrote tracts on arithmetic, geometry, and four mechanical machines.

Moschopulus.   A few years later *Emmanuel Moschopulus*, who died in Italy circ. 1460, wrote a treatise on magic squares. A *magic square*[3] consists of a number of integers arranged in the form of a square so that the sum of the numbers in every row, in every column, and in each diagonal is the same. If the integers be the consecutive numbers from 1 to n^2, the square is said to be of the nth order, and in this case the sum of the numbers in any row, column, or diagonal is equal to $\frac{1}{2}n(n^2 + 1)$. Thus the first 16 integers, arranged in either of the forms given below, form a magic square of the fourth order, the sum of the numbers in every row, every column, and each diagonal being 34.

1	15	14	4
12	6	7	9
8	10	11	5
13	3	2	16

15	10	3	6
4	5	16	9
14	11	2	7
1	8	13	12

In the mystical philosophy then current certain metaphysical ideas were often associated with particular numbers, and thus it was natural that such arrangements of numbers should attract attention and be deemed to possess magical properties. The theory of the formation of magic squares is elegant, and several distinguished mathematicians have written on it, but, though interesting, I need hardly say it is not useful. Moschopulus seems to have been the earliest European writer who attempted to deal with the mathematical theory, but his rules

---

[1] They have been edited by S. P. Tannery, Paris, 1886.

[2] See above, page 95.

[3] On the formation and history of magic squares, see my *Mathematical Recreations*, London, ninth edition, 1920, chap. vii. On the work of Moschopulus, see S. Günther's *Geschichte der mathematischen Wissenschaften*, Leipzig, 1876, chap. iv.

apply only to odd squares. The astrologers of the fifteenth and sixteenth centuries were much impressed by such arrangements. In particular the famous Cornelius Agrippa (1486–1535) constructed magic squares of the orders 3, 4, 5, 6, 7, 8, 9, which were associated respectively with the seven astrological "planets," namely, Saturn, Jupiter, Mars, the Sun, Venus, Mercury, and the Moon. He taught that a square of one cell, in which unity was inserted, represented the unity and eternity of God; while the fact that a square of the second order could not be constructed illustrated the imperfection of the four elements, air, earth, fire, and water; and later writers added that it was symbolic of original sin. A magic square engraved on a silver plate was often prescribed as a charm against the plague, and one (namely, that in the first diagram on the last page) is drawn in the picture of melancholy painted about the year 1500 by Albrecht Dürer. Such charms are still worn in the East.

Constantinople was captured by the Turks in 1453, and the last semblance of a Greek school of mathematics then disappeared. Numerous Greeks took refuge in Italy. In the West the memory of Greek science had vanished, and even the names of all but a few Greek writers were unknown; thus the books brought by these refugees came as a revelation to Europe, and, as we shall see later, gave a considerable stimulus to the study of science.

CHAPTER VII.

SYSTEMS OF NUMERATION AND PRIMITIVE ARITHMETIC.[1]

I HAVE in many places alluded to the Greek method of expressing numbers in writing, and I have thought it best to defer to this chapter the whole of what I wanted to say on the various systems of numerical notation which were displaced by the system introduced by the Arabs.

First, as to symbolism and language. The plan of indicating numbers by the digits of one or both hands is so natural that we find it in universal use among early races, and the members of all tribes now extant are able to indicate by signs numbers at least as high as ten: it is stated that in some languages the names for the first ten numbers are derived from the fingers used to denote them. For larger numbers we soon, however, reach a limit beyond which primitive man is unable to count, while as far as language goes it is well known that many tribes have no word for any number higher than ten, and some have no word for any number beyond four, all higher numbers being expressed by the words plenty or heap: in connection with this it is worth remarking that (as stated above) the Egyptians used the symbol for the word heap to denote an unknown quantity in algebra.

The number five is generally represented by the open hand, and it is said that in almost all languages the words five and hand are derived from the same root. It is possible that in early times men did not readily count beyond five, and things if more numerous were

---

[1]The subject of this chapter has been discussed by Cantor and by Hankel. See also the *Philosophy of Arithmetic* by John Leslie, second edition, Edinburgh, 1820. Besides these authorities the article on *Arithmetic* by George Peacock in the *Encyclopaedia Metropolitana, Pure Sciences,* London, 1845; E. B. Tylor's *Primitive Culture,* London, 1873; *Les signes numéraux et l'arithmétique chez les peuples de l'antiquité* ... by T. H. Martin, Rome, 1864; and *Die Zahlzeichen* ... by G. Friedlein, Erlangen, 1869, should be consulted.

counted by multiples of it. It may be that the Roman symbol X for ten represents two "V"s, placed apex to apex, and, if so, this seems to point to a time when things were counted by fives.[1] In connection with this it is worth noticing that both in Java and among the Aztecs a week consisted of five days.

The members of nearly all races of which we have now any knowledge seem, however, to have used the digits of both hands to represent numbers. They could thus count up to and including ten, and therefore were led to take ten as their radix of notation. In the English language, for example, all the words for numbers higher than ten are expressed on the decimal system: those for 11 and 12, which at first sight seem to be exceptions, being derived from Anglo-Saxon words for one and ten and two and ten respectively.

Some tribes seem to have gone further, and by making use of their toes were accustomed to count by multiples of twenty. The Aztecs, for example, are said to have done so. It may be noticed that we still count some things (for instance, sheep) by scores, the word score signifying a notch or scratch made on the completion of the twenty; while the French also talk of quatrevingts, as though at one time they counted things by multiples of twenty. I am not, however, sure whether the latter argument is worth anything, for I have an impression that I have seen the word *octante* in old French books; and there is no question[2] that *septante* and *nonante* were at one time common words for seventy and ninety, and indeed they are still retained in some dialects.

The only tribes of whom I have read who did not count in terms either of five or of some multiple of five are the Bolans of West Africa who are said to have counted by multiples of seven, and the Maories who are said to have counted by multiples of eleven.

Up to ten it is comparatively easy to count, but primitive people find great difficulty in counting higher numbers; apparently at first this difficulty was only overcome by the method (still in use in South Africa) of getting two men, one to count the units up to ten on his fingers, and the other to count the number of groups of ten so formed. To us it is obvious that it is equally effectual to make a mark of some kind on the completion of each group of ten, but it is alleged that the members of many tribes never succeeded in counting numbers higher than ten

---

[1]See also the *Odyssey*, iv, 413–415, in which apparently reference is made to a similar custom.

[2]See, for example, V. M. de Kempten's *Practique...à ciffrer*, Antwerp, 1556.

unless by the aid of two men.

Most races who shewed any aptitude for civilization proceeded further and invented a way of representing numbers by means of pebbles or counters arranged in sets of ten; and this in its turn developed into the abacus or swan-pan. This instrument was in use among nations so widely separated as the Etruscans, Greeks, Egyptians, Hindoos, Chinese, and Mexicans; and was, it is believed, invented independently at several different centres. It is still in common use in Russia, China, and Japan.

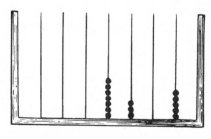

FIGURE 1.

In its simplest form (see Figure 1) the abacus consists of a wooden board with a number of grooves cut in it, or of a table covered with sand in which grooves are made with the fingers. To represent a number, as many counters or pebbles are put on the first groove as there are units, as many on the second as there are tens, and so on. When by its aid a number of objects are counted, for each object a pebble is put on the first groove; and, as soon as there are ten pebbles there, they are taken off and one pebble put on the second groove; and so on. It was sometimes, as in the Aztec *quipus,* made with a number of parallel wires or strings stuck in a piece of wood on which beads could be threaded; and in that form is called a swan-pan. In the number represented in each of the instruments drawn on the next page there are seven thousands, three hundreds, no tens, and five units, that is, the number is 7305. Some races counted from left to right, others from right to left, but this is a mere matter of convention.

The Roman abaci seem to have been rather more elaborate. They contained two marginal grooves or wires, one with four beads to facilitate the addition of fractions whose denominators were four, and one with twelve beads for fractions whose denominators were twelve: but otherwise they do not differ in principle from those described above. They were commonly made to represent numbers up to 100,000,000.

The Greek abaci were similar to the Roman ones. The Greeks and Romans used their abaci as boards on which they played a game something like backgammon.

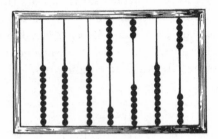

FIGURE 2.

In the Russian *tschotü* (Figure 2) the instrument is improved by having the wires set in a rectangular frame, and ten (or nine) beads are permanently threaded on each of the wires, the wires being considerably longer than is necessary to hold them. If the frame be held horizontal, and all the beads be towards one side, say the lower side of the frame, it is possible to represent any number by pushing towards the other or upper side as many beads on the first wire as there are units in the number, as many beads on the second wire as there are tens in the number, and so on. Calculations can be made somewhat more rapidly if the five beads on each wire next to the upper side be coloured differently to those next to the lower side, and they can be still further facilitated if the first, second, ..., ninth counters in each column be respectively marked with symbols for the numbers 1, 2, ..., 9. Gerbert[1] is said to have introduced the use of such marks, called apices, towards the close of the tenth century.

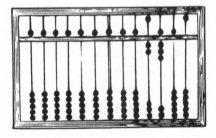

FIGURE 3.

[1]See below, page 114.

Figure 3 represents the form of swan-pan or saroban in common use in China and Japan. There the development is carried one step further, and five beads on each wire are replaced by a single bead of a different form or on a different division, but apices are not used. I am told that an expert Japanese can, by the aid of a swan-pan, add numbers as rapidly as they can be read out to him. It will be noticed that the instrument represented in Figure 3 is made so that two numbers can be expressed at the same time on it.

The use of the abacus in addition and subtraction is evident. It can be used also in multiplication and division; rules for these processes, illustrated by examples, are given in various old works on arithmetic.[1]

The abacus obviously presents a concrete way of representing a number in the decimal system of notation, that is, by means of the local value of the digits. Unfortunately the method of writing numbers developed on different lines, and it was not until about the thirteenth century of our era, when a symbol zero used in conjunction with nine other symbols was introduced, that a corresponding notation in writing was adopted in Europe.

Next, as to the means of representing numbers in writing. In general we may say that in the earliest times a number was (if represented by a sign and not a word) indicated by the requisite number of strokes. Thus in an inscription from Tralles in Caria of the date 398 B.C. the phrase seventh year is represented by $\epsilon\tau\epsilon o\varsigma$ | | | | | | |. These strokes may have been mere marks; or perhaps they originally represented fingers, since in the Egyptian hieroglyphics the symbols for the numbers 1, 2, 3, are one, two, and three fingers respectively, though in the later hieratic writing these symbols had become reduced to straight lines. Additional symbols for 10 and 100 were soon introduced: and the oldest extant Egyptian and Phoenician writings repeat the symbol for unity as many times (up to 9) as was necessary, and then repeat the symbol for ten as many times (up to 9) as was necessary, and so on. No specimens of Greek numeration of a similar kind are in existence, but there is every reason to believe the testimony of Iamblichus who asserts that this was the method by which the Greeks first expressed numbers in writing.

This way of representing numbers remained in current use throughout Roman history; and for greater brevity they or the Etruscans added separate signs for 5, 50, &c. The Roman symbols are generally merely

---

[1] For example in R. Record's *Grounde of Artes*, edition of 1610, London, pp. 225–262.

the initial letters of the names of the numbers; thus C stood for centum or 100, M for mille or 1000. The symbol V for 5 seems to have originally represented an open palm with the thumb extended. The symbols L for 50 and D for 500 are said to represent the upper halves of the symbols used in early times for C and M. The subtractive forms like IV for IIII are probably of a later origin.

Similarly in Attica five was denoted by Π, the first letter of πέντε, or sometimes by Γ; ten by Δ, the initial letter of δέκα; a hundred by H for ἑκατόν; a thousand by X for χίλιοι; while 50 was represented by a Δ written inside a Π; and so on. These Attic symbols continued to be used for inscriptions and formal documents until a late date.

This, if a clumsy, is a perfectly intelligible system; but the Greeks at some time in the third century before Christ abandoned it for one which offers no special advantages in denoting a given number, while it makes all the operations of arithmetic exceedingly difficult. In this, which is known from the place where it was introduced as the Alexandrian system, the numbers from 1 to 9 are represented by the first nine letters of the alphabet; the tens from 10 to 90 by the next nine letters; and the hundreds from 100 to 900 by the next nine letters. To do this the Greeks wanted 27 letters, and as their alphabet contained only 24, they reinserted two letters (the digamma and koppa) which had formerly been in it but had become obsolete, and introduced at the end another symbol taken from the Phoenician alphabet. Thus the ten letters α to ι stood respectively for the numbers from 1 to 10; the next eight letters for the multiples of 10 from 20 to 90; and the last nine letters for 100, 200, etc., up to 900. Intermediate numbers like 11 were represented as the sum of 10 and 1, that is, by the symbol $\iota \alpha'$. This afforded a notation for all numbers up to 999; and by a system of suffixes and indices it was extended so as to represent numbers up to 100,000,000.

There is no doubt that at first the results were obtained by the use of the abacus or some similar mechanical method, and that the signs were only employed to record the result; the idea of operating with the symbols themselves in order to obtain the results is of a later growth, and is one with which the Greeks never became familiar. The non-progressive character of Greek arithmetic may be partly due to their unlucky adoption of the Alexandrian system which caused them for most practical purposes to rely on the abacus, and to supplement it by a table of multiplications which was learnt by heart. The results of the multiplication or division of numbers other than those in the multiplication table might have been obtained by the use of the abacus, but in fact they were generally got by repeated additions and subtractions. Thus, as late as 944, a certain mathematician who in the course of his work wants to multiply 400 by 5 finds the result by addition. The same writer, when he wants to divide 6152 by 15, tries all the multiples of 15 until he gets to 6000, this gives him 400 and a remainder 152; he then begins again with all the multiples of 15 until he gets to 150, and this gives him 10 and a remainder 2. Hence the answer is 410 with a remainder 2.

A few mathematicians, however, such as Hero of Alexandria, Theon, and Eutocius, multiplied and divided in what is essentially the same way as we do. Thus to multiply 18 by 13 they proceeded as follows:—

$$
\begin{aligned}
\iota\gamma + \iota\eta = (\iota + \gamma)(\iota + \eta) \qquad\qquad & 13 \times 18 = (10 + 3)(10 + 8) \\
= \iota(\iota + \eta) + \gamma(\iota + \eta) \qquad\qquad & = 10(10 + 8) + 3(10 + 8) \\
= \rho + \pi + \lambda + \kappa\delta \qquad\qquad & = 100 + 80 + 30 + 24 \\
= \sigma\lambda\delta \qquad\qquad & = 234
\end{aligned}
$$

I suspect that the last step, in which they had to add four numbers together, was obtained by the aid of the abacus.

These, however, were men of exceptional genius, and we must recollect that for all ordinary purposes the art of calculation was performed only by the use of the abacus and the multiplication table, while the term arithmetic was confined to the theories of ratio, proportion, and of numbers.

All the systems here described were more or less clumsy, and they have been displaced among civilized races by the Arabic system in which there are ten digits or symbols, namely, nine for the first nine numbers and another for zero. In this system an integral number is denoted by a succession of digits, each digit representing the product of that digit and a power of ten, and the number being equal to the sum of these products. Thus, by means of the local value attached to nine symbols and a symbol for zero, any number in the decimal scale of notation can be expressed. The history of the development of the science of arithmetic with this notation will be considered below in chapter XI.

SECOND PERIOD.

Mathematics of the Middle Ages and Renaissance.

This period begins about the sixth century, and may be said to end with the invention of analytical geometry and of the infinitesimal calculus. The characteristic feature of this period is the creation or development of modern arithmetic, algebra, and trigonometry.

In this period I consider first, in chapter VIII, the rise of learning in Western Europe, and the mathematics of the middle ages. Next, in chapter IX, I discuss the nature and history of Hindoo and Arabian mathematics, and in chapter X their introduction into Europe. Then, in chapter XI, I trace the subsequent progress of arithmetic to the year 1637. Next, in chapter XII, I treat of the general history of mathematics during the renaissance, from the invention of printing to the beginning of the seventeenth century, say, from 1450 to 1637; this contains an account of the commencement of the modern treatment of arithmetic, algebra, and trigonometry. Lastly, in chapter XIII, I consider the revival of interest in mechanics, experimental methods, and pure geometry which marks the last few years of this period, and serves as a connecting link between the mathematics of the renaissance and the mathematics of modern times.

CHAPTER VIII.

THE RISE OF LEARNING IN WESTERN EUROPE.[1]
CIRC. 600–1200.

Education in the sixth, seventh, and eighth centuries.

THE first few centuries of this second period of our history are singularly barren of interest; and indeed it would be strange if we found science or mathematics studied by those who lived in a condition of perpetual war. Broadly speaking we may say that from the sixth to the eighth centuries the only places of study in western Europe were the Benedictine monasteries. We may find there some slight attempts at a study of literature; but the science usually taught was confined to the use of the abacus, the method of keeping accounts, and a knowledge of the rule by which the date of Easter could be determined. Nor was this unreasonable, for the monk had renounced the world, and there was no reason why he should learn more science than was required for the services of the Church and his monastery. The traditions of Greek and Alexandrian learning gradually died away. Possibly in Rome and a few favoured places copies of the works of the great Greek mathematicians were obtainable though with difficulty, but there were no students, the books were unvalued, and in time became very scarce.

Three authors of the sixth century—Boethius, Cassiodorus, and Isidorus—may be named whose writings serve as a connecting link between the mathematics of classical and of medieval times. As their

---

[1] The mathematics of this period has been discussed by Cantor, by S. Günther, *Geschichte des mathematischen Unterrichtes im deutschen Mittelalter*, Berlin, 1887; and by H. Weissenborn, *Gerbert, Beiträge zur Kenntniss der Mathematik des Mittelalters*, Berlin, 1888; and *Zur Geschichte der Einführung der jetzigen Ziffers*, Berlin, 1892.

works remained standard text-books for some six or seven centuries it is necessary to mention them, but it should be understood that this is the only reason for doing so; they show no special mathematical ability. It will be noticed that these authors were contemporaries of the later Athenian and Alexandrian schools.

Boethius.   *Anicius Manlius Severinus Boethius*, or as the name is sometimes written *Boetius*, born at Rome about 475 and died in 526, belonged to a family which for the two preceding centuries had been esteemed one of the most illustrious in Rome. It was formerly believed that he was educated at Athens: this is somewhat doubtful, but at any rate he was exceptionally well read in Greek literature and science.

Boethius would seem to have wished to devote his life to literary pursuits; but recognizing "that the world would be happy only when kings became philosophers or philosophers kings," he yielded to the pressure put on him and took an active share in politics. He was celebrated for his extensive charities, and, what in those days was very rare, the care that he took to see that the recipients were worthy of them. He was elected consul at an unusually early age, and took advantage of his position to reform the coinage and to introduce the public use of sun-dials, water-clocks, etc. He reached the height of his prosperity in 522 when his two sons were inaugurated as consuls. His integrity and attempts to protect the provincials from the plunder of the public officials brought on him the hatred of the Court. He was sentenced to death while absent from Rome, seized at Ticinum, and in the baptistery of the church there tortured by drawing a cord round his head till the eyes were forced out of the sockets, and finally beaten to death with clubs on October 23, 526. Such at least is the account that has come down to us. At a later time his merits were recognized, and tombs and statues erected in his honour by the state.

Boethius was the last Roman of note who studied the language and literature of Greece, and his works afforded to medieval Europe some glimpse of the intellectual life of the old world. His importance in the history of literature is thus very great, but it arises merely from the accident of the time at which he lived. After the introduction of Aristotle's works in the thirteenth century his fame died away, and he has now sunk into an obscurity which is as great as was once his reputation. He is best known by his *Consolatio*, which was translated by Alfred the Great into Anglo-Saxon. For our purpose it is sufficient to note that the teaching of early medieval mathematics was mainly founded on his geometry and arithmetic.

His *Geometry*[1] consists of the enunciations (only) of the first book of Euclid, and of a few selected propositions in the third and fourth books, but with numerous practical applications to finding areas, etc. He adds an appendix with proofs of the first three propositions to shew that the enunciations may be relied on. His *Arithmetic* is founded on that of Nicomachus.

Cassiodorus.   A few years later another Roman, *Magnus Aurelius Cassiodorus*, who was born about 490 and died in 566, published two works, *De Institutione Divinarum Litterarum* and *De Artibus ac Disciplinis*, in which not only the preliminary trivium of grammar, logic, and rhetoric were discussed, but also the scientific quadrivium of arithmetic, geometry, music, and astronomy. These were considered standard works during the middle ages; the former was printed at Venice in 1598.

Isidorus.   *Isidorus*, bishop of Seville, born in 570 and died in 636, was the author of an encyclopaedic work in twenty volumes called *Origines*, of which the third volume is given up to the quadrivium. It was published at Leipzig in 1833.

The Cathedral and Conventual Schools.[2]

When, in the latter half of the eighth century, Charles the Great had established his empire, he determined to promote learning so far as he was able. He began by commanding that schools should be opened in connection with every cathedral and monastery in his kingdom; an order which was approved and materially assisted by the popes. It is interesting to us to know that this was done at the instance and under the direction of two Englishmen, Alcuin and Clement, who had attached themselves to his court.

Alcuin.[3]   Of these the more prominent was *Alcuin*, who was born in Yorkshire in 735 and died at Tours in 804. He was educated at York under archbishop Egbert, his "beloved master," whom he succeeded as director of the school there. Subsequently he became abbot

---

[1] His works on geometry and arithmetic were edited by G. Friedlein, Leipzig, 1867.

[2] See *The Schools of Charles the Great and the Restoration of Education in the Ninth Century* by J. B. Mullinger, London, 1877.

[3] See the life of Alcuin by F. Lorentz, Halle, 1829, translated by J. M. Slee, London, 1837; *Alcuin und sein Jahrhundert* by K. Werner, Paderborn, 1876; and Cantor, vol. i, pp. 712–721.

of Canterbury, and was sent to Rome by Offa to procure the *pallium* for archbishop Eanbald. On his journey back he met Charles at Parma; the emperor took a great liking to him, and finally induced him to take up his residence at the imperial court, and there teach rhetoric, logic, mathematics, and divinity. Alcuin remained for many years one of the most intimate and influential friends of Charles and was constantly employed as a confidential ambassador; as such he spent the years 791 and 792 in England, and while there reorganized the studies at his old school at York. In 801 he begged permission to retire from the court so as to be able to spend the last years of his life in quiet: with difficulty he obtained leave, and went to the abbey of St. Martin at Tours, of which he had been made head in 796. He established a school in connection with the abbey which became very celebrated, and he remained and taught there till his death on May 19, 804.

Most of the extant writings of Alcuin deal with theology or history, but they include a collection of arithmetical propositions suitable for the instruction of the young. The majority of the propositions are easy problems, either determinate or indeterminate, and are, I presume, founded on works with which he had become acquainted when at Rome. The following is one of the most difficult, and will give an idea of the character of the work. If one hundred bushels of corn be distributed among one hundred people in such a manner that each man receives three bushels, each woman two, and each child half a bushel: how many men, women, and children were there? The general solution is $(20-3n)$ men, $5n$ women, and $(80 - 2n)$ children, where n may have any of the values 1, 2, 3, 4, 5, 6. Alcuin only states the solution for which $n = 3$; that is, he gives as the answer 11 men, 15 women, and 74 children.

This collection however was the work of a man of exceptional genius, and probably we shall be correct in saying that mathematics, if taught at all in a school, was generally confined to the geometry of Boethius, the use of the abacus and multiplication table, and possibly the arithmetic of Boethius; while except in one of these schools or in a Benedictine cloister it was hardly possible to get either instruction or opportunities for study. It was of course natural that the works used should come from Roman sources, for Britain and all the countries included in the empire of Charles had at one time formed part of the western half of the Roman empire, and their inhabitants continued for a long time to regard Rome as the centre of civilization, while the higher clergy kept up a tolerably constant intercourse with Rome.

After the death of Charles many of his schools confined themselves to teaching Latin, music, and theology, some knowledge of which was essential to the worldly success of the higher clergy. Hardly any science or mathematics was taught, but the continued existence of the schools gave an opportunity to any teacher whose learning or zeal exceeded the narrow limits fixed by tradition; and though there were but few who availed themselves of the opportunity, yet the number of those desiring instruction was so large that it would seem as if any one who could teach was sure to attract a considerable audience.

A few schools, where the teachers were of repute, became large and acquired a certain degree of permanence, but even in them the teaching was still usually confined to the trivium and quadrivium. The former comprised the three arts of grammar, logic, and rhetoric, but practically meant the art of reading and writing Latin; nominally the latter included arithmetic and geometry with their applications, especially to music and astronomy, but in fact it rarely meant more than arithmetic sufficient to enable one to keep accounts, music for the church services, geometry for the purpose of land-surveying, and astronomy sufficient to enable one to calculate the feasts and fasts of the church. The seven liberal arts are enumerated in the line, *Lingua, tropus, ratio; numerus, tonus, angulus, astra.* Any student who got beyond the trivium was looked on as a man of great erudition, *Qui tria, qui septem, qui totum scibile novit,* as a verse of the eleventh century runs. The special questions which then and long afterwards attracted the best thinkers were logic and certain portions of transcendental theology and philosophy.

We may sum the matter up by saying that during the ninth and tenth centuries the mathematics taught was still usually confined to that comprised in the two works of Boethius together with the practical use of the abacus and the multiplication table, though during the latter part of the time a wider range of reading was undoubtedly accessible.

Gerbert.[1] In the tenth century a man appeared who would in any age have been remarkable and who gave a great stimulus to learning. This was *Gerbert*, an Aquitanian by birth, who died in 1003 at about the age of fifty. His abilities attracted attention to him even when a boy, and procured his removal from the abbey school at Aurillac to the Spanish march where he received a good education. He was in Rome in

---

[1]Weissenborn, in the works already mentioned, treats Gerbert very fully; see also *La Vie et les Œuvres de Gerbert*, by A. Olleris, Clermont, 1867; *Gerbert von Aurillac*, by K. Werner, second edition, Vienna, 1881; and *Gerberti ... Opera mathematica*, edited by N. Bubnov, Berlin, 1899.

971, where his proficiency in music and astronomy excited considerable interest: but his interests were not confined to these subjects, and he had already mastered all the branches of the trivium and quadrivium, as then taught, except logic; and to learn this he moved to Rheims, which Archbishop Adalbero had made the most famous school in Europe. Here he was at once invited to teach, and so great was his fame that to him Hugh Capet entrusted the education of his son Robert who was afterwards king of France.

Gerbert was especially famous for his construction of abaci and of terrestrial and celestial globes; he was accustomed to use the latter to illustrate his lectures. These globes excited great admiration; and he utilized this by offering to exchange them for copies of classical Latin works, which seem already to have become very scarce; the better to effect this he appointed agents in the chief towns of Europe. To his efforts it is believed we owe the preservation of several Latin works. In 982 he received the abbey of Bobbio, and the rest of his life was taken up with political affairs; he became Archbishop of Rheims in 991, and of Ravenna in 998; in 999 he was elected Pope, when he took the title of Sylvester II.; as head of the Church, he at once commenced an appeal to Christendom to arm and defend the Holy Land, thus forestalling Peter the Hermit by a century, but he died on May 12, 1003, before he had time to elaborate his plans. His library is, I believe, preserved in the Vatican.

So remarkable a personality left a deep impress on his generation, and all sorts of fables soon began to collect around his memory. It seems certain that he made a clock which was long preserved at Magdeburg, and an organ worked by steam which was still at Rheims two centuries after his death. All this only tended to confirm the suspicions of his contemporaries that he had sold himself to the devil; and the details of his interviews with that gentleman, the powers he purchased, and his effort to escape from his bargain when he was dying, may be read in the pages of William of Malmesbury, Orderic Vitalis, and Platina. To these anecdotes the first named writer adds the story of the statue inscribed with the words "strike here," which having amused our ancestors in the *Gesta Romanorum* has been recently told again in the *Earthly Paradise*.

Extensive though his influence was, it must not be supposed that Gerbert's writings shew any great originality. His mathematical works comprise a treatise on arithmetic entitled *De Numerorum Divisione*, and one on *geometry*. An improvement in the abacus, attributed by some writers to Boethius, but which is more likely due to Gerbert, is

the introduction in every column of beads marked by different characters, called *apices*, for each of the numbers from 1 to 9, instead of nine
exactly similar counters or beads. These apices lead to a representation of numbers essentially the same as the Arabic numerals. There
was however no symbol for zero; the step from this concrete system
of denoting numbers by a decimal system on an abacus to the system
of denoting them by similar symbols in writing seems to us to be a
small one, but it would appear that Gerbert did not make it. He found
at Mantua a copy of the geometry of Boethius, and introduced it into
the medieval schools. Gerbert's own work on geometry is of unequal
ability; it includes a few applications to land-surveying and the determination of the heights of inaccessible objects, but much of it seems
to be copied from some Pythagorean text-book. In the course of it he
however solves one problem which was of remarkable difficulty for that
time. The question is to find the sides of a right-angled triangle whose
hypotenuse and area are given. He says, in effect, that if these latter
be denoted respectively by c and h^2, then the lengths of the two sides
will be

$$\tfrac{1}{2}\left\{ \sqrt{c^2 + 4h^2} + \sqrt{c^2 - 4h^2} \right\} \text{ and } \tfrac{1}{2}\left\{ \sqrt{c^2 + 4h^2} - \sqrt{c^2 - 4h^2} \right\}.$$

Bernelinus.   One of Gerbert's pupils, *Bernelinus*, published a
work on the abacus[1] which is, there is very little doubt, a reproduction
of the teaching of Gerbert. It is valuable as indicating that the Arabic
system of writing numbers was still unknown in Europe.

The Early Medieval Universities.[2]

At the end of the eleventh century or the beginning of the twelfth a
revival of learning took place at several of these cathedral or monastic
schools; and in some cases, at the same time, teachers who were not
members of the school settled in its vicinity and, with the sanction of
the authorities, gave lectures which were in fact always on theology,
logic, or civil law. As the students at these centres grew in numbers,
it became desirable to act together whenever any interest common to
all was concerned. The association thus formed was a sort of guild or

---

[1]It is reprinted in Olleris's edition of Gerbert's works, pp. 311–326.

[2]See the *Universities of Europe in the Middle Ages* by H. Rashdall, Oxford, 1895;
Die Universitäten des Mittelalters bis 1400 by P. H. Denifle, 1885; and vol. i of the
University of Cambridge by J. B. Mullinger, Cambridge, 1873.

trades union, or in the language of the time a *universitas magistrorum et scholarium*. This was the first stage in the development of the earliest medieval universities. In some cases, as at Paris, the governing body of the university was formed by the teachers alone, in others, as at Bologna, by both teachers and students; but in all cases precise rules for the conduct of business and the regulation of the internal economy of the guild were formulated at an early stage in its history. The municipalities and numerous societies which existed in Italy supplied plenty of models for the construction of such rules, but it is possible that some of the regulations were derived from those in force in the Mohammedan schools at Cordova.

We are, almost inevitably, unable to fix the exact date of the commencement of these voluntary associations, but they existed at Paris, Bologna, Salerno, Oxford, and Cambridge before the end of the twelfth century: these may be considered the earliest universities in Europe. The instruction given at Salerno and Bologna was mainly technical—at Salerno in medicine, and at Bologna in law—and their claim to recognition as universities, as long as they were merely technical schools, has been disputed.

Although the organization of these early universities was independent of the neighbouring church and monastic schools they seem in general to have been, at any rate originally, associated with such schools, and perhaps indebted to them for the use of rooms, etc. The universities or guilds (self-governing and formed by teachers and students), and the adjacent schools (under the direct control of church or monastic authorities), continued to exist side by side, but in course of time the latter diminished in importance, and often ended by becoming subject to the rule of the university authorities. Nearly all the medieval universities grew up under the protection of a bishop (or abbot), and were in some matters subject to his authority or to that of his chancellor, from the latter of whom the head of the university subsequently took his title. The universities, however, were not ecclesiastical organizations, and, though the bulk of their members were ordained, their direct connection with the Church arose chiefly from the fact that clerks were then the only class of the community who were left free by the state to pursue intellectual studies.

A *universitas magistrorum et scholarium*, if successful in attracting students and acquiring permanency, always sought special legal privileges, such as the right to fix the price of provisions and the power to try legal actions in which its members were concerned. These privileges generally led to a recognition of its power to grant degrees which conferred a right of teaching anywhere within the kingdom. The university was frequently incorporated at or about the same time. Paris received its charter in 1200, and probably was the earliest university in Europe thus officially recognized. Legal privileges were conferred on Oxford in 1214, and on Cambridge in 1231: the development of Oxford and Cambridge followed closely the precedent of Paris on which their organization was modelled. In the course of the thirteenth century universities were founded at (among other places) Naples, Orleans, Padua, and Prague; and in the course of the fourteenth century at Pavia and Vienna. The title of university was generally accredited to any teaching body as soon as it was recognized as a *studium generale*.

The most famous medieval universities aspired to a still wider recognition, and the final step in their evolution was an acknowledgment by the pope or emperor of their degrees as a title to teach throughout Christendom—such universities were closely related one with the other. Paris was thus recognized in 1283, Oxford in 1296, and Cambridge in 1318.

The standard of education in mathematics has been largely fixed by the universities, and most of the mathematicians of subsequent times have been closely connected with one or more of them; and therefore I may be pardoned for adding a few words on the general course of studies[1] in a university in medieval times.

The students entered when quite young, sometimes not being more than eleven or twelve years old when first coming into residence. It is misleading to describe them as undergraduates, for their age, their studies, the discipline to which they were subjected, and their position in the university shew that they should be regarded as schoolboys. The first four years of their residence were supposed to be spent in the study of the trivium, that is, Latin grammar, logic, and rhetoric. In quite early times, a considerable number of the students did not progress beyond the study of Latin grammar—they formed an inferior faculty and were eligible only for the degree of master of grammar or master of rhetoric—but the more advanced students (and in later times all students) spent these years in the study of the trivium.

The title of bachelor of arts was conferred at the end of this course,

---

[1] For fuller details as to their organization of studies, their system of instruction, and their constitution, see my *History of the Study of Mathematics at Cambridge*, Cambridge, 1889.

and signified that the student was no longer a schoolboy and therefore in pupilage. The average age of a commencing bachelor may be taken as having been about seventeen or eighteen. Thus at Cambridge in the presentation for a degree the technical term still used for an undergraduate is *juvenis*, while that for a bachelor is *vir*. A bachelor could not take pupils, could teach only under special restrictions, and probably occupied a position closely analogous to that of an undergraduate nowadays. Some few bachelors proceeded to the study of civil or canon law, but it was assumed in theory that they next studied the quadrivium, the course for which took three years, and which included about as much science as was to be found in the pages of Boethius and Isidorus.

The degree of master of arts was given at the end of this course. In the twelfth and thirteenth centuries it was merely a license to teach: no one sought it who did not intend to use it for that purpose and to reside in the university, and only those who had a natural aptitude for such work were likely to enter a profession so ill-paid as that of a teacher. The degree was obtainable by any student who had gone through the recognized course of study, and shewn that he was of good moral character. Outsiders were also admitted, but not as a matter of course. I may here add that towards the end of the fourteenth century students began to find that a degree had a pecuniary value, and most universities subsequently conferred it only on condition that the new master should reside and teach for at least a year. Somewhat later the universities took a further step and began to refuse degrees to those who were not intellectually qualified. This power was assumed on the precedent of a case which arose in Paris in 1426, when the university declined to confer a degree on a student—a Slavonian, one Paul Nicholas—who had performed the necessary exercises in a very indifferent manner: he took legal proceedings to compel the university to grant the degree, but their right to withhold it was established. Nicholas accordingly has the distinction of being the first student who under modern conditions was "plucked."

Although science and mathematics were recognised as the standard subjects of study for a bachelor, it is probable that, until the renaissance, the majority of the students devoted most of their time to logic, philosophy, and theology. The subtleties of scholastic philosophy were dreary and barren, but it is only just to say that they provided a severe intellectual training.

We have now arrived at a time when the results of Arab and Greek science became known in Europe. The history of Greek mathematics has been already discussed; I must now temporarily leave the subject of medieval mathematics, and trace the development of the Arabian schools to the same date; and I must then explain how the schoolmen became acquainted with the Arab and Greek text-books, and how their introduction affected the progress of European mathematics.

CHAPTER IX.

THE MATHEMATICS OF THE ARABS.[1]

THE story of Arab mathematics is known to us in its general outlines, but we are as yet unable to speak with certainty on many of its
details. It is, however, quite clear that while part of the early knowledge of the Arabs was derived from Greek sources, part was obtained
from Hindoo works; and that it was on those foundations that Arab
science was built. I will begin by considering in turn the extent of
mathematical knowledge derived from these sources.

Extent of Mathematics obtained from Greek Sources.

According to their traditions, in themselves very probable, the scientific knowledge of the Arabs was at first derived from the Greek doctors who attended the caliphs at Bagdad. It is said that when the Arab
conquerors settled in towns they became subject to diseases which had
been unknown to them in their life in the desert. The study of medicine
was then confined mainly to Greeks and Jews, and many of these, encouraged by the caliphs, settled at Bagdad, Damascus, and other cities;
their knowledge of all branches of learning was far more extensive and
accurate than that of the Arabs, and the teaching of the young, as has

---

[1]The subject is discussed at length by Cantor, chaps. xxxii–xxxv; by Hankel,
pp. 172–293; by A. von Kremer in *Kulturgeschichte des Orientes unter den Chalifen*,
Vienna, 1877; and by H. Suter in his "Die Mathematiker und Astronomen der
Araber und ihre Werke," *Zeitschrift für Mathematik und Physik, Abhandlungen zur
Geschichte der Mathematik*, Leipzig, vol. xlv, 1900. See also *Matériaux pour servir
à l'histoire comparée des sciences mathématiques chez les Grecs et les Orientaux*,
by L. A. Sédillot, Paris, 1845–9; and the following articles by Fr. Woepcke, *Sur
l'introduction de l'arithmétique Indienne en Occident*, Rome, 1859; *Sur l'histoire
des sciences mathématiques chez les Orientaux*, Paris, 1860; and *Mémoire sur la
propagation des chiffres Indiens*, Paris, 1863.

often happened in similar cases, fell into their hands. The introduction of European science was rendered the more easy as various small Greek schools existed in the countries subject to the Arabs: there had for many years been one at Edessa among the Nestorian Christians, and there were others at Antioch, Emesa, and even at Damascus, which had preserved the traditions and some of the results of Greek learning.

The Arabs soon remarked that the Greeks rested their medical science on the works of Hippocrates, Aristotle, and Galen; and these books were translated into Arabic by order of the caliph Haroun Al Raschid about the year 800. The translation excited so much interest that his successor Al Mamun (813–833) sent a commission to Constantinople to obtain copies of as many scientific works as was possible, while an embassy for a similar purpose was also sent to India. At the same time a large staff of Syrian clerks was engaged, whose duty it was to translate the works so obtained into Arabic and Syriac. To disarm fanaticism these clerks were at first termed the caliph's doctors, but in 851 they were formed into a college, and their most celebrated member, Honein ibn Ishak, was made its first president by the caliph Mutawakkil (847–861). Honein and his son Ishak ibn Honein revised the translations before they were finally issued. Neither of them knew much mathematics, and several blunders were made in the works issued on that subject, but another member of the college, Tabit ibn Korra, shortly published fresh editions which thereafter became the standard texts.

In this way before the end of the ninth century the Arabs obtained translations of the works of Euclid, Archimedes, Apollonius, Ptolemy, and others; and in some cases these editions are the only copies of the books now extant. It is curious, as indicating how completely Diophantus had dropped out of notice, that as far as we know the Arabs got no manuscript of his great work till 150 years later, by which time they were already acquainted with the idea of algebraic notation and processes.

Extent of Mathematics obtained from Hindoo Sources.

The Arabs had considerable commerce with India, and a knowledge of one or both of the two great original Hindoo works on algebra had been thus obtained in the caliphate of Al Mansur (754–775), though it was not until fifty or sixty years later that they attracted much attention. The algebra and arithmetic of the Arabs were largely founded on these treatises, and I therefore devote this section to the consideration of Hindoo mathematics.

The Hindoos, like the Chinese, have pretended that they are the most ancient people on the face of the earth, and that to them all sciences owe their creation. But it is probable that these pretensions have no foundation; and in fact no science or useful art (except a rather fantastic architecture and sculpture) can be definitely traced back to the inhabitants of the Indian peninsula prior to the Aryan invasion. This invasion seems to have taken place at some time in the latter half of the fifth century or in the sixth century, when a tribe of Aryans entered India by the north-west frontier, and established themselves as rulers over a large part of the country. Their descendants, wherever they have kept their blood pure, may still be recognised by their superiority over the races they originally conquered; but as is the case with the modern Europeans, they found the climate trying and gradually degenerated. For the first two or three centuries they, however, retained their intellectual vigour, and produced one or two writers of great ability.

Arya-Bhata.   The earliest of these, of whom we have definite information, is *Arya-Bhata*,[1] who was born at Patna in the year 476. He is frequently quoted by Brahmagupta, and in the opinion of many commentators he created algebraic analysis, though it has been suggested that he may have seen Diophantus's *Arithmetic*. The chief work of Arya-Bhata with which we are acquainted is his *Aryabhathiya*, which consists of mnemonic verses embodying the enunciations of various rules and propositions. There are no proofs, and the language is so obscure and concise that it long defied all efforts to translate it.

The book is divided into four parts: of these three are devoted to astronomy and the elements of spherical trigonometry; the remaining part contains the enunciations of thirty-three rules in arithmetic, algebra, and plane trigonometry. It is probable that Arya-Bhata regarded himself as an astronomer, and studied mathematics only so far as it was useful to him in his astronomy.

In algebra Arya-Bhata gives the sum of the first, second, and third powers of the first n natural numbers; the general solution of a quadratic

---

[1]The subject of prehistoric Indian mathematics has been discussed by G. Thibaut, Von Schroeder, and H. Vogt. A Sanskrit text of the *Aryabhathiya*, edited by H. Kern, was published at Leyden in 1874; there is also an article on it by the same editor in the *Journal of the Asiatic Society*, London, 1863, vol. xx, pp. 371–387; a French translation by L. Rodet of that part which deals with algebra and trigonometry is given in the *Journal Asiatique*, 1879, Paris, series 7, vol. xiii, pp. 393–434.

equation; and the solution in integers of certain indeterminate equations of the first degree. His solutions of numerical equations have been supposed to imply that he was acquainted with the decimal system of enumeration.

In trigonometry he gives a table of natural sines of the angles in the first quadrant, proceeding by multiples of $3\frac{3}{4}°$, defining a sine as the semi-chord of double the angle. Assuming that for the angle $3\frac{3}{4}°$ the sine is equal to the circular measure, he takes for its value 225, *i.e.* the number of minutes in the angle. He then enunciates a rule which is nearly unintelligible, but probably is the equivalent of the statement

$$\sin(n+1)\alpha - \sin n\alpha = \sin n\alpha - \sin(n-1)\alpha - \sin n\alpha \operatorname{cosec} \alpha,$$

where α stands for $3\frac{3}{4}°$; and working with this formula he constructs a table of sines, and finally finds the value of $\sin 90°$ to be 3438. This result is correct if we take 3.1416 as the value of π, and it is interesting to note that this is the number which in another place he gives for π. The correct trigonometrical formula is

$$\sin(n+1)\alpha - \sin n\alpha = \sin n\alpha - \sin(n-1)\alpha - 4\sin n\alpha \sin^2 \tfrac{1}{2}\alpha.$$

Arya-Bhata, therefore, took $4\sin^2 \frac{1}{2}\alpha$ as equal to $\operatorname{cosec} \alpha$, that is, he supposed that $2\sin\alpha = 1 + \sin 2\alpha$: using the approximate values of $\sin\alpha$ and $\sin 2\alpha$ given in his table, this reduces to $2(225) = 1 + 449$, and hence to that degree of approximation his formula is correct. A considerable proportion of the geometrical propositions which he gives is wrong.

Brahmagupta.   The next Hindoo writer of note is *Brahmagupta*, who is said to have been born in 598, and probably was alive about 660. He wrote a work in verse entitled *Brahma-Sphuta-Siddhanta*, that is, the *Siddhanta*, or system of Brahma in astronomy. In this, two chapters are devoted to arithmetic, algebra, and geometry.[1]

The arithmetic is entirely rhetorical. Most of the problems are worked out by the rule of three, and a large proportion of them are on the subject of interest.

In his algebra, which is also rhetorical, he works out the fundamental propositions connected with an arithmetical progression, and solves a quadratic equation (but gives only the positive value to the radical). As

---

[1] These two chapters (chaps. xii and xviii) were translated by H. T. Colebrooke, and published at London in 1817.

an illustration of the problems given I may quote the following, which was reproduced in slightly different forms by various subsequent writers, but I replace the numbers by letters. "Two apes lived at the top of a cliff of height h, whose base was distant mh from a neighbouring village. One descended the cliff and walked to the village, the other flew up a height x and then flew in a straight line to the village. The distance traversed by each was the same. Find x." Brahmagupta gave the correct answer, namely $x = mh/(m+2)$. In the question as enunciated originally $h = 100$, $m = 2$.

Brahmagupta finds solutions in integers of several indeterminate equations of the first degree, using the same method as that now practised. He states one indeterminate equation of the second degree, namely, $nx^2 + 1 = y^2$, and gives as its solution $x = 2t/(t^2 - n)$ and $y = (t^2 + n)/(t^2 - n)$. To obtain this general form he proved that, if one solution either of that or of certain allied equations could be guessed, the general solution could be written down; but he did not explain how one solution could be obtained. Curiously enough this equation was sent by Fermat as a challenge to Wallis and Lord Brouncker in the seventeenth century, and the latter found the same solutions as Brahmagupta had previously done. Brahmagupta also stated that the equation $y^2 = nx^2 - 1$ could not be satisfied by integral values of x and y unless n could be expressed as the sum of the squares of two integers. It is perhaps worth noticing that the early algebraists, whether Greeks, Hindoos, Arabs, or Italians, drew no distinction between the problems which led to determinate and those which led to indeterminate equations. It was only after the introduction of syncopated algebra that attempts were made to give general solutions of equations, and the difficulty of giving such solutions of indeterminate equations other than those of the first degree has led to their practical exclusion from elementary algebra.

In geometry Brahmagupta proved the Pythagorean property of a right-angled triangle (Euc. I, 47). He gave expressions for the area of a triangle and of a quadrilateral inscribable in a circle in terms of their sides; and shewed that the area of a circle was equal to that of a rectangle whose sides were the radius and semiperimeter. He was less successful in his attempt to rectify a circle, and his result is equivalent to taking $\sqrt{10}$ for the value of π. He also determined the surface and volume of a pyramid and cone; problems over which Arya-Bhata had blundered badly. The next part of his geometry is almost unintelligible, but it seems to be an attempt to find expressions for several magnitudes connected with a quadrilateral inscribed in a circle in terms of its sides: much of this is wrong.

It must not be supposed that in the original work all the propositions which deal with any one subject are collected together, and it is only for convenience that I have tried to arrange them in that way. It is impossible to say whether the whole of Brahmagupta's results given above are original. He knew of Arya-Bhata's work, for he reproduces the table of sines there given; it is likely also that some progress in mathematics had been made by Arya-Bhata's immediate successors, and that Brahmagupta was acquainted with their works; but there seems no reason to doubt that the bulk of Brahmagupta's algebra and arithmetic is original, although perhaps influenced by Diophantus's writings: the origin of the geometry is more doubtful, probably some of it is derived from Hero's works, and maybe some represents indigenous Hindoo work.

Bhaskara.   To make this account of Hindoo mathematics complete I may depart from the chronological arrangement and say that the only remaining Indian mathematician of exceptional eminence of whose works we know anything was *Bhaskara,* who was born in 1114. He is said to have been the lineal successor of Brahmagupta as head of an astronomical observatory at Ujein. He wrote an astronomy, of which four chapters have been translated. Of these one termed *Lilavati* is on arithmetic; a second termed *Bija Ganita* is on algebra; the third and fourth are on astronomy and the sphere;[1] some of the other chapters also involve mathematics. This work was, I believe, known to the Arabs almost as soon as it was written, and influenced their subsequent writings, though they failed to utilize or extend most of the discoveries contained in it. The results thus became indirectly known in the West before the end of the twelfth century, but the text itself was not introduced into Europe till within recent times.

The treatise is in verse, but there are explanatory notes in prose. It is not clear whether it is original or whether it is merely an exposition of the results then known in India; but in any case it is most probable that Bhaskara was acquainted with the Arab works which had been written in the tenth and eleventh centuries, and with the results of Greek mathematics as transmitted through Arabian sources. The algebra is

[1]See the article *Viga Ganita* in the *Penny Cyclopaedia*, London, 1843; and the translations of the *Lilavati* and the *Bija Ganita* issued by H. T. Colebrooke, London, 1817. The chapters on astronomy and the sphere were edited by L. Wilkinson, Calcutta, 1842.

syncopated and almost symbolic, which marks a great advance over that of Brahmagupta and of the Arabs. The geometry is also superior to that of Brahmagupta, but apparently this is due to the knowledge of various Greek works obtained through the Arabs.

The first book or *Lilavati* commences with a salutation to the god of wisdom. The general arrangement of the work may be gathered from the following table of contents. Systems of weights and measures. Next decimal numeration, briefly described. Then the eight operations of arithmetic, namely, addition, subtraction, multiplication, division, square, cube, square-root, and cube-root. Reduction of fractions to a common denominator, fractions of fractions, mixed numbers, and the eight rules applied to fractions. The "rules of cipher," namely, $a \pm 0 = a$, $0^2 = 0$, $\sqrt{0} = 0$, $a \div 0 = \infty$. The solution of some simple equations which are treated as questions of arithmetic. The rule of false assumption. Simultaneous equations of the first degree with applications. Solution of a few quadratic equations. Rule of three and compound rule of three, with various cases. Interest, discount, and partnership. Time of filling a cistern by several fountains. Barter. Arithmetical progressions, and sums of squares and cubes. Geometrical progressions. Problems on triangles and quadrilaterals. Approximate value of π. Some trigonometrical formulae. Contents of solids. Indeterminate equations of the first degree. Lastly, the book ends with a few questions on combinations.

This is the earliest known work which contains a systematic exposition of the decimal system of numeration. It is possible that Arya-Bhata was acquainted with it, and it is most likely that Brahmagupta was so, but in Bhaskara's arithmetic we meet with the Arabic or Indian numerals and a sign for zero as part of a well-recognised notation. It is impossible at present to definitely trace these numerals farther back than the eighth century, but there is no reason to doubt the assertion that they were in use at the beginning of the seventh century. Their origin is a difficult and disputed question. I mention below[1] the view which on the whole seems most probable, and perhaps is now generally accepted, and I reproduce there some of the forms used in early times.

To sum the matter up briefly, it may be said that the *Lilavati* gives the rules now current for addition, subtraction, multiplication, and division, as well as for the more common processes in arithmetic; while the greater part of the work is taken up with the discussion of the

---

[1]See below, page 152.

rule of three, which is divided into direct and inverse, simple and compound, and is used to solve numerous questions chiefly on interest and exchange—the numerical questions being expressed in the decimal system of notation with which we are familiar.

Bhaskara was celebrated as an astrologer no less than as a mathematician. He learnt by this art that the event of his daughter Lilavati marrying would be fatal to himself. He therefore declined to allow her to leave his presence, but by way of consolation he not only called the first book of his work by her name, but propounded many of his problems in the form of questions addressed to her. For example, "Lovely and dear Lilavati, whose eyes are like a fawn's, tell me what are the numbers resulting from 135 multiplied by 12. If thou be skilled in multiplication, whether by whole or by parts, whether by division or by separation of digits, tell me, auspicious damsel, what is the quotient of the product when divided by the same multiplier."

I may add here that the problems in the Indian works give a great deal of interesting information about the social and economic condition of the country in which they were written. Thus Bhaskara discusses some questions on the price of slaves, and incidentally remarks that a female slave was generally supposed to be most valuable when 16 years old, and subsequently to decrease in value in inverse proportion to the age; for instance, if when 16 years old she were worth 32 nishkas, her value when 20 would be represented by $(16 \times 32) \div 20$ nishkas. It would appear that, as a rough average, a female slave of 16 was worth about 8 oxen which had worked for two years. The interest charged for money in India varied from $3\frac{1}{2}$ to 5 per cent per month. Amongst other data thus given will be found the prices of provisions and labour.

The chapter termed *Bija Ganita* commences with a sentence so ingeniously framed that it can be read as the enunciation of a religious, or a philosophical, or a mathematical truth. Bhaskara after alluding to his *Lilavati*, or arithmetic, states that he intends in this book to proceed to the general operations of analysis. The idea of the notation is as follows. Abbreviations and initials are used for symbols; subtraction is indicated by a dot placed above the coefficient of the quantity to be subtracted; addition by juxtaposition merely; but no symbols are used for multiplication, equality, or inequality, these being written at length. A product is denoted by the first syllable of the word subjoined to the factors, between which a dot is sometimes placed. In a quotient or fraction the divisor is written under the dividend without a line of separation. The two sides of an equation are written one under the other, confusion being prevented by the recital in words of all the steps which accompany the operation. Various symbols for the unknown quantity are used, but most of them are the initials of names of colours, and the word colour is often used as synonymous with unknown quantity; its Sanskrit equivalent also signifies a letter, and letters are sometimes used either from the alphabet or from the initial syllables of subjects of the problem. In one or two cases symbols are used for the given as well as for the unknown quantities. The initials of the words square and solid denote the second and third powers, and the initial syllable of square root marks a surd. Polynomials are arranged in powers, the absolute quantity being always placed last and distinguished by an initial syllable denoting known quantity. Most of the equations have numerical coefficients, and the coefficient is always written after the unknown quantity. Positive or negative terms are indiscriminately allowed to come first; and every power is repeated on both sides of an equation, with a zero for the coefficient when the term is absent. After explaining his notation, Bhaskara goes on to give the rules for addition, subtraction, multiplication, division, squaring, and extracting the square root of algebraical expressions; he then gives the rules of cipher as in the *Lilavati*; solves a few equations; and lastly concludes with some operations on surds. Many of the problems are given in a poetical setting with allusions to fair damsels and gallant warriors.

Fragments of other chapters, involving algebra, trigonometry, and geometrical applications, have been translated by Colebrooke. Amongst the trigonometrical formulae is one which is equivalent to the equation $d(\sin\theta) = \cos\theta\, d\theta$.

I have departed from the chronological order in treating here of Bhaskara, but I thought it better to mention him at the same time as I was discussing his compatriots. It must be remembered, however, that he flourished subsequently to all the Arab mathematicians considered in the next section. The works with which the Arabs first became acquainted were those of Arya-Bhata and Brahmagupta, and perhaps of their successors Sridhara and Padmanabha; it is doubtful if they ever made much use of the great treatise of Bhaskara.

It is probable that the attention of the Arabs was called to the works of the first two of these writers by the fact that the Arabs adopted the Indian system of arithmetic, and were thus led to look at the mathematical text-books of the Hindoos. The Arabs had always had considerable commerce with India, and with the establishment of their empire the amount of trade naturally increased; at that time, about the year 700, they found the Hindoo merchants beginning to use the system of numeration with which we are familiar, and adopted it at once. This immediate acceptance of it was made the easier, as they had no works of science or literature in which another system was used, and it is doubtful whether they then possessed any but the most primitive system of notation for expressing numbers. The Arabs, like the Hindoos, seem also to have made little or no use of the abacus, and therefore must have found Greek and Roman methods of calculation extremely laborious. The earliest definite date assigned for the use in Arabia of the decimal system of numeration is 773. In that year some Indian astronomical tables were brought to Bagdad, and it is almost certain that in these Indian numerals (including a zero) were employed.

The Development of Mathematics in Arabia.[1]

In the preceding sections of this chapter I have indicated the two sources from which the Arabs derived their knowledge of mathematics, and have sketched out roughly the amount of knowledge obtained from each. We may sum the matter up by saying that before the end of the eighth century the Arabs were in possession of a good numerical notation and of Brahmagupta's work on arithmetic and algebra; while before the end of the ninth century they were acquainted with the masterpieces of Greek mathematics in geometry, mechanics, and astronomy. I have now to explain what use they made of these materials.

Alkarismi.   The first and in some respects the most illustrious of the Arabian mathematicians was *Mohammed ibn Musa Abu Djefar Al-Khwārizmī*. There is no common agreement as to which of these names is the one by which he is to be known: the last of them refers to the place where he was born, or in connection with which he was best known, and I am told that it is the one by which he would have been usually known among his contemporaries. I shall therefore refer to him by that name; and shall also generally adopt the corresponding titles to designate the other Arabian mathematicians. Until recently, this was almost always written in the corrupt form *Alkarismi*, and, though this way of spelling it is incorrect, it has been sanctioned by so many writers that I shall make use of it.

---

[1] A work by B. Baldi on the lives of several of the Arab mathematicians was printed in Boncompagni's *Bulletino di bibliografia*. 1872, vol. v, pp. 427–534.

We know nothing of Alkarismi's life except that he was a native of Khorassan and librarian of the caliph Al Mamun; and that he accompanied a mission to Afghanistan, and possibly came back through India. On his return, about 830, he wrote an algebra,[1] which is founded on that of Brahmagupta, but in which some of the proofs rest on the Greek method of representing numbers by lines. He also wrote a treatise on arithmetic: an anonymous tract termed *Algoritmi De Numero Indorum*, which is in the university library at Cambridge, is believed to be a Latin translation of this treatise.[2] Besides these two works he compiled some astronomical tables, with explanatory remarks; these included results taken from both Ptolemy and Brahmagupta.

The algebra of Alkarismi holds a most important place in the history of mathematics, for we may say that the subsequent Arab and the early medieval works on algebra were founded on it, and also that through it the Arabic or Indian system of decimal numeration was introduced into the West. The work is termed *Al-gebr we' l mukabala: al-gebr*, from which the word algebra is derived, means *the restoration*, and refers to the fact that any the same magnitude may be added to or subtracted from both sides of an equation; *al mukabala* means the process of simplification, and is generally used in connection with the combination of like terms into a single term. The unknown quantity is termed either "the thing" or "the root" (that is, of a plant), and from the latter phrase our use of the word root as applied to the solution of an equation is derived. The square of the unknown is called "the power." All the known quantities are numbers.

The work is divided into five parts. In the first Alkarismi gives rules for the solution of quadratic equations, divided into five classes of the forms $ax^2 = bx$, $ax^2 = c$, $ax^2 + bx = c$, $ax^2 + c = bx$, and $ax^2 = bx + c$, where a, b, c are positive numbers, and in all the applications $a = 1$. He considers only real and positive roots, but he recognises the existence of two roots, which as far as we know was never done by the Greeks. It is somewhat curious that when both roots are positive he generally takes only that root which is derived from the negative value of the radical.

He next gives geometrical proofs of these rules in a manner analogous to that of Euclid II, 4. For example, to solve the equation $x^2 + 10x = 39$, or any equation of the form $x^2 + px = q$, he gives

---

[1] It was published by F. Rosen, with an English translation, London, 1831.

[2] It was published by B. Boncompagni, Rome, 1857.

two methods of which one is as follows. Let AB represent the value of x, and construct on it the square $ABCD$ (see figure below). Produce DA to H and DC to F so that $AH = CF = 5$ (or $\frac{1}{2}p$); and complete the figure as drawn below. Then the areas AC, HB, and BF represent the magnitudes x^2, $5x$, and $5x$. Thus the left-hand side of the equation is represented by the sum of the areas AC, HB, and BF, that is, by the gnomon HCG. To both sides of the equation add the square KG, the area of which is 25 (or $\frac{1}{4}p^2$), and we shall get a new square whose area is by hypothesis equal to 39+25, that is, to 64 (or $q + \frac{1}{4}p^2$) and whose side therefore is 8. The side of this square DH, which is equal to 8, will exceed AH, which is equal to 5, by the value of the unknown required, which, therefore, is 3.

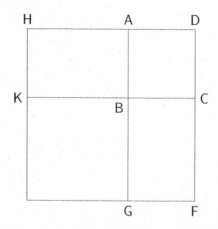

In the third part of the book Alkarismi considers the product of $(x \pm a)$ and $(x \pm b)$. In the fourth part he states the rules for addition and subtraction of expressions which involve the unknown, its square, or its square root; gives rules for the calculation of square roots; and concludes with the theorems that $a\sqrt{b} = \sqrt{a^2b}$ and $\sqrt{a}\sqrt{b} = \sqrt{ab}$. In the fifth and last part he gives some problems, such, for example, as to find two numbers whose sum is 10 and the difference of whose squares is 40.

In all these early works there is no clear distinction between arithmetic and algebra, and we find the account and explanation of arithmetical processes mixed up with algebra and treated as part of it. It was from this book then that the Italians first obtained not only the ideas of algebra, but also of an arithmetic founded on the decimal system. This arithmetic was long known as *algorism*, or the art of Alkarismi, which served to distinguish it from the arithmetic of Boethius; this name remained in use till the eighteenth century.

Tabit ibn Korra. The work commenced by Alkarismi was carried on by *Tabit ibn Korra,* born at Harran in 836, and died in 901, who was one of the most brilliant and accomplished scholars produced by the Arabs. As I have already stated, he issued translations of the chief works of Euclid, Apollonius, Archimedes, and Ptolemy. He also wrote several original works, all of which are lost with the exception of a fragment on algebra, consisting of one chapter on cubic equations, which are solved by the aid of geometry in somewhat the same way as that given later.[1]

Algebra continued to develop very rapidly, but it remained entirely rhetorical. The problems with which the Arabs were chiefly concerned were solution of equations, problems leading to equations, or properties of numbers. The two most prominent algebraists of a later date were Alkayami and Alkarki, both of whom flourished at the beginning of the eleventh century.

Alkayami. The first of these, *Omar Alkayami,* is noticeable for his geometrical treatment of cubic equations by which he obtained a root as the abscissa of a point of intersection of a conic and a circle.[2] The equations he considers are of the following forms, where a and c stand for positive integers, (i) $x^3 + b^2 x = b^2 c$, whose root he says is the abscissa of a point of intersection of $x^2 = by$ and $y^2 = x(c-x)$; (ii) $x^3 + ax^2 = c^3$, whose root he says is the abscissa of a point of intersection of $xy = c^2$ and $y^2 = c(x + a)$; (iii) $x^3 \pm ax^2 + b^2 x = b^2 c$, whose root he says is the abscissa of a point of intersection of $y^2 = (x \pm a)(c - x)$ and $x(b \pm y) = bc$. He gives one biquadratic, namely, $(100 - x^2)(10 - x)^2 = 8100$, the root of which is determined by the point of intersection of $(10 - x)y = 90$ and $x^2 + y^2 = 100$. It is sometimes said that he stated that it was impossible to solve the equation $x^3 + y^3 = z^3$ in positive integers, or in other words that the sum of two cubes can never be a cube; though whether he gave an accurate proof, or whether, as is more likely, the proposition (if enunciated at all) was the result of a wide induction, it is now impossible to say; but the fact that such a theorem is attributed to him will serve to illustrate the extraordinary progress the Arabs had made in algebra.

Alkarki. The other mathematician of this time (circ. 1000) whom

---

[1]See below, page 185.

[2]His treatise on algebra was published by Fr. Woepcke, Paris, 1851.

I mentioned was *Alkarki*.[1] He gave expressions for the sums of the first, second, and third powers of the first n natural numbers; solved various equations, including some of the forms $ax^{2p} \pm bx^p \pm c = 0$ and discussed surds, shewing, for example, that $\sqrt{8} + \sqrt{18} = \sqrt{50}$.

Even where the methods of Arab algebra are quite general the applications are confined in all cases to numerical problems, and the algebra is so arithmetical that it is difficult to treat the subjects apart. From their books on arithmetic and from the observations scattered through various works on algebra, we may say that the methods used by the Arabs for the four fundamental processes were analogous to, though more cumbrous than, those now in use; but the problems to which the subject was applied were similar to those given in modern books, and were solved by similar methods, such as rule of three, &c. Some minor improvements in notation were introduced, such, for instance, as the introduction of a line to separate the numerator from the denominator of a fraction; and hence a line between two symbols came to be used as a symbol of division.[2] Alhossein (980–1037) used a rule for testing the correctness of the results of addition and multiplication by "casting out the nines." Various forms of this rule have been given, but they all depend on the proposition that, if each number in the question be replaced by the remainder when it is divided by 9, and if these remainders be added or multiplied as directed in the question, then this result when divided by 9 will leave the same remainder as the answer whose correctness it is desired to test when divided by 9: if these remainders differ, there is an error. The selection of 9 as a divisor was due to the fact that the remainder when a number is divided by 9 can be obtained by adding the digits of the number and dividing the sum by 9.

I am not concerned with the views of Arab writers on astronomy or the value of their observations, but I may remark in passing that they accepted the theory as laid down by Hipparchus and Ptolemy, and did not materially alter or advance it. I may, however, add that Al Mamun caused the length of a degree of latitude to be measured, and he, as well as the two mathematicians to be next named, determined the obliquity of the ecliptic.

Albategni.   Albuzjani.   Like the Greeks, the Arabs rarely, if ever, employed trigonometry except in connection with astronomy; but

---

[1]His algebra was published by Fr. Woepcke, 1853, and his arithmetic was translated into German by Ad. Hochheim, Halle, 1878.

[2]See below, page 199.

in effect they used the trigonometrical ratios which are now current, and worked out the plane trigonometry of a single angle. They are also acquainted with the elements of spherical trigonometry. *Albategni*, born at Batan in Mesopotamia, in 877, and died at Bagdad in 929, was among the earliest of the many distinguished Arabian astronomers. He wrote the *Science of the Stars*,[1] which is worthy of note from its containing a mention of the motion of the sun's apogee. In this work angles are determined by "the semi-chord of twice the angle," that is, by the sine of the angle (taking the radius vector as unity). It is doubtful whether he was acquainted with the previous introduction of sines by Arya-Bhata and Brahmagupta; Hipparchus and Ptolemy, it will be remembered, had used the chord. Albategni was also acquainted with the fundamental formula in spherical trigonometry, giving the side of a triangle in terms of the other sides and the angle included by them. Shortly after the death of Albategni, *Albuzjani*, who is also known as *Abul-Wafa*, born in 940, and died in 998, introduced certain trigonometrical functions, and constructed tables of tangents and cotangents. He was celebrated as a geometrician as well as an astronomer.

Alhazen. Abd-al-gehl.   The Arabs were at first content to take the works of Euclid and Apollonius for their text-books in geometry without attempting to comment on them, but *Alhazen*, born at Bassora in 987 and died at Cairo in 1038, issued in 1036 a collection[2] of problems something like the *Data* of Euclid. Besides commentaries on the definitions of Euclid and on the *Almagest*, Alhazen also wrote a work on optics,[3] which includes the earliest scientific account of atmospheric refraction. It also contains some ingenious geometry, amongst other things, a geometrical solution of the problem to find at what point of a concave mirror a ray from a given point must be incident so as to be reflected to another given point. Another geometrician of a slightly later date was *Abd-al-gehl* (circ. 1100), who wrote on conic sections, and was also the author of three small geometrical tracts.

It was shortly after the last of the mathematicians mentioned above that Bhaskara, the third great Hindoo mathematician, flourished; there is every reason to believe that he was familiar with the works of the Arab school as described above, and also that his writings were at once known in Arabia.

---

[1] It was edited by Regiomontanus, Nuremberg, 1537.
[2] It was translated by L. A. Sédillot, and published at Paris in 1836.
[3] It was published at Bâle in 1572.

The Arab schools continued to flourish until the fifteenth century. But they produced no other mathematician of any exceptional genius, nor was there any great advance on the methods indicated above, and it is unnecessary for me to crowd my pages with the names of a number of writers who did not materially affect the progress of the science in Europe.

From this rapid sketch it will be seen that the work of the Arabs (including therein writers who wrote in Arabia and lived under Eastern Mohammedan rule) in arithmetic, algebra, and trigonometry was of a high order of excellence. They appreciated geometry and the applications of geometry to astronomy, but they did not extend the bounds of the science. It may be also added that they made no special progress in statics, or optics, or hydrostatics; though there is abundant evidence that they had a thorough knowledge of practical hydraulics.

The general impression left is that the Arabs were quick to appreciate the work of others—notably of the Greek masters and of the Hindoo mathematicians—but, like the ancient Chinese and Egyptians, they did not systematically develop a subject to any considerable extent. Their schools may be taken to have lasted in all for about 650 years, and if the work produced be compared with that of Greek or modern European writers it is, as a whole, second-rate both in quantity and quality.

CHAPTER X.

THE INTRODUCTION OF ARAB WORKS INTO EUROPE.
CIRC. 1150–1450.

IN the last chapter but one I discussed the development of European mathematics to a date which corresponds roughly with the end of the "dark ages"; and in the last chapter I traced the history of the mathematics of the Indians and Arabs to the same date. The mathematics of the two or three centuries that follow and are treated in this chapter are characterised by the introduction of the Arab mathematical text-books and of Greek books derived from Arab sources, and the assimilation of the new ideas thus presented.

It was, however, from Spain, and not from Arabia, that a knowledge of eastern mathematics first came into western Europe. The Moors had established their rule in Spain in 747, and by the tenth or eleventh century had attained a high degree of civilisation. Though their political relations with the caliphs at Bagdad were somewhat unfriendly, they gave a ready welcome to the works of the great Arab mathematicians. In this way the Arab translations of the writings of Euclid, Archimedes, Apollonius, Ptolemy, and perhaps of other Greek authors, together with the works of the Arabian algebraists, were read and commented on at the three great Moorish schools of Granada, Cordova, and Seville. It seems probable that these works indicate the full extent of Moorish learning, but, as all knowledge was jealously guarded from Christians, it is impossible to speak with certainty either on this point or on that of the time when the Arab books were first introduced into Spain.

The eleventh century. The earliest Moorish writer of distinction of whom I find mention is **Geber ibn Aphla**, who was born at Seville and died towards the latter part of the eleventh century at Cordova. He wrote on astronomy and trigonometry, and was acquainted with the theorem that the sines of the angles of a spherical triangle are proportional to the sines of the opposite sides.[1]

Arzachel.[2]   Another Arab of about the same date was *Arzachel*, who was living at Toledo in 1080. He suggested that the planets moved in ellipses, but his contemporaries with scientific intolerance declined to argue about a statement which was contrary to Ptolemy's conclusions in the *Almagest*.

The twelfth century.   During the course of the twelfth century copies of the books used in Spain were obtained in western Christendom. The first step towards procuring a knowledge of Arab and Moorish science was taken by an English monk, **Adelhard of Bath**,[3] who, under the disguise of a Mohammedan student, attended some lectures at Cordova about 1120 and obtained a copy of Euclid's *Elements*. This copy, translated into Latin, was the foundation of all the editions known in Europe till 1533, when the Greek text was recovered. How rapidly a knowledge of the work spread we may judge when we recollect that before the end of the thirteenth century Roger Bacon was familiar with it, while before the close of the fourteenth century the first five books formed part of the regular curriculum at many universities. The enunciations of Euclid seem to have been known before Adelhard's time, and possibly as early as the year 1000, though copies were rare. Adelhard also issued a text-book on the use of the abacus.

Ben Ezra.[3]   During the same century other translations of the Arab text-books or commentaries on them were obtained. Amongst those who were most influential in introducing Moorish learning into Europe I may mention *Abraham Ben Ezra*. Ben Ezra was born at Toledo in 1097, and died at Rome in 1167. He was one of the most distinguished Jewish rabbis who had settled in Spain, where it must be recollected that they were tolerated and even protected by the Moors on account of their medical skill. Besides some astronomical tables and an astrology, Ben Ezra wrote an arithmetic;[4] in this he explains the Arab system of numeration with nine symbols and a zero, gives the fundamental processes of arithmetic, and explains the rule of three.

---

[1] Geber's works were translated into Latin by Gerard, and published at Nuremberg in 1533.

[2] See a memoir by M. Steinschneider in Boncompagni's *Bulletino di Bibliografia*, 1887, vol xx.

[3] On the influence of Adelhard and Ben Ezra, see the "Abhandlungen zur Geschichte der Mathematik" in the *Zeitschrift für Mathematik*, vol. xxv, 1880.

[4] An analysis of it was published by O. Terquem in Liouville's *Journal* for 1841.

Gerard.[1]   Another European who was induced by the reputation of the Arab schools to go to Toledo was *Gerard*, who was born at Cremona in 1114 and died in 1187. He translated the Arab edition of the *Almagest*, the works of Alhazen, and the works of Alfarabius, whose name is otherwise unknown to us: it is believed that the Arabic numerals were used in this translation, made in 1136, of Ptolemy's work. Gerard also wrote a short treatise on algorism which exists in manuscript in the Bodleian Library at Oxford. He was acquainted with one of the Arab editions of Euclid's *Elements*, which he translated into Latin.

John Hispalensis.   Among the contemporaries of Gerard was *John Hispalensis* of Seville, originally a rabbi, but converted to Christianity and baptized under the name given above. He made translations of several Arab and Moorish works, and also wrote an algorism which contains the earliest examples of the extraction of the square roots of numbers by the aid of the decimal notation.

The thirteenth century.   During the thirteenth century there was a revival of learning throughout Europe, but the new learning was, I believe, confined to a very limited class. The early years of this century are memorable for the development of several universities, and for the appearance of three remarkable mathematicians—Leonardo of Pisa, Jordanus, and Roger Bacon, the Franciscan monk of Oxford. Henceforward it is to Europeans that we have to look for the development of mathematics, but until the invention of printing the knowledge was confined to a very limited class.

Leonardo.[2]   *Leonardo Fibonacci* (*i.e.* filius Bonaccii) generally known as *Leonardo of Pisa*, was born at Pisa about 1175. His father Bonacci was a merchant, and was sent by his fellow-townsmen to control the custom-house at Bugia in Barbary; there Leonardo was educated, and he thus became acquainted with the Arabic or decimal system of numeration, as also with Alkarismi's work on Algebra, which was described in the last chapter. It would seem that Leonardo was entrusted with some duties, in connection with the custom-house, which required him to travel. He returned to Italy about 1200, and in 1202

---

[1]See Boncompagni's *Della vita e delle opere di Gherardo Cremonese*, Rome, 1851.

[2]See the *Leben und Schriften Leonardos da Pisa*, by J. Giesing, Döbeln, 1886; Cantor, chaps. xli, xlii; and an article by V. Lazzarini in the *Bollettino di Bibliografia e Storia*, Rome, 1904, vol. vii. Most of Leonardo's writings were edited and published by B. Boncompagni, Rome, vol. i, 1857, and vol. ii, 1862.

published a work called *Algebra et almuchabala* (the title being taken from Alkarismi's work), but generally known as the *Liber Abaci*. He there explains the Arabic system of numeration, and remarks on its great advantages over the Roman system. He then gives an account of algebra, and points out the convenience of using geometry to get rigid demonstrations of algebraical formulae. He shews how to solve simple equations, solves a few quadratic equations, and states some methods for the solution of indeterminate equations; these rules are illustrated by problems on numbers. The algebra is rhetorical, but in one case letters are employed as algebraical symbols. This work had a wide circulation, and for at least two centuries remained a standard authority from which numerous writers drew their inspiration.

The *Liber Abaci* is especially interesting in the history of arithmetic, since practically it introduced the use of the Arabic numerals into Christian Europe. The language of Leonardo implies that they were previously unknown to his countrymen; he says that having had to spend some years in Barbary he there learnt the Arabic system, which he found much more convenient than that used in Europe; he therefore published it "in order that the Latin[1] race might no longer be deficient in that knowledge." Now Leonardo had read very widely, and had travelled in Greece, Sicily, and Italy; there is therefore every presumption that the system was not then commonly employed in Europe.

Though Leonardo introduced the use of Arabic numerals into commercial affairs, it is probable that a knowledge of them as current in the East was previously not uncommon among travellers and merchants, for the intercourse between Christians and Mohammedans was sufficiently close for each to learn something of the language and common practices of the other. We can also hardly suppose that the Italian merchants were ignorant of the method of keeping accounts used by some of their best customers; and we must recollect, too, that there were numerous Christians who had escaped or been ransomed after serving the Mohammedans as slaves. It was, however, Leonardo who brought the Arabic system into general use, and by the middle of the thirteenth century a large proportion of the Italian merchants employed it by the side of the old system.

---

[1]Dean Peacock says that the earliest known application of the word Italians to describe the inhabitants of Italy occurs about the middle of the thirteenth century; by the end of that century it was in common use.

The majority of mathematicians must have already known of the system from the works of Ben Ezra, Gerard, and John Hispalensis. But shortly after the appearance of Leonardo's book Alfonso of Castile (in 1252) published some astronomical tables, founded on observations made in Arabia, which were computed by Arabs, and which, it is generally believed, were expressed in Arabic notation. Alfonso's tables had a wide circulation among men of science, and probably were largely instrumental in bringing these numerals into universal use among mathematicians. By the end of the thirteenth century it was generally assumed that all scientific men would be acquainted with the system: thus Roger Bacon writing in that century recommends algorism (that is, the arithmetic founded on the Arab notation) as a necessary study for theologians who ought, he says, "to abound in the power of numbering." We may then consider that by the year 1300, or at the latest 1350, these numerals were familiar both to mathematicians and to Italian merchants.

So great was Leonardo's reputation that the Emperor Frederick II. stopped at Pisa in 1225 in order to hold a sort of mathematical tournament to test Leonardo's skill, of which he had heard such marvellous accounts. The competitors were informed beforehand of the questions to be asked, some or all of which were composed by John of Palermo, who was one of Frederick's suite. This is the first time that we meet with an instance of those challenges to solve particular problems which were so common in the sixteenth and seventeenth centuries. The first question propounded was to find a number of which the square, when either increased or decreased by 5, would remain a square. Leonardo gave an answer, which is correct, namely 41/12. The next question was to find by the methods used in the tenth book of Euclid a line whose length x should satisfy the equation $x^3 + 2x^2 + 10x = 20$. Leonardo showed by geometry that the problem was impossible, but he gave an approximate value of the root of this equation, namely, $1 \cdot 22' \, 7'' \, 42''' \, 33'''' \, 4^v \, 40^{vi}$, which is equal to $1.3688081075\ldots$, and is correct to nine places of decimals.[1] Another question was as follows. Three men, A, B, C, possess a sum of money u, their shares being in the ratio $3 : 2 : 1$. A takes away x, keeps half of it, and deposits the remainder with D; B takes away y, keeps two-thirds of it, and deposits the remainder with D; C takes away all that is left, namely z, keeps five-sixths of it, and deposits the remainder with D. This deposit with D is found to belong to A, B, and C in equal

---

[1] See Fr. Woepcke in Liouville's *Journal* for 1854, p. 401.

proportions. Find u, x, y, and z. Leonardo showed that the problem was indeterminate, and gave as one solution $u = 47$, $x = 33$, $y = 13$, $z = 1$. The other competitors failed to solve any of these questions.

The chief work of Leonardo is the *Liber Abaci* alluded to above. This work contains a proof of the well-known result

$$(a^2 + b^2)(c^2 + d^2) = (ac + bd)^2 + (bc - ad)^2 = (ad + bc)^2 + (bd - ac)^2.$$

He also wrote a geometry, termed *Practica Geometriae*, which was issued in 1220. This is a good compilation, and some trigonometry is introduced; among other propositions and examples he finds the area of a triangle in terms of its sides. Subsequently he published a *Liber Quadratorum* dealing with problems similar to the first of the questions propounded at the tournament.[1] He also issued a tract dealing with determinate algebraical problems: these are all solved by the rule of false assumption in the manner explained above.

Frederick II.   The Emperor *Frederick II.*, who was born in 1194, succeeded to the throne in 1210, and died in 1250, was not only interested in science, but did as much as any other single man of the thirteenth century to disseminate a knowledge of the works of the Arab mathematicians in western Europe. The university of Naples remains as a monument of his munificence. I have already mentioned that the presence of the Jews had been tolerated in Spain on account of their medical skill and scientific knowledge, and as a matter of fact the titles of physician and algebraist[2] were for a long time nearly synonymous; thus the Jewish physicians were admirably fitted both to get copies of the Arab works and to translate them. Frederick II. made use of this fact to engage a staff of learned Jews to translate the Arab works which he obtained, though there is no doubt that he gave his patronage to them the more readily because it was singularly offensive to the pope, with whom he was then engaged in a quarrel. At any rate, by the end of the thirteenth century copies of the works of Euclid, Archimedes, Apollonius, Ptolemy, and of several Arab authors were obtainable from this source, and by the end of the next century were not uncommon. From this time, then, we may say that the development of science in Europe was independent of the aid of the Arabian schools.

---

[1] Fr. Woepcke in Liouville's *Journal* for 1855, p. 54, has given an analysis of Leonardo's method of treating problems on square numbers.

[2] For instance the reader may recollect that in *Don Quixote* (part ii, ch. 15), when Samson Carasco is thrown by the knight from his horse and has his ribs broken, an *algebrista* is summoned to bind up his wounds.

Jordanus.[1]   Among Leonardo's contemporaries was a German mathematician, whose works were until the last few years almost unknown. This was *Jordanus Nemorarius*, sometimes called *Jordanus de Saxonia* or *Teutonicus*. Of the details of his life we know but little, save that he was elected general of the Dominican order in 1222. The works enumerated in the footnote[2] hereto are attributed to him, and if we assume that these works have not been added to or improved by subsequent annotators, we must esteem him one of the most eminent mathematicians of the middle ages.

His knowledge of geometry is illustrated by his *De Triangulis* and *De Isoperimetris*. The most important of these is the *De Triangulis*, which is divided into four books. The first book, besides a few definitions, contains thirteen propositions on triangles which are based on Euclid's *Elements*. The second book contains nineteen propositions, mainly on the ratios of straight lines and the comparison of the areas of triangles; for example, one problem is to find a point inside a triangle so that the lines joining it to the angular points may divide the triangle into three equal parts. The third book contains twelve propositions mainly concerning arcs and chords of circles. The fourth book contains twenty-eight propositions, partly on regular polygons and partly on miscellaneous questions such as the duplication and trisection problems.

The *Algorithmus Demonstratus* contains practical rules for the four fundamental processes, and Arabic numerals are generally (but not always) used. It is divided into ten books dealing with properties of numbers, primes, perfect numbers, polygonal numbers, ratios, powers, and the progressions. It would seem from it that Jordanus knew the general expression for the square of any algebraic multinomial.

The *De Numeris Datis* consists of four books containing solutions

---

[1] See Cantor, chaps. xliii, xliv, where references to the authorities on Jordanus are collected.

[2] Prof. Curtze, who has made a special study of the subject, considers that the following works are due to Jordanus. "Geometria vel de Triangulis," published by M. Curtze in 1887 in vol. vi of the *Mitteilungen des Copernicus-Vereins zu Thorn*; *De Isoperimetris*; *Arithmetica Demonstrata*, published by Faber Stapulensis at Paris in 1496, second edition, 1514; *Algorithmus Demonstratus*, published by J. Schöner at Nuremberg in 1534; *De Numeris Datis*, published by P. Treutlein in 1879 and edited in 1891 with comments by M. Curtze in vol. xxxvi of the *Zeitschrift für Mathematik und Physik*; *De Ponderibus*, published by P. Apian at Nuremberg in 1533, and reissued at Venice in 1565; and, lastly, two or three tracts on Ptolemaic astronomy.

of one hundred and fifteen problems. Some of these lead to simple or quadratic equations involving more than one unknown quantity. He shews a knowledge of proportion; but many of the demonstrations of his general propositions are only numerical illustrations of them.

In several of the propositions of the *Algorithmus* and *De Numeris Datis* letters are employed to denote both known and unknown quantities, and they are used in the demonstrations of the rules of arithmetic as well as of algebra. As an example of this I quote the following proposition,[1] the object of which is to determine two quantities whose sum and product are known.

Dato numero per duo diuiso si, quod ex ductu unius in alterum producitur, datum fuerit, et utrumque eorum datum esse necesse est.

Sit numerus datus *abc* diuisus in *ab* et *c*, atque ex *ab* in *c* fiat *d* datus, itemque ex *abc* in se fiat *e*. Sumatur itaque quadruplum *d*, qui fit *f*, quo dempto de *e* remaneat *g*, et ipse erit quadratum differentiae *ab* ad *c*. Extrahatur ergo radix ex *g*, et sit *h*, eritque *h* differentia *ab* ad *c*. cumque sic *h* datum, erit et *c* et *ab* datum.

Huius operatio facile constabit hoc modo. Verbi gratia sit x diuisus in numeros duos, atque ex ductu unius eorum in alium fiat XXI; cuius quadruplum et ipsum est LXXXIIII, tollatur de quadrato X, hoc est C, et remanent XVI, cuius radix extrahatur, quae erit quatuor, et ipse est differentia. Ipsa tollatur de X et reliquum, quod est VI, dimidietur, eritque medietas III, et ipse est minor portio et maior VII.

It will be noticed that Jordanus, like Diophantus and the Hindoos, denotes addition by juxtaposition. Expressed in modern notation his argument is as follows. Let the numbers be $a + b$ (which I will denote by γ) and c. Then $\gamma + c$ is given; hence $(\gamma + c)^2$ is known; denote it by e. Again γc is given; denote it by d; hence $4\gamma c$, which is equal to $4d$, is known; denote it by f. Then $(\gamma - c)^2$ is equal to $e - f$, which is known; denote it by g. Therefore $\gamma - c = \sqrt{g}$, which is known; denote it by h. Hence $\gamma + c$ and $\gamma - c$ are known, and therefore γ and c can be at once found. It is curious that he should have taken a sum like $a + b$ for one of his unknowns. In his numerical illustration he takes the sum to be 10 and the product 21.

Save for one instance in Leonardo's writings, the above works are the earliest instances known in European mathematics of syncopated algebra in which letters are used for algebraical symbols. It is probable that the *Algorithmus* was not generally known until it was printed in

---

[1] From the *De Numeris Datis*, book i, prop. 3.

1534, and it is doubtful how far the works of Jordanus exercised any considerable influence on the development of algebra. In fact it constantly happens in the history of mathematics that improvements in notation or method are made long before they are generally adopted or their advantages realized. Thus the same thing may be discovered over and over again, and it is not until the general standard of knowledge requires some such improvement, or it is enforced by some one whose zeal or attainments compel attention, that it is adopted and becomes part of the science. Jordanus in using letters or symbols to represent any quantities which occur in analysis was far in advance of his contemporaries. A similar notation was tentatively introduced by other and later mathematicians, but it was not until it had been thus independently discovered several times that it came into general use.

It is not necessary to describe in detail the mechanics, optics, or astronomy of Jordanus. The treatment of mechanics throughout the middle ages was generally unintelligent.

No mathematicians of the same ability as Leonardo and Jordanus appear in the history of the subject for over two hundred years. Their individual achievements must not be taken to imply the standard of knowledge then current, but their works were accessible to students in the following two centuries, though there were not many who seem to have derived much benefit therefrom, or who attempted to extend the bounds of arithmetic and algebra as there expounded.

During the thirteenth century the most famous centres of learning in western Europe were Paris and Oxford, and I must now refer to the more eminent members of those schools.

Holywood.[1] I will begin by mentioning *John de Holywood*, whose name is often written in the latinized form of *Sacrobosco*. Holywood was born in Yorkshire and educated at Oxford; but after taking his master's degree he moved to Paris, and taught there till his death in 1244 or 1246. His lectures on algorism and algebra are the earliest of which I can find mention. His work on arithmetic was for many years a standard authority; it contains rules, but no proofs; it was printed at Paris in 1496. He also wrote a treatise on the sphere, which was made public in 1256: this had a wide and long-continued circulation, and indicates how rapidly a knowledge of mathematics was spreading. Besides these, two pamphlets by him, entitled respectively *De Computo Ecclesiastico* and *De Astrolabio*, are still extant.

---

[1]See Cantor, chap. xlv.

Roger Bacon.[1]   Another contemporary of Leonardo and Jordanus was Roger Bacon, who for physical science did work somewhat analogous to what they did for arithmetic and algebra. *Roger Bacon* was born near Ilchester in 1214, and died at Oxford on June 11, 1294. He was the son of royalists, most of whose property had been confiscated at the end of the civil wars: at an early age he was entered as a student at Oxford, and is said to have taken orders in 1233. In 1234 he removed to Paris, then the intellectual capital of western Europe, where he lived for some years devoting himself especially to languages and physics; and there he spent on books and experiments all that remained of his family property and his savings. He returned to Oxford soon after 1240, and there for the following ten or twelve years he laboured incessantly, being chiefly occupied in teaching science. His lecture room was crowded, but everything that he earned was spent in buying manuscripts and instruments. He tells us that altogether at Paris and Oxford he spent over £2000 in this way—a sum which represents at least £20,000 nowadays.

Bacon strove hard to replace logic in the university curriculum by mathematical and linguistic studies, but the influences of the age were too strong for him. His glowing eulogy on "divine mathematics" which should form the foundation of a liberal education, and which "alone can purge the intellect and fit the student for the acquirement of all knowledge," fell on deaf ears. We can judge how small was the amount of geometry which was implied in the quadrivium, when he tells us that in geometry few students at Oxford read beyond Euc. I, 5; though we might perhaps have inferred as much from the character of the work of Boethius.

At last worn out, neglected, and ruined, Bacon was persuaded by his friend Grosseteste, the great Bishop of Lincoln, to renounce the world and take the Franciscan vows. The society to which he now found himself confined was singularly uncongenial to him, and he beguiled the time by writing on scientific questions and perhaps lecturing. The superior of the order heard of this, and in 1257 forbade him to lecture or publish anything under penalty of the most severe punishments, and at the same time directed him to take up his residence at Paris, where he could be more closely watched.

[1]See *Roger Bacon, sa vie, ses ouvrages* ... by E. Charles, Paris, 1861; and the memoir by J. S. Brewer, prefixed to the *Opera Inedita*, Rolls Series, London, 1859: a somewhat depreciatory criticism of the former of these works is given in *Roger Bacon, eine Monographie*, by L. Schneider, Augsburg, 1873.

Clement IV., when in England, had heard of Bacon's abilities, and in 1266 when he became Pope he invited Bacon to write. The Franciscan order reluctantly permitted him to do so, but they refused him any assistance. With difficulty Bacon obtained sufficient money to get paper and the loan of books, and in the short space of fifteen months he produced in 1267 his *Opus Majus* with two supplements which summarized what was then known in physical science, and laid down the principles on which it, as well as philosophy and literature, should be studied. He stated as the fundamental principle that the study of natural science must rest solely on experiment; and in the fourth part he explained in detail how astronomy and physical sciences rest ultimately on mathematics, and progress only when their fundamental principles are expressed in a mathematical form. Mathematics, he says, should be regarded as the alphabet of all philosophy.

The results that he arrived at in this and his other works are nearly in accordance with modern ideas, but were too far in advance of that age to be capable of appreciation or perhaps even of comprehension, and it was left for later generations to rediscover his works, and give him that credit which he never experienced in his lifetime. In astronomy he laid down the principles for a reform of the calendar, explained the phenomena of shooting stars, and stated that the Ptolemaic system was unscientific in so far as it rested on the assumption that circular motion was the natural motion of a planet, while the complexity of the explanations required made it improbable that the theory was true. In optics he enunciated the laws of reflexion and in a general way of refraction of light, and used them to give a rough explanation of the rainbow and of magnifying glasses. Most of his experiments in chemistry were directed to the transmutation of metals, and led to no useful results. He gave the composition of gunpowder, but there is no doubt that it was not his own invention, though it is the earliest European mention of it. On the other hand, some of his statements appear to be guesses which are more or less ingenious, while some of them are certainly erroneous.

In the years immediately following the publication of his *Opus Majus* he wrote numerous works which developed in detail the principles there laid down. Most of these have now been published, but I do not know of the existence of any complete edition. They deal only with applied mathematics and physics.

Clement took no notice of the great work for which he had asked, except to obtain leave for Bacon to return to England. On the death of Clement, the general of the Franciscan order was elected Pope and took the title of Nicholas IV. Bacon's investigations had never been approved of by his superiors, and he was now ordered to return to Paris, where we are told he was immediately accused of magic; he was condemned in 1280 to imprisonment for life, but was released about a year before his death.

Campanus. The only other mathematician of this century whom I need mention is *Giovanni Campano*, or in the latinized form *Campanus*, a canon of Paris. A copy of Adelhard's translation of Euclid's *Elements* fell into the hands of Campanus, who added a commentary thereon in which he discussed the properties of a regular re-entrant pentagon.[1] He also, besides some minor works, wrote the *Theory of the Planets*, which was a free translation of the *Almagest*.

The fourteenth century. The history of the fourteenth century, like that of the one preceding it, is mostly concerned with the assimilation of Arab mathematical text-books and of Greek books derived from Arab sources.

Bradwardine.[2] A mathematician of this time, who was perhaps sufficiently influential to justify a mention here, is *Thomas Bradwardine*, Archbishop of Canterbury. Bradwardine was born at Chichester about 1290. He was educated at Merton College, Oxford, and subsequently lectured in that university. From 1335 to the time of his death he was chiefly occupied with the politics of the church and state; he took a prominent part in the invasion of France, the capture of Calais, and the victory of Cressy. He died at Lambeth in 1349. His mathematical works, which were probably written when he was at Oxford, are the *Tractatus de Proportionibus*, printed at Paris in 1495; the *Arithmetica Speculativa*, printed at Paris in 1502; the *Geometria Speculativa*, printed at Paris in 1511; and the *De Quadratura Circuli*, printed at Paris in 1495. They probably give a fair idea of the nature of the mathematics then read at an English university.

Oresmus.[3] *Nicholas Oresmus* was another writer of the fourteenth century. He was born at Caen in 1323, became the confidential adviser of Charles V., by whom he was made tutor to Charles VI., and

---

[1] This edition of Euclid was printed by Ratdolt at Venice in 1482, and was formerly believed to be due to Campanus. On this work see J. L. Heiberg in the *Zeitschrift für Mathematik*, vol. xxxv, 1890.

[2] See Cantor, vol. ii, p. 102 *et seq.*

[3] See *Die mathematischen Schriften des Nicole Oresme*, by M. Curtze, Thorn, 1870.

subsequently was appointed bishop of Lisieux, at which city he died on July 11, 1382. He wrote the *Algorismus Proportionum*, in which the idea of fractional indices is introduced. He also issued a treatise dealing with questions of coinage and commercial exchange; from the mathematical point of view it is noticeable for the use of vulgar fractions and the introduction of symbols for them.

By the middle of this century Euclidean geometry (as expounded by Campanus) and algorism were fairly familiar to all professed mathematicians, and the Ptolemaic astronomy was also generally known. About this time the almanacks began to add to the explanation of the Arabic symbols the rules of addition, subtraction, multiplication, and division, "de algorismo." The more important calendars and other treatises also inserted a statement of the rules of proportion, illustrated by various practical questions.

In the latter half of this century there was a general revolt of the universities against the intellectual tyranny of the schoolmen. This was largely due to Petrarch, who in his own generation was celebrated as a humanist rather than as a poet, and who exerted all his power to destroy scholasticism and encourage scholarship. The result of these influences on the study of mathematics may be seen in the changes then introduced in the study of the quadrivium. The stimulus came from the university of Paris, where a statute to that effect was passed in 1366, and a year or two later similar regulations were made at other universities; unfortunately no text-books are mentioned. We can, however, form a reasonable estimate of the range of mathematical reading required, by looking at the statutes of the universities of Prague, of Vienna, and of Leipzig.

By the statutes of Prague, dated 1384, candidates for the bachelor's degree were required to have read Holywood's treatise on the sphere, and candidates for the master's degree to be acquainted with the first six books of Euclid, optics, hydrostatics, the theory of the lever, and astronomy. Lectures were actually delivered on arithmetic, the art of reckoning with the fingers, and the algorism of integers; on almanacks, which probably meant elementary astrology; and on the *Almagest*, that is, on Ptolemaic astronomy. There is, however, some reason for thinking that mathematics received far more attention here than was then usual at other universities.

At Vienna, in 1389, a candidate for a master's degree was required to have read five books of Euclid, common perspective, proportional parts, the measurement of superficies, and the *Theory of the Planets*. The book last named is the treatise by Campanus which was founded on that by Ptolemy. This was a fairly respectable mathematical standard, but I would remind the reader that there was no such thing as "plucking" in a medieval university. The student had to keep an act or give a lecture on certain subjects, but whether he did it well or badly he got his degree, and it is probable that it was only the few students whose interests were mathematical who really mastered the subjects mentioned above.

The fifteenth century.   A few facts gleaned from the history of the fifteenth century tend to shew that the regulations about the study of the quadrivium were not seriously enforced. The lecture lists for the years 1437 and 1438 of the university of Leipzig (founded in 1409, the statutes of which are almost identical with those of Prague as quoted above) are extant, and shew that the only lectures given there on mathematics in those years were confined to astrology. The records of Bologna, Padua, and Pisa seem to imply that there also astrology was the only scientific subject taught in the fifteenth century, and even as late as 1598 the professor of mathematics at Pisa was required to lecture on the *Quadripartitum*, an astrological work purporting (probably falsely) to have been written by Ptolemy. The only mathematical subjects mentioned in the registers of the university of Oxford as read there between the years 1449 and 1463 were Ptolemy's astronomy, or some commentary on it, and the first two books of Euclid. Whether most students got as far as this is doubtful. It would seem, from an edition of Euclid's *Elements* published at Paris in 1536, that after 1452 candidates for the master's degree at that university had to take an oath that they had attended lectures on the first six books of that work.

Beldomandi.   The only writer of this time that I need mention here is *Prodocimo Beldomandi* of Padua, born about 1380, who wrote an algoristic arithmetic, published in 1410, which contains the summation of a geometrical series; and some geometrical works.[1]

By the middle of the fifteenth century printing had been introduced, and the facilities it gave for disseminating knowledge were so great as to revolutionize the progress of science. We have now arrived at a time when the results of Arab and Greek science were known in Europe; and this perhaps, then, is as good a date as can be fixed for the close of this period, and the commencement of that of the renaissance. The mathematical history of the renaissance begins with the career of Regiomontanus; but before proceeding with the general history it will

---

[1]For further details see Boncompagni's *Bulletino di bibliografia*, vols. xii, xviii.

be convenient to collect together the chief facts connected with the development of arithmetic during the middle ages and the renaissance. To this the next chapter is devoted.

CHAPTER XI.

THE DEVELOPMENT OF ARITHMETIC.[1]
CIRC. 1300–1637.

WE have seen in the last chapter that by the end of the thirteenth century the Arabic arithmetic had been fairly introduced into Europe and was practised by the side of the older arithmetic which was founded on the work of Boethius. It will be convenient to depart from the chronological arrangement and briefly to sum up the subsequent history of arithmetic, but I hope, by references in the next chapter to the inventions and improvements in arithmetic here described, that I shall be able to keep the order of events and discoveries clear.

The older arithmetic consisted of two parts: practical arithmetic or the art of calculation which was taught by means of the abacus and possibly the multiplication table; and theoretical arithmetic, by which was meant the ratios and properties of numbers taught according to Boethius—a knowledge of the latter being confined to professed mathematicians. The theoretical part of this system continued to be taught till the middle of the fifteenth century, and the practical part of it was used by the smaller tradesmen in England,[2] Germany, and France till the beginning of the seventeenth century.

[1] See the article on Arithmetic by G. Peacock in the *Encyclopaedia Metropolitana*, vol. i, London, 1845; *Arithmetical Books* by A. De Morgan, London, 1847; and an article by P. Treutlein of Karlsruhe, in the *Zeitschrift für Mathematik*, 1877, vol. xxii, supplement, pp. 1–100.

[2] See, for instance, Chaucer, *The Miller's Tale*, v, 22–25; Shakespeare, *The Winter's Tale*, Act IV, Sc. 2; *Othello*, Act I, Sc. 1. There are similar references in French and German literature; notably by Montaigne and Molière. I believe that the Exchequer division of the High Court of Justice derives its name from the table before which the judges and officers of the court originally sat: this was covered with black cloth divided into squares or chequers by white lines, and apparently was used as an abacus.

The new Arabian arithmetic was called *algorism* or the art of Alka-rismi, to distinguish it from the old or Boethian arithmetic. The text-books on algorism commenced with the Arabic system of notation, and began by giving rules for addition, subtraction, multiplication, and division; the principles of proportion were then applied to various practical problems, and the books usually concluded with general rules for many of the common problems of commerce. Algorism was in fact a mercantile arithmetic, though at first it also included all that was then known as algebra.

Thus algebra has its origin in arithmetic; and to most people the term *universal arithmetic*, by which it was sometimes designated, conveys a more accurate impression of its objects and methods than the more elaborate definitions of modern mathematicians—certainly better than the definition of Sir William Hamilton as the science of pure time, or that of De Morgan as the calculus of succession. No doubt logically there is a marked distinction between arithmetic and algebra, for the former is the theory of discrete magnitude, while the latter is that of continuous magnitude; but a scientific distinction such as this is of comparatively recent origin, and the idea of continuity was not introduced into mathematics before the time of Kepler.

Of course the fundamental rules of this algorism were not at first strictly proved—that is the work of advanced thought—but until the middle of the seventeenth century there was some discussion of the principles involved; since then very few arithmeticians have attempted to justify or prove the processes used, or to do more than enunciate rules and illustrate their use by numerical examples.

I have alluded frequently to the Arabic system of numerical notation. I may therefore conveniently begin by a few notes on the history of the symbols now current.

Their origin is obscure and has been much disputed.[1] On the whole it seems probable that the symbols for the numbers 4, 5, 6, 7, and 9 (and possibly 8 too) are derived from the initial letters of the corresponding words in the Indo-Bactrian alphabet in use in the north of India perhaps 150 years before Christ; that the symbols for the numbers 2 and 3 are derived respectively from two and three parallel penstrokes written

---

[1]See A. L'Esprit, *Histoire des chiffres*, Paris, 1893; A. P. Pihan, *Signes de numération*, Paris, 1860; Fr. Woepcke, *La propagation des chiffres Indiens*, Paris, 1863; A. C. Burnell, *South Indian Palaeography*, Mangalore, 1874; Is. Taylor, *The Alphabet*, London, 1883; and Cantor.

cursively; and similarly that the symbol for the number 1 represents a single penstroke. Numerals of this type were in use in India before the end of the second century of our era. The origin of the symbol for zero is unknown; it is not impossible that it was originally a dot inserted to indicate a blank space, or it may represent a closed hand, but these are mere conjectures; there is reason to believe that it was introduced in India towards the close of the fifth century of our era, but the earliest writing now extant in which it occurs is assigned to the eighth century.

Devanagari (Indian) numerals, *circ.* 950.	⟩
Gobar Arabic numerals, *circ.* 1100 (?).	⟩
From a missal, *circ.* 1385, of German origin.	⟩
European (probably Italian) numerals, *circ.* 1400.	⟩
From the *Mirrour of the World*, printed by Caxton in 1480.	⟩
From a Scotch calendar for 1482, probably of French origin.	⟩

The numerals used in India in the eighth century and for a long time afterwards are termed Devanagari numerals, and their forms are shewn in the first line of the table given above. These forms were slightly modified by the eastern Arabs, and the resulting symbols were again slightly modified by the western Arabs or Moors. It is perhaps probable that at first the Spanish Arabs discarded the use of the symbol for zero, and only reinserted it when they found how inconvenient the omission proved. The symbols ultimately adopted by the Arabs are termed Gobar numerals, and an idea of the forms most commonly used may be gathered from those printed in the second line of the table given above. From Spain or Barbary the Gobar numerals passed into western Europe, and they occur on a Sicilian coin as early as 1138. The further evolution of the forms of the symbols to those with which we are familiar is indicated below by facsimiles[1] of the numerals used at different times. All the sets of numerals here represented are written from left to right and in the order 1, 2, 3, 4, 5, 6, 7, 8, 9, 10. From 1500 onwards the symbols employed are practically the same as those now in use.[2]

$$ \text{١, ٢, ٣, ٤, ٥, ٦, ٧, ٨, ٩, ١٠} $$

The further evolution in the East of the Gobar numerals proceeded almost independently of European influence. There are minute differences in the forms used by various writers, and in some cases alternative forms; without, however, entering into these details we may say that the numerals they commonly employed finally took the form shewn above, but the symbol there given for 4 is at the present time generally written cursively.

Leaving now the history of the symbols I proceed to discuss their introduction into general use and the development of algoristic arithmetic. I have already explained how men of science, and particularly astronomers, had become acquainted with the Arabic system by the middle of the thirteenth century. The trade of Europe during the thirteenth and fourteenth centuries was mostly in Italian hands, and the obvious advantages of the algoristic system led to its general adoption in Italy for mercantile purposes. This change was not effected, however, without considerable opposition; thus, an edict was issued at Florence in 1299 forbidding bankers to use Arabic numerals, and in 1348 the authorities of the university of Padua directed that a list should be kept of books for sale with the prices marked "non per cifras sed per literas claras."

The rapid spread of the use of Arabic numerals and arithmetic through the rest of Europe seems to have been as largely due to the makers of almanacks and calendars as to merchants and men of science. These calendars had a wide circulation in medieval times. Some of them were composed with special reference to ecclesiastical purposes,

---

[1]The first, second, and fourth examples are taken from Is. Taylor's *Alphabet*, London, 1883, vol. ii, p. 266; the others are taken from Leslie's *Philosophy of Arithmetic*, 2nd ed., Edinburgh, 1820, pp. 114, 115.

[2]See, for example, Tonstall's *De Arte Supputandi*, London, 1522; or Record's *Grounde of Artes*, London, 1540, and *Whetstone of Witte*, London, 1557.

and contained the dates of the different festivals and fasts of the church
for a period of some seven or eight years in advance, as well as notes on
church ritual. Nearly every monastery and church of any pretensions
possessed one of these. Others were written specially for the use of astrologers and physicians, and some of them contained notes on various
scientific subjects, especially medicine and astronomy. Such almanacks
were not then uncommon, but, since it was only rarely that they found
their way into any corporate library, specimens are now rather scarce.
It was the fashion to use the Arabic symbols in ecclesiastical works;
while their occurrence in all astronomical tables and their Oriental origin (which savoured of magic) secured their use in calendars intended
for scientific purposes. Thus the symbols were generally employed in
both kinds of almanacks, and there are but few specimens of calendars
issued after the year 1300 in which an explanation of the Arabic numerals is not included. Towards the middle of the fourteenth century
the rules of arithmetic *de algorismo* were also sometimes added, and
by the year 1400 we may consider that the Arabic symbols were generally known throughout Europe, and were used in most scientific and
astronomical works.

Outside Italy most merchants continued, however, to keep their accounts in Roman numerals till about 1550, and monasteries and colleges
till about 1650; though in both cases it is probable that in and after
the fifteenth century the processes of arithmetic were performed in the
algoristic manner. Arabic numerals are used in the pagination of some
books issued at Venice in 1471 and 1482. No instance of a date or
number being written in Arabic numerals is known to occur in any
English parish register or the court rolls of any English manor before
the sixteenth century; but in the rent-roll of the St Andrews Chapter,
Scotland, the Arabic numerals were used in 1490. The Arabic numerals
were used in Constantinople by Planudes[1] in the fourteenth century.

The history of modern mercantile arithmetic in Europe begins then
with its use by Italian merchants, and it is especially to the Florentine
traders and writers that we owe its early development and improvement. It was they who invented the system of book-keeping by double
entry. In this system every transaction is entered on the credit side in
one ledger, and on the debtor side in another; thus, if cloth be sold
to *A*, *A*'s account is debited with the price, and the stock-book, containing the transactions in cloth, is credited with the amount sold. It

---

[1] See above, p. 98.

was they, too, who arranged the problems to which arithmetic could be applied in different classes, such as rule of three, interest, profit and loss, &c. They also reduced the fundamental operations of arithmetic "to seven, in reverence," says Pacioli, "of the seven gifts of the Holy Spirit: namely, numeration, addition, subtraction, multiplication, division, raising to powers, and extraction of roots." Brahmagupta had enumerated twenty processes, besides eight subsidiary ones, and had stated that "a distinct and several knowledge of these" was "essential to all who wished to be calculators"; and, whatever may be thought of Pacioli's reason for the alteration, the consequent simplification of the elementary processes was satisfactory. It may be added that arithmetical schools were founded in various parts of Germany, especially in and after the fourteenth century, and did much towards familiarizing traders in northern and western Europe with commercial algoristic arithmetic.

The operations of algoristic arithmetic were at first very cumbersome. The chief improvements subsequently introduced into the early Italian algorism were (i) the simplification of the four fundamental processes; (ii) the introduction of signs for addition, subtraction, equality, and (though not so important) for multiplication and division; (iii) the invention of logarithms; and (iv) the use of decimals. I will consider these in succession.

(i) In addition and subtraction the Arabs usually worked from left to right. The modern plan of working from right to left is said to have been introduced by an Englishman named Garth, of whose life I can find no account. The old plan continued in partial use till about 1600; even now it would be more convenient in approximations where it is necessary to keep only a certain number of places of decimals.

The Indians and Arabs had several systems of multiplication. These were all somewhat laborious, and were made the more so as multiplication tables, if not unknown, were at any rate used but rarely. The operation was regarded as one of considerable difficulty, and the test of the accuracy of the result by "casting out the nines" was invented as a check on the correctness of the work. Various other systems of multiplication were subsequently employed in Italy, of which several examples are given by Pacioli and Tartaglia; and the use of the multiplication table—at least as far as 5×5—became common. From this limited table the resulting product of the multiplication of all numbers up to 10×10 can be deduced by what was termed the *regula ignavi*. This is a statement of the identity $(5+a)(5+b) = (5-a)(5-b) + 10(a+b)$. The rule was usually enunciated in the following form. Let the number five be represented by the open hand; the number six by the hand with one finger closed; the number seven by the hand with two fingers closed; the number eight by the hand with three fingers closed; and the number nine by the hand with four fingers closed. To multiply one number by another let the multiplier be represented by one hand, and the number multiplied by the other, according to the above convention. Then the required answer is the product of the number of fingers (counting the thumb as a finger) open in the one hand by the number of fingers open in the other together with ten times the total number of fingers closed. The system of multiplication now in use seems to have been first introduced at Florence.

Figure 1.

1	2	3	4	5	6	7	8	9	0
2	4	6	8	10	12	14	16	18	0
3	6	9	12	15	18	21	24	27	0
4	8	12	16	20	24	28	32	36	0
5	10	15	20	25	30	35	40	45	0
6	12	18	24	30	36	42	48	54	0
7	14	21	28	35	42	49	56	63	0
8	16	24	32	40	48	56	64	72	0
9	18	27	36	45	54	63	72	81	0

Figure 2.

7
14
21
28
35
42
49
56
63

Figure 3.

2	9	8	5
4	18	16	10
6	27	24	15
8	36	32	20
10	45	40	25
12	54	48	30
14	63	56	35
16	72	64	40
18	81	72	45

The difficulty which all but professed mathematicians experienced in the multiplication of large numbers led to the invention of several mechanical ways of effecting the process. Of these the most celebrated is that of Napier's rods invented in 1617. In principle it is the same as a method which had been long in use both in India and Persia, and which has been described in the diaries of several travellers, and notably in the *Travels of Sir John Chardin in Persia*, London, 1686. To use the method a number of rectangular slips of bone, wood, metal, or cardboard are prepared, and each of them divided by cross lines into nine little squares, a slip being generally about three inches long and a third of an inch across. In the top square one of the digits is engraved, and the results of multiplying it by 2, 3, 4, 5, 6, 7, 8, and 9 are respectively entered in the eight lower squares; where the result is a number of two digits, the ten-digit is written above and to the left of the unit-digit and separated from it by a diagonal line. The slips are usually arranged in a box. Figure 1 above represents nine such slips side by side; figure 2 shews the seventh slip, which is supposed to be taken out of the box and put by itself. Suppose we wish to multiply 2985 by 317. The process as effected by the use of these slips is as follows. The slips headed 2, 9, 8, and 5 are taken out of the box and put side by side as shewn in figure 3 above. The result of multiplying 2985 by 7 may be written thus—

$$
\begin{array}{r}
2985 \\
7 \\
\hline
35 \\
56 \\
63 \\
14 \\
\hline
20895 \\
\hline\hline
\end{array}
$$

Now if the reader will look at the seventh line in figure 3, he will see that the upper and lower rows of figures are respectively 1653 and 4365; moreover, these are arranged by the diagonals so that roughly the 4 is under the 6, the 3 under the 5, and the 6 under the 3; thus

$$
\begin{array}{cccc}
1 & 6 & 5 & 3 \\
  & 4 & 3 & 6 & 5
\end{array}
$$

The addition of these two numbers gives the required result. Hence the result of multiplying by 7, 1, and 3 can be successively determined in this way, and the required answer (namely, the product of 2985 and 317) is then obtained by addition.

The whole process was written as follows:

$$
\begin{array}{ll}
2985 \\
\hline
20895 & /7 \\
2985 & /1 \\
8955 & /3 \\
\hline
946245
\end{array}
$$

The modification introduced by Napier in his *Rabdologia*, published in 1617, consisted merely in replacing each slip by a prism with square ends, which he called "a rod," each lateral face being divided and marked in the same way as one of the slips above described. These rods not only economized space, but were easier to handle, and were arranged in such a way as to facilitate the operations required.

1	7	9	7	8
4	7	2		
		0		

Figure 1.

1	7	9	7	8
1	2			
	5	9	7	8
	2	1		
	3	8	7	8
			6	
	3	8	1	8
	4	7	2	
4̸	7̸	2̸		
		0	3	

Figure 2.

1	7	9	7	8
1	2			
	5	9	7	8
	2	1		
	3	8	7	8
			6	
	3	8	1	8
	3	2		
		6	1	8
		5	6	
			5	8
			1	6
			4	2
	4	7	2	
4	7	2		
		0	3	8

Figure 3.

If multiplication was considered difficult, division was at first regarded as a feat which could be performed only by skilled mathematicians. The method commonly employed by the Arabs and Persians for the division of one number by another will be sufficiently illustrated by a concrete instance. Suppose we require to divide 17978 by 472. A sheet of paper is divided into as many vertical columns as there are figures in the number to be divided. The number to be divided is written at the top and the divisor at the bottom; the first digit of each number being placed at the left-hand side of the paper. Then, taking the left-hand column, 4 will go into 1 no times, hence the first figure in the dividend is 0, which is written under the last figure of the divisor. This is represented in figure 1 above. Next (see figure 2) rewrite the 472 immediately above its former position, but shifted one place to the right, and cancel the old figures. Then 4 will go into 17 four times; but, as on trial it is found that 4 is too big for the first digit of the dividend, 3 is selected; 3 is therefore written below the last digit of the divisor and next to the digit of the dividend last found. The process of multiplying the divisor by 3 and subtracting from the number to be divided is indicated in figure 2, and shews that the remainder is 3818. A similar process is then repeated, that is, 472 is divided into 3818, shewing that the quotient is 38 and the remainder 42. This is represented in figure 3, which shews the whole operation.

The method described above never found much favour in Italy. The present system was in use there as early as the beginning of the fourteenth century, but the method generally employed was that known as the *galley* or *scratch* system. The following example from Tartaglia, in which it is required to divide 1330 by 84, will serve to illustrate this method: the arithmetic given by Tartaglia is shewn below, where numbers in thin type are supposed to be scratched out in the course of the work.

$$
\begin{array}{l}
\mathbf{0\ 7} \\
4\ 9 \\
0\ 5\ 9\ \mathbf{0} \\
1\ 3\ 3\ 0\ (\ \mathbf{15} \\
8\ 4\ 4 \\
\ \ \ 8
\end{array}
$$

The process is as follows. First write the 84 beneath the 1330, as indicated below, then 84 will go into 133 once, hence the first figure in the quotient is 1. Now $1 \times 8 = 8$, which subtracted from 13 leaves 5. Write this above the 13, and cancel the 13 and the 8, and we have as the result of the first step

$$
\begin{array}{l}
\mathbf{5} \\
1\ 3\ \mathbf{3\ 0}\ (\ \mathbf{1} \\
8\ 4
\end{array}
$$

Next, $1 \times 4 = 4$, which subtracted from 53 leaves 49. Insert the 49, and cancel the 53 and the 4, and we have as the next step

$$
\begin{array}{l}
\mathbf{4} \\
5\ 9 \\
1\ 3\ 3\ 0\ (\ \mathbf{1} \\
8\ 4
\end{array}
$$

which shews a remainder 490.

We have now to divide 490 by 84. Hence the next figure in the quotient will be 5, and re-writing the divisor we have

$$
\begin{array}{l}
\mathbf{4} \\
5\ 9 \\
1\ 3\ 3\ 0\ (\ \mathbf{15} \\
8\ 4\ 4 \\
\ \ \ \mathbf{8}
\end{array}
$$

Then $5 \times 8 = 40$, and this subtracted from 49 leaves 9. Insert the 9, and cancel the 49 and the 8, and we have the following result

$$4\ \mathbf{9}$$
$$5\ 9$$
$$1\ 3\ 3\ \mathbf{0}\ (\ \mathbf{15}$$
$$8\ 4\ \mathbf{4}$$
$$8$$

Next $5 \times 4 = 20$, and this subtracted from 90 leaves 70. Insert the 70, and cancel the 90 and the 4, and the final result, shewing a remainder 70, is

$$\mathbf{7}$$
$$4\ 9$$
$$5\ 9\ \mathbf{0}$$
$$1\ 3\ 3\ 0\ (\ \mathbf{15}$$
$$8\ 4\ 4$$
$$8$$

The three extra zeros inserted in Tartaglia's work are unnecessary, but they do not affect the result, as it is evident that a figure in the dividend may be shifted one or more places up in the same vertical column if it be convenient to do so.

The medieval writers were acquainted with the method now in use, but considered the scratch method more simple. In some cases the latter is very clumsy, as may be illustrated by the following example taken from Pacioli. The object is to divide 23400 by 100. The result is obtained thus

$$0$$
$$0\ 4\ 0$$
$$0\ 3\ 4\ 0\ 0$$
$$2\ 3\ 4\ 0\ 0\ (\ \mathbf{234}$$
$$1\ 0\ 0\ 0\ 0$$
$$1\ 0\ 0$$
$$1$$

The galley method was used in India, and the Italians may have derived it thence. In Italy it became obsolete somewhere about 1600; but it continued in partial use for at least another century in other countries. I should add that Napier's rods can be, and sometimes were used to obtain the result of dividing one number by another.

(ii) The signs + and − to indicate addition and subtraction[1] occur in Widman's arithmetic published in 1489, but were first brought into general notice, at any rate as symbols of operation, by Stifel in 1544. They occur, however, in a work by G. V. Hoecke, published at Antwerp in 1514. I believe I am correct in saying that Vieta in 1591 was the first well-known writer who used these signs consistently throughout his work, and that it was not until the beginning of the seventeenth century that they became recognized as well-known symbols. The sign = to denote equality[2] was introduced by Record in 1557.

(iii) The invention of logarithms,[3] without which many of the numerical calculations which have constantly to be made would be practically impossible, was due to Napier of Merchiston. The first public announcement of the discovery was made in his *Mirifici Logarithmorum Canonis Descriptio*, published in 1614, and of which an English translation was issued in the following year; but he had privately communicated a summary of his results to Tycho Brahe as early as 1594. In this work Napier explains the nature of logarithms by a comparison between corresponding terms of an arithmetical and geometrical progression. He illustrates their use, and gives tables of the logarithms of the sines and tangents of all angles in the first quadrant, for differences of every minute, calculated to seven places of decimals. His definition of the logarithm of a quantity n was what we should now express by $10^7 \log_e(10^7/n)$. This work is the more interesting to us as it is the first valuable contribution to the progress of mathematics which was made by any British writer. The method by which the logarithms were calculated was explained in the *Constructio*, a posthumous work issued in 1619: it seems to have been very laborious, and depended either on direct involution and evolution, or on the formation of geometrical means. The method by finding the approximate value of a convergent series was introduced by Newton, Cotes, and Euler. Napier had determined to change the base to one which was a power of 10, but died before he could effect it.

The rapid recognition throughout Europe of the advantages of using

---

[1]See below, pp. 171, 172, 177, 179.

[2]See below, p. 177.

[3]See the article on *Logarithms* in the *Encyclopaedia Britannica*, ninth edition; see also below, pp. 195, 196.

logarithms in practical calculations was mainly due to Briggs, who was one of the earliest to recognize the value of Napier's invention. Briggs at once realized that the base to which Napier's logarithms were calculated was inconvenient; he accordingly visited Napier in 1616, and urged the change to a decimal base, which was recognized by Napier as an improvement. On his return Briggs immediately set to work to calculate tables to a decimal base, and in 1617 he brought out a table of logarithms of the numbers from 1 to 1000 calculated to fourteen places of decimals.

It would seem that J. Bürgi, independently of Napier, had constructed before 1611 a table of antilogarithms of a series of natural numbers: this was published in 1620. In the same year a table of the logarithms, to seven places of decimals, of the sines and tangents of angles in the first quadrant was brought out by Edmund Gunter, one of the Gresham lecturers. Four years later the latter mathematician introduced a "line of numbers," which provided a mechanical method for finding the product of two numbers: this was the precursor of the slide-rule, first described by Oughtred in 1632. In 1624, Briggs published tables of the logarithms of some additional numbers and of various trigonometrical functions. His logarithms of the natural numbers are equal to those to the base 10 when multiplied by 10^8, and of the sines of angles to those to the base 10 when multiplied by 10^{12}. The calculation of the logarithms of 70,000 numbers which had been omitted by Briggs from his tables of 1624 was performed by Adrian Vlacq and published in 1628: with this addition the table gave the logarithms of all numbers from 1 to 101,000.

The *Arithmetica Logarithmica* of Briggs and Vlacq are substantially the same as the existing tables: parts have at different times been recalculated, but no tables of an equal range and fulness entirely founded on fresh computations have been published since. These tables were supplemented by Briggs's *Trigonometrica Britannica*, which contains tables not only of the logarithms of the trigonometrical functions, but also of their natural values: it was published posthumously in 1633. A table of logarithms to the base e of the numbers from 1 to 1000 and of the sines, tangents, and secants of angles in the first quadrant was published by John Speidell at London as early as 1619, but of course these were not so useful in practical calculations as those to the base 10. By 1630 tables of logarithms were in general use.

(iv) The introduction of the decimal notation for fractions is due to Pitiscus, in whose Tables, 1608 and 1612, it appears; it was employed in 1616 in the English translation of Napier's *Descriptio*, and occurs in the logarithmic Tables published by Briggs in 1617, after which date its use may be taken to be established. The idea was not new, for Stevinus had in 1585 used a somewhat similar notation, writing a number such as 25·379 either in the form 25, 3′ 7″ 9‴, or in the form 25 ⓪ 3 ① 7 ② 9 ③. This latter notation was also used by Napier in 1617 in his essay on Rods, and by Rudolff. These writers employed it only as a concise way of stating results, and made no use of it as an operative form; probably Briggs did more than any other writer to establish its use as an operative form. The subject is one of much interest, and a considerable body of literature has grown up about it. Some of the facts are in dispute, and the above statement must only be taken to represent my general conclusions. The reader interested in the subject may consult the Napier Tercentenary Volume issued by the Edinburgh Royal Society in 1915. Before the sixteenth century fractions were commonly written in the sexagesimal notation.[1]

In Napier's work of 1619 the point is written in the form now adopted in England. Witt in 1613 and Napier in 1617 used a solidus to separate the integral from the fractional part. Briggs underlined the decimal figures, and would have printed a number such as 25·379 in the form 25379. Subsequent writers added another line, and would have written it as 25⌊379; nor was it till the beginning of the eighteenth century that the current notation was generally employed. Even now the notation varies slightly in different countries: thus the fraction $\frac{1}{4}$ would in the decimal notation be written in England as 0·25, in America as 0.25, and in Germany and France as 0,25. A knowledge of the decimal notation became general among practical men with the introduction of the French decimal standards.

---

[1] For examples, see above, pp. 81, 84, 140.

CHAPTER XII.

THE MATHEMATICS OF THE RENAISSANCE.[1]
CIRC. 1450–1637.

THE last chapter is a digression from the chronological arrangement to which, as far as possible, I have throughout adhered, but I
trust by references in this chapter to keep the order of events and discoveries clear. I return now to the general history of mathematics in
western Europe. Mathematicians had barely assimilated the knowledge
obtained from the Arabs, including their translations of Greek writers,
when the refugees who escaped from Constantinople after the fall of the
eastern empire brought the original works and the traditions of Greek
science into Italy. Thus by the middle of the fifteenth century the chief
results of Greek and Arabian mathematics were accessible to European
students.

The invention of printing about that time rendered the dissemination of discoveries comparatively easy. It is almost a truism to remark
that until printing was introduced a writer appealed to a very limited
class of readers, but we are perhaps apt to forget that when a medieval
writer "published" a work the results were known to only a few of his
contemporaries. This had not been the case in classical times, for then
and until the fourth century of our era Alexandria was the recognized
centre for the reception and dissemination of new works and discoveries. In medieval Europe, on the other hand, there was no common
centre through which men of science could communicate with one another, and to this cause the slow and fitful development of medieval
mathematics may be partly ascribed.

---

[1] Where no other references are given, see parts xii, xiii, xiv, and the early chapters of part xv of Cantor's *Vorlesungen*; on the Italian mathematicians of this period
see also G. Libri, *Histoire des sciences mathématiques en Italie*, 4 vols., Paris, 1838–
1841.

The introduction of printing marks the beginning of the modern world in science as in politics; for it was contemporaneous with the assimilation by the indigenous European school (which was born from scholasticism, and whose history was traced in chapter VIII) of the results of the Indian and Arabian schools (whose history and influence were traced in chapters IX and X), and of the Greek schools (whose history was traced in chapters II to V).

The last two centuries of this period of our history, which may be described as the renaissance, were distinguished by great mental activity in all branches of learning. The creation of a fresh group of universities (including those in Scotland), of a somewhat less complex type than the medieval universities above described, testify to the general desire for knowledge. The discovery of America in 1492 and the discussions that preceded the Reformation flooded Europe with new ideas which, by the invention of printing, were widely disseminated; but the advance in mathematics was at least as well marked as that in literature and that in politics.

During the first part of this time the attention of mathematicians was to a large extent concentrated on syncopated algebra and trigonometry; the treatment of these subjects is discussed in the first section of this chapter, but the relative importance of the mathematicians of this period is not very easy to determine. The middle years of the renaissance were distinguished by the development of symbolic algebra: this is treated in the second section of this chapter. The close of the sixteenth century saw the creation of the science of dynamics: this forms the subject of the first section of chapter XIII. About the same time and in the early years of the seventeenth century considerable attention was paid to pure geometry: this forms the subject of the second section of chapter XIII.

The development of syncopated algebra and trigonometry.

Regiomontanus.[1]   Amongst the many distinguished writers of this time *Johann Regiomontanus* was the earliest and one of the most able. He was born at Königsberg on June 6, 1436, and died at Rome

---

[1] His life was written by P. Gassendi, The Hague, second edition, 1655. His letters, which afford much valuable information on the mathematics of his time, were collected and edited by C. G. von Murr, Nuremberg, 1786. An account of his works will be found in *Regiomontanus, ein geistiger Vorläufer des Copernicus*, by A. Ziegler, Dresden, 1874; see also Cantor, chap. lv.

on July 6, 1476. His real name was *Johannes Müller*, but, following the custom of that time, he issued his publications under a Latin pseudonym which in his case was taken from his birthplace. To his friends, his neighbours, and his tradespeople he may have been Johannes Müller, but the literary and scientific world knew him as Regiomontanus, just as they knew Zepernik as Copernicus, and Schwarzerd as Melanchthon. It seems as pedantic as it is confusing to refer to an author by his actual name when he is universally recognized under another: I shall therefore in all cases as far as possible use that title only, whether latinized or not, by which a writer is generally known.

Regiomontanus studied mathematics at the university of Vienna, then one of the chief centres of mathematical studies in Europe, under Purbach who was professor there. His first work, done in conjunction with Purbach, consisted of an analysis of the *Almagest*. In this the trigonometrical functions *sine* and *cosine* were used and a table of natural sines was introduced. Purbach died before the book was finished: it was finally published at Venice, but not till 1496. As soon as this was completed Regiomontanus wrote a work on astrology, which contains some astronomical tables and a table of natural tangents: this was published in 1490.

Leaving Vienna in 1462, Regiomontanus travelled for some time in Italy and Germany; and at last in 1471 settled for a few years at Nuremberg, where he established an observatory, opened a printing-press, and probably lectured. Three tracts on astronomy by him were written here. A mechanical eagle, which flapped its wings and saluted the Emperor Maximilian I. on his entry into the city, bears witness to his mechanical ingenuity, and was reckoned among the marvels of the age. Thence Regiomontanus moved to Rome on an invitation from Sixtus IV. who wished him to reform the calendar. He was assassinated, shortly after his arrival, at the age of 40.

Regiomontanus was among the first to take advantage of the recovery of the original texts of the Greek mathematical works in order to make himself acquainted with the methods of reasoning and results there used; the earliest notice in modern Europe of the algebra of Diophantus is a remark of his that he had seen a copy of it at the Vatican. He was also well read in the works of the Arab mathematicians.

The fruit of his study was shewn in his *De Triangulis* written in 1464. This is the earliest modern systematic exposition of trigonometry, plane and spherical, though the only trigonometrical functions introduced are those of the sine and cosine. It is divided into five books.

The first four are given up to plane trigonometry, and in particular to determining triangles from three given conditions. The fifth book is devoted to spherical trigonometry. The work was printed at Nuremberg in 1533, nearly a century after the death of Regiomontanus.

As an example of the mathematics of this time I quote one of his propositions at length. It is required to determine a triangle when the difference of two sides, the perpendicular on the base, and the difference between the segments into which the base is thus divided are given [book II, prop. 23]. The following is the solution given by Regiomontanus.

Sit talis triangulus *ABG*, cujus duo latera *AB* et *AG* differentia habeant nota *HG*, ductaque perpendiculari *AD* duorum casuum *BD* et *DG*, differentia sit *EG*: hae duae differentiae sint datae, et ipsa perpendicularis *AD* data. Dico quod omnia latera trianguli nota concludentur. Per artem rei et census hoc problema absolvemus. Detur ergo differentia laterum ut 3, differentia casuum 12, et perpendicularis 10. Pono pro basi unam rem, et pro aggregato laterum 4 res, nae proportio basis ad congeriem laterum est ut *HG* ad *GE*, scilicet unius ad 4. Erit ergo *BD* $\frac{1}{2}$ rei minus 6, sed *AB* erit 2 res demptis $\frac{3}{2}$. Duco *AB* in se, producuntur 4 census et $2\frac{1}{4}$ demptis 6 rebus. Item *BD* in se facit $\frac{1}{4}$ census et 36 minus 6 rebus: huic addo quadratum de 10 qui est 100. Colliguntur $\frac{1}{4}$ census et 136 minus 6 rebus aequales videlicet 4 censibus et $2\frac{1}{4}$ demptis 6 rebus. Restaurando itaque defectus et auferendo utrobique aequalia, quemadmodum ars ipsa praecipit, habemus census aliquot aequales numero, unde cognitio rei patebit, et inde tria latera trianguli more suo innotescet.

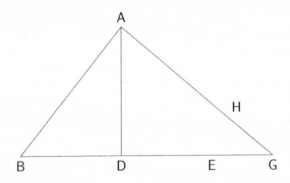

To explain the language of the proof I should add that Regiomontanus calls the unknown quantity *res*, and its square *census* or *zensus*; but though he uses these technical terms he writes the words in full. He commences by saying that he will solve the problem by means of a quadratic equation (per artem rei et census); and that he will suppose the difference of the sides of the triangle to be 3, the difference of the segments of the base to be 12, and the altitude of the triangle to be 10. He then takes for his unknown quantity (unam rem or x) the base of the triangle, and therefore the sum of the sides will be $4x$. Therefore BD will be equal to $\frac{1}{2}x - 6$ ($\frac{1}{2}$ rei minus 6), and AB will be equal to $2x - \frac{3}{2}$ (2 res demptis $\frac{3}{2}$); hence AB^2 (AB in se) will be $4x^2 + 2\frac{1}{4} - 6x$ (4 census et $2\frac{1}{4}$ demptis 6 rebus), and BD^2 will be $\frac{1}{4}x^2 + 36 - 6x$. To BD^2 he adds AD^2 (quadratum de 10) which is 100, and states that the sum of the two is equal to AB^2. This he says will give the value of x^2 (census), whence a knowledge of x (cognitio rei) can be obtained, and the triangle determined.

To express this in the language of modern algebra we have

$$AG^2 - DG^2 = AB^2 - DB^2,$$
$$\therefore AG^2 - AB^2 = DG^2 - DB^2,$$

but by the given numerical conditions

$$AG - AB = 3 = \frac{1}{4}(DG - DB),$$
$$\therefore AG + AB = 4(DG + DB) = 4x.$$

Therefore $AB = 2x - \frac{3}{2}$, and $BD = \frac{1}{2}x - 6$.

Hence $(2x - \frac{3}{2})^2 = (\frac{1}{2}x - 6)^2 + 100$.

From which x can be found, and all the elements of the triangle determined.

It is worth noticing that Regiomontanus merely aimed at giving a general method, and the numbers are not chosen with any special reference to the particular problem. Thus in his diagram he does not attempt to make GE anything like four times as long as GH, and, since x is ultimately found to be equal to $\frac{1}{3}\sqrt{321}$, the point D really falls outside the base. The order of the letters ABG, used to denote the triangle, is of course derived from the Greek alphabet.

Some of the solutions which he gives are unnecessarily complicated, but it must be remembered that algebra and trigonometry were still only in the rhetorical stage of development, and when every step of the argument is expressed in words at full length it is by no means easy to realize all that is contained in a formula.

It will be observed from the above example that Regiomontanus did not hesitate to apply algebra to the solution of geometrical problems.

Another illustration of this is to be found in his discussion of a question which appears in Brahmagupta's *Siddhanta*. The problem was to construct a quadrilateral, having its sides of given lengths, which should be inscribable in a circle. The solution[1] given by Regiomontanus was effected by means of algebra and trigonometry.

The *Algorithmus Demonstratus* of Jordanus, described above, which was first printed in 1534, was formerly attributed to Regiomontanus.

Regiomontanus was one of the most prominent mathematicians of his generation, and I have dealt with his works in some detail as typical of the most advanced mathematics of the time. Of his contemporaries I shall do little more than mention the names of a few of those who are best known; none were quite of the first rank, and I should sacrifice the proportion of the parts of the subject were I to devote much space to them.

Purbach.[2]   I may begin by mentioning *George Purbach*, first the tutor and then the friend of Regiomontanus, born near Linz on May 30, 1423, and died at Vienna on April 8, 1461, who wrote a work on planetary motions which was published in 1460; an arithmetic, published in 1511; a table of eclipses, published in 1514; and a table of natural sines, published in 1541.

Cusa.[3]   Next I may mention *Nicolas de Cusa*, who was born in 1401 and died in 1464. Although the son of a poor fisherman and without influence, he rose rapidly in the church, and in spite of being "a reformer before the reformation" became a cardinal. His mathematical writings deal with the reform of the calendar and the quadrature of the circle; in the latter problem his construction is equivalent to taking $\frac{3}{4}(\sqrt{3} + \sqrt{6})$ as the value of π. He argued in favour of the diurnal rotation of the earth.

Chuquet.   I may also here notice a treatise on arithmetic, known as *Le Triparty*,[4] by *Nicolas Chuquet*, a bachelor of medicine in the university of Paris, which was written in 1484. This work indicates that the extent of mathematics then taught was somewhat greater than was generally believed a few years ago. It contains the earliest known use of the radical sign with indices to mark the root taken, 2 for a square-root,

[1]It was published by C. G. von Murr at Nuremberg in 1786.

[2]Purbach's life was written by P. Gassendi, The Hague, second edition, 1655.

[3]Cusa's life was written by F. A. Scharpff, Tübingen, 1871; and his collected works, edited by H. Petri, were published at Bâle in 1565.

[4]See an article by A. Marre in Boncompagni's *Bulletino di bibliografia* for 1880, vol. xiii, pp. 555–659.

3 for a cube-root, and so on; and also a definite statement of the rule of signs. The words plus and minus are denoted by the contractions $\bar{p}$, $\bar{m}$. The work is in French.

Introduction[1] of signs + and −. In England and Germany algorists were less fettered by precedent and tradition than in Italy, and introduced some improvements in notation which were hardly likely to occur to an Italian. Of these the most prominent were the introduction, if not the invention, of the current symbols for addition, subtraction, and equality.

The earliest instances of the regular use of the signs + and − of which we have any knowledge occur in the fifteenth century. *Johannes Widman* of Eger, born about 1460, matriculated at Leipzig in 1480, and probably by profession a physician, wrote a *Mercantile Arithmetic*, published at Leipzig in 1489 (and modelled on a work by Wagner printed some six or seven years earlier): in this book these signs are used merely as marks signifying excess or deficiency; the corresponding use of the word surplus or overplus[2] was once common and is still retained in commerce.

It is noticeable that the signs generally occur only in practical mercantile questions: hence it has been conjectured that they were originally warehouse marks. Some kinds of goods were sold in a sort of wooden chest called a *lagel*, which when full was apparently expected to weigh roughly either three or four *centners*; if one of these cases were a little lighter, say 5 lbs., than four centners, Widman describes it as weighing $4c - 5$ lbs.: if it were 5 lbs. heavier than the normal weight it is described as weighing $4c \mathbin{\underline{+}} 5$ lbs. The symbols are used as if they would be familiar to his readers; and there are some slight reasons for thinking that these marks were chalked on the chests as they came into the warehouses. We infer that the more usual case was for a chest to weigh a little less than its reputed weight, and, as the sign − placed between two numbers was a common symbol to signify some connection between them, that seems to have been taken as the standard case, while the vertical bar was originally a small mark super-added on the sign − to distinguish the two symbols. It will be observed that the vertical line in the symbol for excess, printed above, is somewhat shorter

---

[1] Recently new light has been thrown on the history of the subject by the researches of J. W. L. Glaisher, *Messenger of Mathematics*, Cambridge, vol. li, pp. 1 *et seq.* The account in the text is based on the earlier investigations of P. Treutlein, A. de Morgan, and Boncompagni.

[2] See *passim* Levit. xxv, verse 27, and 1 Maccab. x, verse 41.

than the horizontal line. This is also the case with Stifel and most of the early writers who used the symbol: some presses continued to print it in this, its earliest form, till the end of the seventeenth century. Xylander, on the other hand, in 1575 has the vertical bar much longer than the horizontal line, and the symbol is something like $+$.

Another conjecture is that the symbol for *plus* is derived from the Latin abbreviation & for *et*; while that for *minus* is obtained from the bar which is often used in ancient manuscripts to indicate an omission, or which is written over the contracted form of a word to signify that certain letters have been left out. This view has been often supported on a priori grounds, but it has recently found powerful advocates in Professors Zangmeister and Le Paige who also consider that the introduction of these symbols for plus and minus may be referred to the fourteenth century.

These explanations of the origin of our symbols for *plus* and *minus* are the most plausible that have been yet advanced, but the question is difficult and cannot be said to be solved. Another suggested derivation is that $+$ is a contraction of ⅊ the initial letter in Old German of plus, while $-$ is the limiting form of m (for minus) when written rapidly. De Morgan[1] proposed yet another derivation: the Hindoos sometimes used a dot to indicate subtraction, and this dot might, he thought, have been elongated into a bar, and thus give the sign for *minus*; while the origin of the sign for *plus* was derived from it by a super-added bar as explained above; but I take it that at a later time he abandoned this theory for what has been called the warehouse explanation.

I should perhaps here add that till the close of the sixteenth century the sign $+$ connecting two quantities like a and b was also used in the sense that if a were taken as the answer to some question one of the given conditions would be too little by b. This was a relation which constantly occurred in solutions of questions by the rule of false assumption.

Lastly, I would repeat again that these signs in Widman are only abbreviations and not symbols of operation; he attached little or no importance to them, and no doubt would have been amazed if he had been told that their introduction was preparing the way for a revolution of the processes used in algebra.

The *Algorithmus* of Jordanus was not published till 1534; Widman's work was hardly known outside Germany; and it is to Pacioli that we

---

[1] See his *Arithmetical Books*, London, 1847, p. 19.

owe the introduction into general use of syncopated algebra; that is, the use of abbreviations for certain of the more common algebraical quantities and operations, but where in using them the rules of syntax are observed.

Pacioli.[1]   *Lucas Pacioli*, sometimes known as *Lucas di Burgo*, and sometimes, but more rarely, as *Lucas Paciolus*, was born at Burgo in Tuscany about the middle of the fifteenth century. We know little of his life except that he was a Franciscan friar; that he lectured on mathematics at Rome, Pisa, Venice, and Milan; and that at the last-named city he was the first occupant of a chair of mathematics founded by Sforza: he died at Florence about the year 1510.

His chief work was printed at Venice in 1494 and is termed *Summa de arithmetica, geometria, proporzioni e proporzionalita*. It is divided into two parts, the first dealing with arithmetic and algebra, the second with geometry. This was the earliest printed book on arithmetic and algebra. It is mainly based on the writings of Leonardo of Pisa, and its importance in the history of mathematics is largely due to its wide circulation.

In the arithmetic Pacioli gives rules for the four simple processes, and a method for extracting square roots. He deals pretty fully with all questions connected with mercantile arithmetic, in which he works out numerous examples, and in particular discusses at great length bills of exchange and the theory of book-keeping by double entry. This part was the first systematic exposition of algoristic arithmetic, and has been already alluded to in chapter XI. It and the similar work by Tartaglia are the two standard authorities on the subject.

Many of his problems are solved by "the method of false assumption," which consists in assuming any number for the unknown quantity, and if on trial the given conditions be not satisfied, altering it by a simple proportion as in rule of three. As an example of this take the problem to find the original capital of a merchant who spent a quarter of it in Pisa and a fifth of it in Venice, who received on these transactions 180 ducats, and who has in hand 224 ducats. Suppose that we assume that he had originally 100 ducats. Then if he spent $25 + 20$ ducats at Pisa and Venice, he would have had 55 ducats left. But by the enunciation he then had $224 - 180$, that is, 44 ducats. Hence the ratio of his original capital to 100 ducats is as 44 to 55. Thus his original

---

[1] See H. Staigmüller in the *Zeitschrift für Mathematik*, 1889, vol. xxxiv; also Libri, vol. iii, pp. 133–145; and Cantor, chap. lvii.

capital was 80 ducats.

The following example will serve as an illustration of the kind of arithmetical problems discussed.

I buy for 1440 ducats at Venice 2400 sugar loaves, whose nett weight is 7200 lire; I pay as a fee to the agent 2 per cent.; to the weighers and porters on the whole, 2 ducats; I afterwards spend in boxes, cords, canvas, and in fees to the ordinary packers in the whole, 8 ducats; for the tax or octroi duty on the first amount, 1 ducat per cent.; afterwards for duty and tax at the office of exports, 3 ducats per cent.; for writing directions on the boxes and booking their passage, 1 ducat; for the bark to Rimini, 13 ducats; in compliments to the captains and in drink for the crews of armed barks on several occasions, 2 ducats; in expenses for provisions for myself and servant for one month, 6 ducats; for expenses for several short journeys over land here and there, for barbers, for washing of linen, and of boots for myself and servant, 1 ducat; upon my arrival at Rimini I pay to the captain of the port for port dues in the money of that city, 3 lire; for porters, disembarkation on land, and carriage to the magazine, 5 lire; as a tax upon entrance, 4 soldi a load which are in number 32 (such being the custom); for a booth at the fair, 4 soldi per load; I further find that the measures used at the fair are different to those used at Venice, and that 140 lire of weight are there equivalent to 100 at Venice, and that 4 lire of their silver coinage are equal to a ducat of gold. I ask, therefore, at how much I must sell a hundred lire Rimini in order that I may gain 10 per cent. upon my whole adventure, and what is the sum which I must receive in Venetian money?

In the algebra he discusses in some detail simple and quadratic equations, and problems on numbers which lead to such equations. He mentions the Arabic classification of cubic equations, but adds that their solution appears to be as impossible as the quadrature of the circle. The following is the rule he gives[1] for solving a quadratic equation of the form $x^2 + x = a$: it is rhetorical and not syncopated, and will serve to illustrate the inconvenience of that method.

> "Si res et census numero coaequantur, a rebus
> dimidio sumpto censum producere debes,
> addereque numero, cujus a radice totiens
> tolle semis rerum, census latusque redibit."

He confines his attention to the positive roots of equations.

[1]Edition of 1494, p. 145.

Though much of the matter described above is taken from Leonardo's *Liber Abaci*, yet the notation in which it is expressed is superior to that of Leonardo. Pacioli follows Leonardo and the Arabs in calling the unknown quantity the *thing*, in Italian *cosa*—hence algebra was sometimes known as the cossic art—or in Latin *res*, and sometimes denotes it by *co* or *R* or *Rj*. He calls the square of it *census* or *zensus*, and sometimes denotes it by *ce* or *Z*; similarly the cube of it, or *cuba*, is sometimes represented by *cu* or *C*; the fourth power, or *censo di censo*, is written either at length or as *ce di ce* or as *ce ce*. It may be noticed that all his equations are numerical, so that he did not rise to the conception of representing known quantities by letters as Jordanus had done and as is the case in modern algebra; but Libri gives two instances in which in a proportion he represents a number by a letter. He indicates addition by p or $\bar{p}$, the initial letter of the word *plus*, but he generally evades the introduction of a symbol for *minus* by writing his quantities on that side of the equation which makes them positive, though in a few places he denotes it by $\bar{m}$ for *minus* or by *de* for *demptus*. Similarly, equality is sometimes indicated by *ae* for *aequalis*. This is a commencement of syncopated algebra.

There is nothing striking in the results he arrives at in the second or geometrical part of the work; nor in two other tracts on geometry which he wrote and which were printed at Venice in 1508 and 1509. It may be noticed, however, that, like Regiomontanus, he applied algebra to aid him in investigating the geometrical properties of figures.

The following problem will illustrate the kind of geometrical questions he attacked. The radius of the inscribed circle of a triangle is 4 inches, and the segments into which one side is divided by the point of contact are 6 inches and 8 inches respectively. Determine the other sides. To solve this it is sufficient to remark that $rs = \Delta = \sqrt{s(s-a)(s-b)(s-c)}$ which gives $4s = \sqrt{s \times (s-14) \times 6 \times 8}$, hence $s = 21$; therefore the required sides are $21 - 6$ and $21 - 8$, that is, 15 and 13. But Pacioli makes no use of these formulae (with which he was acquainted), but gives an elaborate geometrical construction, and then uses algebra to find the lengths of various segments of the lines he wants. The work is too long for me to reproduce here, but the following analysis of it will afford sufficient materials for its reproduction. Let ABC be the triangle, D, E, F the points of contact of the sides, and O the centre of the given circle. Let H be the point of intersection of OB and DF, and K that of OC and DE. Let L and M be the feet of the perpendiculars drawn from E and F on BC. Draw EP parallel to

AB and cutting *BC* in *P*. Then Pacioli determines in succession the magnitudes of the following lines: (i) *OB*, (ii) *OC*, (iii) *FD*, (iv) *FH*, (v) *ED*, (vi) *EK*. He then forms a quadratic equation, from the solution of which he obtains the values of *MB* and *MD*. Similarly he finds the values of *LC* and *LD*. He now finds in succession the values of *EL*, *FM*, *EP*, and *LP*; and then by similar triangles obtains the value of *AB*, which is 13. This proof was, even sixty years later, quoted by Cardan as "incomparably simple and excellent, and the very crown of mathematics." I cite it as an illustration of the involved and inelegant methods then current. The problems enunciated are very similar to those in the *De Triangulis* of Regiomontanus.

Leonardo da Vinci.   The fame of *Leonardo da Vinci* as an artist has overshadowed his claim to consideration as a mathematician, but he may be said to have prepared the way for a more accurate conception of mechanics and physics, while his reputation and influence drew some attention to the subject; he was an intimate friend of Pacioli. Leonardo was the illegitimate son of a lawyer of Vinci in Tuscany, was born in 1452, and died in France in 1519 while on a visit to Francis I. Several manuscripts by him were seized by the French revolutionary armies at the end of the last century, and Venturi, at the request of the Institute, reported on those concerned with physical or mathematical subjects.[1]

Leaving out of account Leonardo's numerous and important artistic works, his mathematical writings are concerned chiefly with mechanics, hydraulics, and optics—his conclusions being usually based on experiments. His treatment of hydraulics and optics involves but little mathematics. The mechanics contain numerous and serious errors; the best portions are those dealing with the equilibrium of a lever under any forces, the laws of friction, the stability of a body as affected by the position of its centre of gravity, the strength of beams, and the orbit of a particle under a central force; he also treated a few easy problems by virtual moments. A knowledge of the triangle of forces is occasionally attributed to him, but it is probable that his views on the subject were somewhat indefinite. Broadly speaking, we may say that his mathematical work is unfinished, and consists largely of suggestions which he did not discuss in detail and could not (or at any rate did not) verify.

---

[1] *Essai sur les ouvrages physico-mathématiques de Léonard de Vinci*, by J.-B. Venturi, Paris, 1797.

Dürer.   *Albrecht Dürer*[1] was another artist of the same time who was also known as a mathematician. He was born at Nuremberg on May 21, 1471, and died there on April 6, 1528. His chief mathematical work was issued in 1525, and contains a discussion of perspective, some geometry, and certain graphical solutions; Latin translations of it were issued in 1532, 1555, and 1605.

Copernicus.   An account of *Nicolaus Copernicus*, born at Thorn on Feb. 19, 1473, and died at Frauenberg on May 7, 1543, and his conjecture that the earth and planets all revolved round the sun, belong to astronomy rather than to mathematics. I may, however, add that Copernicus wrote on trigonometry, his results being published as a text-book at Wittenberg in 1542; it is clear though it contains nothing new. It is evident from this and his astronomy that he was well read in the literature of mathematics, and was himself a mathematician of considerable power. I describe his statement as to the motion of the earth as a conjecture, because he advocated it only on the ground that it gave a simple explanation of natural phenomena. Galileo in 1632 was the first to try to supply a proof of this hypothesis.

By the beginning of the sixteenth century the printing-press began to be active, and many of the works of the earlier mathematicians became now for the first time accessible to all students. This stimulated inquiry, and before the middle of the century numerous works were issued which, though they did not include any great discoveries, introduced a variety of small improvements all tending to make algebra more analytical.

Record.   The sign now used to denote equality was introduced by *Robert Record.*[2] Record was born at Tenby in Pembrokeshire about 1510, and died at London in 1558. He entered at Oxford, and obtained a fellowship at All Souls College in 1531; thence he migrated to Cambridge, where he took a degree in medicine in 1545. He then returned to Oxford and lectured there, but finally settled in London and became physician to Edward VI. and to Mary. His prosperity must have been short-lived, for at the time of his death he was confined in the King's Bench prison for debt.

In 1540 he published an arithmetic, termed the *Grounde of Artes*, in which he employed the signs + to indicate excess and − to indi-

---

[1] See *Dürer als Mathematiker*, by H. Staigmüller, Stuttgart, 1891.

[2] On the life and career of Robert Record, see D. E. Smith in *The American Mathematical Monthly*, vol. 28, 1921, p. 296 *et seq.*

cate deficiency; "+ whyche betokeneth too muche, as this line − plaine
without a crosse line betokeneth too little." In this book the equality
of two ratios is indicated by two equal and parallel lines whose opposite ends are joined diagonally, *ex. gr.* by ⊐. A few years later, in
1557, he wrote an algebra under the title of the *Whetstone of Witte*.
This is interesting as it contains the earliest introduction of the sign =
for equality, and he says he selected that particular symbol because[1]
than two parallel straight lines "noe 2 thynges can be moare equalle."
M. Charles Henry has, however, asserted that this sign is a recognized
abbreviation for the word *est* in medieval manuscripts; and, if this be
established, it would seem to indicate a more probable origin. In this
work Record shewed how the square root of an algebraic expression
could be extracted. He also wrote an astronomy. These works give a
clear view of the knowledge of the time.

Rudolff. Riese.   About the same time in Germany, Rudolff and
Riese took up the subjects of algebra and arithmetic. Their investigations form the basis of Stifel's well-known work. *Christoff Rudolff*[2]
published his algebra in 1525; it is entitled *Die Coss*, and is founded on
the writings of Pacioli, and perhaps of Jordanus. Rudolff introduced
the sign of $\sqrt{}$ for the square root, the symbol being a corruption of
the initial letter of the word *radix*, similarly $\sqrt{}\sqrt{}\sqrt{}$ denoted the cube
root, and $\sqrt{}\sqrt{}$ the fourth root. *Adam Riese*[3] was born near Bamberg,
Bavaria, in 1489, of humble parentage, and after working for some years
as a miner set up a school; he died at Annaberg on March 30, 1559. He
wrote a treatise on practical geometry, but his most important book
was his well-known arithmetic (which may be described as algebraical),
issued in 1536, and founded on Pacioli's work. Riese used the symbols
+ and −.

Stifel.[4]   The methods used by Rudolff and Riese and their results
were brought into general notice through Stifel's work, which had a
wide circulation. *Michael Stifel*, sometimes known by the Latin name
of *Stiffelius*, was born at Esslingen in 1486, and died at Jena on April 19,
1567. He was originally an Augustine monk, but he accepted the doctrines of Luther, of whom he was a personal friend. He tells us in his

---

[1]See *Whetstone of Witte*, f. Ff, j. v.

[2]See E. Wappler, *Geschichte der deutschen Algebra im xv. Jahrhunderte*,
Zwickau, 1887.

[3]See two works by B. Berlet, *Ueber Adam Riese*, Annaberg, 1855; and *Die Coss
von Adam Riese*, Annaberg, 1860.

[4]The authorities on Stifel are given by Cantor chap. lxii.

algebra that his conversion was finally determined by noticing that the pope Leo X. was the beast mentioned in the Revelation. To shew this, it was only necessary to add up the numbers represented by the letters in Leo decimus (the m had to be rejected since it clearly stood for *mysterium*) and the result amounts to exactly ten less than 666, thus distinctly implying that it was Leo the tenth. Luther accepted his conversion, but frankly told him he had better clear his mind of any nonsense about the number of the beast.

Unluckily for himself Stifel did not act on this advice. Believing that he had discovered the true way of interpreting the biblical prophecies, he announced that the world would come to an end on October 3, 1533. The peasants of Holzdorf, of which place he was pastor, aware of his scientific reputation, accepted his assurance on this point. Some gave themselves up to religious exercises, others wasted their goods in dissipation, but all abandoned their work. When the day foretold had passed, many of the peasants found themselves ruined. Furious at having been deceived, they seized the unfortunate prophet, and he was lucky in finding a refuge in the prison at Wittenberg, from which he was after some time released by the personal intercession of Luther.

Stifel wrote a small treatise on algebra, but his chief mathematical work is his *Arithmetica Integra*, published at Nuremberg in 1544, with a preface by Melanchthon.

The first two books of the *Arithmetica Integra* deal with surds and incommensurables, and are Euclidean in form. The third book is on algebra, and is noticeable for having called general attention to the German practice of using the signs $+$ and $-$ to denote addition and subtraction. There are traces of these signs being occasionally employed by Stifel as symbols of operation and not only as abbreviations; in this use of them he seems to have followed G. V. Hoecke. He not only employed the usual abbreviations for the Italian words which represent the unknown quantity and its powers, but in at least one case when there were several unknown quantities he represented them respectively by the letters A, B, C, &c.; thus re-introducing the general algebraic notation which had fallen into disuse since the time of Jordanus. It used to be said that Stifel was the real inventor of logarithms, but it is now certain that this opinion was due to a misapprehension of a passage in which he compares geometrical and arithmetical progressions. Stifel is said to have indicated a formula for writing down the coefficients of the various terms in the expansion of $(1 + x)^n$ if those in the expansion of $(1 + x)^{n-1}$ were known.

In 1553 Stifel brought out an edition of Rudolff's *Die Coss*, in which he introduced an improvement in the algebraic notation then current. The symbols at that time ordinarily used for the unknown quantity and its powers were letters which stood for abbreviations of the words. Among those frequently adopted were R or Rj for *radix* or *res* (x), Z or C for *zensus* or *census* (x^2), C or K for *cubus* (x^3), &c. Thus $x^2 + 5x - 4$ would have been written

$$1 \; Z \; \text{p.} \; 5 \; R \; \text{m.} \; 4;$$

where p stands for plus and m for minus. Other letters and symbols were also employed: thus Xylander (1575) would have denoted the above expression by

$$1Q + 5N - 4;$$

a notation similar to this was sometimes used by Vieta and even by Fermat. The advance made by Stifel was that he introduced the symbols $1A$, $1AA$, $1AAA$, for the unknown quantity, its square, and its cube, which shewed at a glance the relation between them.

Tartaglia.   *Niccolo Fontana*, generally known as *Nicholas Tartaglia*, that is, Nicholas the stammerer, was born at Brescia in 1500, and died at Venice on December 14, 1557. After the capture of the town by the French in 1512, most of the inhabitants took refuge in the cathedral, and were there massacred by the soldiers. His father, who was a postal messenger at Brescia, was amongst the killed. The boy himself had his skull split through in three places, while his jaws and his palate were cut open; he was left for dead, but his mother got into the cathedral, and finding him still alive managed to carry him off. Deprived of all resources she recollected that dogs when wounded always licked the injured place, and to that remedy he attributed his ultimate recovery, but the injury to his palate produced an impediment in his speech, from which he received his nickname. His mother managed to get sufficient money to pay for his attendance at school for fifteen days, and he took advantage of it to steal a copy-book from which he subsequently taught himself how to read and write; but so poor were they that he tells us he could not afford to buy paper, and was obliged to make use of the tombstones as slates on which to work his exercises.

He commenced his public life by lecturing at Verona, but he was appointed at some time before 1535 to a chair of mathematics at Venice, where he was living, when he became famous through his acceptance of a challenge from a certain *Antonio del Fiore* (or *Florido*). Fiore had learnt from his master, one *Scipione Ferro* (who died at Bologna in 1526), an empirical solution of a cubic equation of the form $x^3 + qx = r$. This solution was previously unknown in Europe, and it is possible that Ferro had found the result in an Arab work. Tartaglia, in answer to a request from Colla in 1530, stated that he could effect the solution of a numerical equation of the form $x^3 + px^2 = r$. Fiore, believing that Tartaglia was an impostor, challenged him to a contest. According to this challenge each of them was to deposit a certain stake with a notary, and whoever could solve the most problems out of a collection of thirty propounded by the other was to get the stakes, thirty days being allowed for the solution of the questions proposed. Tartaglia was aware that his adversary was acquainted with the solution of a cubic equation of some particular form, and suspecting that the questions proposed to him would all depend on the solution of such cubic equations, set himself the problem to find a general solution, and certainly discovered how to obtain a solution of some if not all cubic equations. His solution is believed to have depended on a geometrical construction,[1] but led to the formula which is often, but unjustly, described as Cardan's.

When the contest took place, all the questions proposed to Tartaglia were, as he had suspected, reducible to the solution of a cubic equation, and he succeeded within two hours in bringing them to particular cases of the equation $x^3 + qx = r$, of which he knew the solution. His opponent failed to solve any of the problems proposed to him, most of which were, as a matter of fact, reducible to numerical equations of the form $x^3 + px^2 = r$. Tartaglia was therefore the conqueror; he subsequently composed some verses commemorative of his victory.

The chief works of Tartaglia are as follows: (i) His *Nova scienza*, published in 1537: in this he investigated the fall of bodies under gravity; and he determined the range of a projectile, stating that it was a maximum when the angle of projection was 45°, but this seems to have been a lucky guess. (ii) His *Inventioni*, published in 1546, and containing, *inter alia*, his solution of cubic equations. (iii) His *Trattato di numeri e misuri*, consisting of an arithmetic, published in 1556, and a treatise on numbers, published in 1560; in this he shewed how the coefficients of x in the expansion of $(1 + x)^n$ could be calculated, by the use of an arithmetical triangle,[2] from those in the expansion of $(1 + x)^{n-1}$ for the cases when n is equal to 2, 3, 4, 5, or 6. His works

[1]See below, p. 185.
[2]See below, pp. 234, 235.

were collected into a single edition and republished at Venice in 1606.

The treatise on arithmetic and numbers is one of the chief authorities for our knowledge of the early Italian algorism. It is verbose, but gives a clear account of the arithmetical methods then in use, and has numerous historical notes which, as far as we can judge, are reliable, and are ultimately the authorities for many of the statements in the last chapter. It contains an immense number of questions on every kind of problem which would be likely to occur in mercantile arithmetic, and there are several attempts to frame algebraical formulae suitable for particular problems.

These problems give incidentally a good deal of information as to the ordinary life and commercial customs of the time. Thus we find that the interest demanded on first-class security in Venice ranged from 5 to 12 per cent. a year; while the interest on commercial transactions ranged from 20 per cent. a year upwards. Tartaglia illustrates the evil effects of the law forbidding usury by the manner in which it was evaded in farming. Farmers who were in debt were forced by their creditors to sell all their crops immediately after the harvest; the market being thus glutted, the price obtained was very low, and the money-lenders purchased the corn in open market at an extremely cheap rate. The farmers then had to borrow their seed-corn on condition that they replaced an equal quantity, or paid the then price of it, in the month of May, when corn was dearest. Again, Tartaglia, who had been asked by the magistrates at Verona to frame for them a sliding scale by which the price of bread would be fixed by that of corn, enters into a discussion on the principles which it was then supposed should regulate it. In another place he gives the rules at that time current for preparing medicines.

Pacioli had given in his arithmetic some problems of an amusing character, and Tartaglia imitated him by inserting a large collection of mathematical puzzles. He half apologizes for introducing them by saying that it was not uncommon at dessert to propose arithmetical questions to the company by way of amusement, and he therefore adds some suitable problems. He gives several questions on how to guess a number thought of by one of the company, or the relationships caused by the marriage of relatives, or difficulties arising from inconsistent bequests. Other puzzles are similar to the following. "Three beautiful ladies have for husbands three men, who are young, handsome, and gallant, but also jealous. The party are travelling, and find on the bank of a river, over which they have to pass, a small boat which can hold no more than two persons. How can they pass, it being agreed that, in order to avoid scandal, no woman shall be left in the society of a man unless her husband is present?" "A ship, carrying as passengers fifteen Turks and fifteen Christians, encounters a storm; and the pilot declares that in order to save the ship and crew one-half of the passengers must be thrown into the sea. To choose the victims, the passengers are placed in a circle, and it is agreed that every ninth man shall be cast overboard, reckoning from a certain point. In what manner must they be arranged, so that the lot may fall exclusively upon the Turks?"   "Three men robbed a gentleman of a vase containing 24 ounces of balsam. Whilst running away they met in a wood with a glass-seller of whom in a great hurry they purchased three vessels. On reaching a place of safety they wish to divide the booty, but they find that their vessels contain 5, 11, and 13 ounces respectively. How can they divide the balsam into equal portions?"

These problems—some of which are of oriental origin—form the basis of the collections of mathematical recreations by Bachet de Méziriac, Ozanam, and Montucla.[1]

Cardan.[2]   The life of Tartaglia was embittered by a quarrel with his contemporary Cardan, who published Tartaglia's solution of a cubic equation which he had obtained under a pledge of secrecy. *Girolamo Cardan* was born at Pavia on September 24, 1501, and died at Rome on September 21, 1576. His career is an account of the most extraordinary and inconsistent acts. A gambler, if not a murderer, he was also an ardent student of science, solving problems which had long baffled all investigation; at one time of his life he was devoted to intrigues which were a scandal even in the sixteenth century, at another he did nothing

---

[1]Solutions of these and other similar problems are given in my *Mathematical Recreations*, chaps. i, ii. On Bachet, see below, p. 252. *Jacques Ozanam*, born at Bouligneux in 1640, and died in 1717, left numerous works of which one, worth mentioning here, is his *Récréations mathématiques et physiques*, two volumes, Paris, 1696. *Jean Étienne Montucla*, born at Lyons in 1725, and died in Paris in 1799, edited and revised Ozanam's mathematical recreations. His history of attempts to square the circle, 1754, and history of mathematics to the end of the seventeenth century, in two volumes, 1758, are interesting and valuable works.

[2]There is an admirable account of Cardan's life in the *Nouvelle biographie générale*, by V. Sardou. Cardan left an autobiography of which an analysis by H. Morley was published in two volumes in London in 1854. All Cardan's printed works were collected by Sponius, and published in ten volumes, Lyons, 1663; the works on arithmetic and geometry are contained in the fourth volume. It is said that there are in the Vatican several manuscript note-books of his which have not been yet edited.

but rave on astrology, and yet at another he declared that philosophy was the only subject worthy of man's attention. His was the genius that was closely allied to madness.

He was the illegitimate son of a lawyer of Milan, and was educated at the universities of Pavia and Padua. After taking his degree he commenced life as a doctor, and practised his profession at Sacco and Milan from 1524 to 1550; it was during this period that he studied mathematics and published his chief works. After spending a year or so in France, Scotland, and England, he returned to Milan as professor of science, and shortly afterwards was elected to a chair at Pavia. Here he divided his time between debauchery, astrology, and mechanics. His two sons were as wicked and passionate as himself: the elder was in 1560 executed for poisoning his wife, and about the same time Cardan in a fit of rage cut off the ears of the younger who had committed some offence; for this scandalous outrage he suffered no punishment, as the Pope Gregory XIII. granted him protection. In 1562 Cardan moved to Bologna, but the scandals connected with his name were so great that the university took steps to prevent his lecturing, and only gave way under pressure from Rome. In 1570 he was imprisoned for heresy on account of his having published the horoscope of Christ, and when released he found himself so generally detested that he determined to resign his chair. At any rate he left Bologna in 1571, and shortly afterwards moved to Rome. Cardan was the most distinguished astrologer of his time, and when he settled at Rome he received a pension in order to secure his services as astrologer to the papal court. This proved fatal to him for, having foretold that he should die on a particular day, he felt obliged to commit suicide in order to keep up his reputation—so at least the story runs.

The chief mathematical work of Cardan is the *Ars Magna* published at Nuremberg in 1545. Cardan was much interested in the contest between Tartaglia and Fiore, and as he had already begun writing this book he asked Tartaglia to communicate his method of solving a cubic equation. Tartaglia refused, whereupon Cardan abused him in the most violent terms, but shortly afterwards wrote saying that a certain Italian nobleman had heard of Tartaglia's fame and was most anxious to meet him, and begged him to come to Milan at once. Tartaglia came, and though he found no nobleman awaiting him at the end of his journey, he yielded to Cardan's importunity, and gave him the rule, Cardan on his side taking a solemn oath that he would never reveal it. Cardan asserts that he was given merely the result, and that he obtained the proof himself, but this is doubtful. He seems to have at once taught the method, and one of his pupils Ferrari reduced the equation of the fourth degree to a cubic and so solved it.

When the *Ars Magna* was published in 1545 the breach of faith was made manifest.[1] Tartaglia not unnaturally was very angry, and after an acrimonious controversy he sent a challenge to Cardan to take part in a mathematical duel. The preliminaries were settled, and the place of meeting was to be a certain church in Milan, but when the day arrived Cardan failed to appear, and sent Ferrari in his stead. Both sides claimed the victory, though I gather that Tartaglia was the more successful; at any rate his opponents broke up the meeting, and he deemed himself fortunate in escaping with his life. Not only did Cardan succeed in his fraud, but modern writers have often attributed the solution to him, so that Tartaglia has not even that posthumous reputation which at least is his due.

The *Ars Magna* is a great advance on any algebra previously published. Hitherto algebraists had confined their attention to those roots of equations which were positive. Cardan discussed negative and even complex roots, and proved that the latter would always occur in pairs, though he declined to commit himself to any explanation as to the meaning of these "sophistic" quantities which he said were ingenious though useless. Most of his analysis of cubic equations seems to have been original; he shewed that if the three roots were real, Tartaglia's solution gave them in a form which involved imaginary quantities. Except for the somewhat similar researches of Bombelli a few years later, the theory of imaginary quantities received little further attention from mathematicians until John Bernoulli and Euler took up the matter after the lapse of nearly two centuries. Gauss first put the subject on a systematic and scientific basis, introduced the notation of complex variables, and used the symbol i, which had been introduced by Euler in 1777, to denote the square root of (-1): the modern theory is chiefly based on his researches.

Cardan established the relations connecting the roots with the coefficients of an equation. He was also aware of the principle that underlies Descartes's "rule of signs," but as he followed the custom, then general, of writing his equations as the equality of two expressions in each

---

[1] The history of the subject and of the doings of Fiore, Tartaglia, and Cardan are given in an Appendix to the 2nd edition of the French translation of my *Mathematical Recreations*, Paris, 1908, vol. ii, p. 322 *et seq.*

of which all the terms were positive he was unable to express the rule concisely. He gave a method of approximating to the root of a numerical equation, founded on the fact that, if a function have opposite signs when two numbers are substituted in it, the equation obtained by equating the function to zero will have a root between these two numbers.

Cardan's solution of a quadratic equation is geometrical and substantially the same as that given by Alkarismi. His solution of a cubic equation is also geometrical, and may be illustrated by the following case which he gives in chapter XI. To solve the equation $x^3 + 6x = 20$ (or any equation of the form $x^3 + qx = r$), take two cubes such that the rectangle under their respective edges is 2 (or $\frac{1}{3}q$) and the difference of their volumes is 20 (or r). Then x will be equal to the difference between the edges of the cubes. To verify this he first gives a geometrical lemma to shew that, if from a line AC a portion CB be cut off, then the cube on AB will be less than the difference between the cubes on AC and BC by three times the right parallelepiped whose edges are respectively equal to AC, BC, and AB—this statement is equivalent to the algebraical identity $(a - b)^3 = a^3 - b^3 - 3ab(a - b)$—and the fact that x satisfies the equation is then obvious. To obtain the lengths of the edges of the two cubes he has only to solve a quadratic equation for which the geometrical solution previously given sufficed.

Like all previous mathematicians he gives separate proofs of his rule for the different forms of equations which can fall under it. Thus he proves the rule independently for equations of the form $x^3 + px = q$, $x^3 = px + q$, $x^3 + px + q = 0$, and $x^3 + q = px$. It will be noticed that with geometrical proofs this was the natural course, but it does not appear that he was aware that the resulting formulae were general. The equations he considers are numerical.

Shortly after Cardan came a number of mathematicians who did good work in developing the subject, but who are hardly of sufficient importance to require detailed mention here. Of these the most celebrated are perhaps Ferrari and Rheticus.

Ferrari. *Ludovico Ferraro*, usually known as *Ferrari*, whose name I have already mentioned in connection with the solution of a biquadratic equation, was born at Bologna on Feb. 2, 1522, and died on Oct. 5, 1565. His parents were poor and he was taken into Cardan's service as an errand boy, but was allowed to attend his master's lectures, and subsequently became his most celebrated pupil. He is described as "a neat rosy little fellow, with a bland voice, a cheerful face, and an agreeable short nose, fond of pleasure, of great natural powers," but "with the temper of a fiend." His manners and numerous accomplishments procured him a place in the service of the Cardinal Ferrando Gonzago, where he managed to make a fortune. His dissipations told on his health, and he retired in 1565 to Bologna where he began to lecture on mathematics. He was poisoned the same year either by his sister, who seems to have been the only person for whom he had any affection, or by her paramour.

Such work as Ferrari produced is incorporated in Cardan's *Ars Magna* or Bombelli's *Algebra*, but nothing can be definitely assigned to him except the solution of a biquadratic equation. Colla had proposed the solution of the equation $x^4 + 6x^2 + 36 = 60x$ as a challenge to mathematicians: this particular equation had probably been found in some Arabic work. Nothing is known about the history of this problem except that Ferrari succeeded where Tartaglia and Cardan had failed.

Rheticus.   *Georg Joachim Rheticus,* born at Feldkirch on Feb. 15, 1514, and died at Kaschau on Dec. 4, 1576, was professor at Wittenberg, and subsequently studied under Copernicus whose works were produced under the direction of Rheticus. Rheticus constructed various trigonometrical tables, some of which were published by his pupil Otho in 1596. These were subsequently completed and extended by Vieta and Pitiscus, and are the basis of those still in use. Reticus also found the values of $\sin 2\theta$ and $\sin 3\theta$ in terms of $\sin \theta$ and $\cos \theta$, and was aware that trigonometrical ratios might be defined by means of the ratios of the sides of a right-angled triangle without introducing a circle.

I add here the names of some other celebrated mathematicians of about the same time, though their works are now of little value to any save antiquarians. **Franciscus Maurolycus,** born at Messina of Greek parents in 1494, and died in 1575, translated numerous Latin and Greek mathematical works, and discussed the conics regarded as sections of a cone: his works were published at Venice in 1575. **Jean Borrel,** born in 1492 and died at Grenoble in 1572, wrote an algebra, founded on that of Stifel; and a history of the quadrature of the circle: his works were published at Lyons in 1559. **Wilhelm Xylander,** born at Augsburg on Dec. 26, 1532, and died on Feb. 10, 1576, at Heidelberg, where since 1558 he had been professor, brought out an edition of the works of Psellus in 1556; an edition of Euclid's *Elements* in 1562; an edition of the *Arithmetic* of Diophantus in 1575; and some minor works which were collected and published in 1577. **Frederigo**

Commandino, born at Urbino in 1509, and died there on Sept. 3, 1575, published a translation of the works of Archimedes in 1558; selections from Apollonius and Pappus in 1566; an edition of Euclid's *Elements* in 1572; and selections from Aristarchus, Ptolemy, Hero, and Pappus in 1574: all being accompanied by commentaries. **Jacques Peletier,** born at le Mans on July 25, 1517, and died at Paris in July 1582, wrote text-books on algebra and geometry: most of the results of Stifel and Cardan are included in the former. **Adrian Romanus,** born at Louvain on Sept. 29, 1561, and died on May 4, 1625, professor of mathematics and medicine at the university of Louvain, was the first to prove the usual formula for $\sin(A+B)$. And lastly, **Bartholomäus Pitiscus,** born on Aug. 24, 1561, and died at Heidelberg, where he was professor of mathematics, on July 2, 1613, published his *Trigonometry* in 1599: this contains the expressions for $\sin(A \pm B)$ and $\cos(A \pm B)$ in terms of the trigonometrical ratios of A and B.

About this time also several text-books were produced which if they did not extend the boundaries of the subject systematized it. In particular I may mention those by Ramus and Bombelli.

Ramus.[1]   *Peter Ramus* was born at Cuth in Picardy in 1515, and was killed at Paris in the massacre of St. Bartholomew on Aug. 24, 1572. He was educated at the university of Paris, and on taking his degree he astonished and charmed the university with the brilliant declamation he delivered on the thesis that everything Aristotle had taught was false. He lectured—for it will be remembered that in early days there were no professors—first at le Mans, and afterwards at Paris; at the latter he founded the first chair of mathematics. Besides some works on philosophy he wrote treatises on arithmetic, algebra, geometry (founded on Euclid), astronomy (founded on the works of Copernicus), and physics, which were long regarded on the Continent as the standard text-books in these subjects. They are collected in an edition of his works published at Bâle in 1569.

Bombelli.   Closely following the publication of Cardan's great work, *Rafaello Bombelli* published in 1572 an algebra which is a systematic exposition of the knowledge then current on the subject. In the preface he traces the history of the subject, and alludes to Diophantus who, in spite of the notice of Regiomontanus, was still unknown in Europe. He discusses radicals, real and complex. He also treats the theory

---

[1] See the monographs by Ch. Waddington, Paris, 1855; and by C. Desmaze, Paris, 1864.

of equations, and shews that in the irreducible case of a cubic equation the roots are all real; and he remarks that the problem to trisect a given angle is the same as that of the solution of a cubic equation. Finally he gave a large collection of problems.

Bombelli's work is noticeable for his use of symbols which indicate an approach to index notation. Following in the steps of Stifel, he introduced a symbol ⌣₁⌣ for the unknown quantity, ⌣₂⌣ for its square, ⌣₃⌣ for its cube, and so on, and therefore wrote $x^2 + 5x - 4$ as

$$1 \cup_2 \cup \text{ p. } 5 \cup_1 \cup \text{ m. } 4$$

Stevinus in 1586 employed ①, ②, ③, ... in a similar way; and suggested, though he did not use, a corresponding notation for fractional indices. He would have written the above expression as

$$1 \, ② + 5 \, ① - 4 \, ⓪.$$

But whether the symbols were more or less convenient they were still only abbreviations for words, and were subject to all the rules of syntax. They merely afforded a sort of shorthand by which the various steps and results could be expressed concisely. The next advance was the creation of symbolic algebra, and the chief credit of that is due to Vieta.

The development of symbolic algebra.

We have now reached a point beyond which any considerable development of algebra, so long as it was strictly syncopated, could hardly proceed. It is evident that Stifel and Bombelli and other writers of the sixteenth century had introduced or were on the point of introducing some of the ideas of symbolic algebra. But so far as the credit of inventing symbolic algebra can be put down to any one man we may perhaps assign it to Vieta, while we may say that Harriot and Descartes did more than any other writers to bring it into general use. It must be remembered, however, that it took time before all these innovations became generally known, and they were not familiar to mathematicians until the lapse of some years after they had been published.

Vieta.[1]   *Franciscus Vieta (François Viète)* was born in 1540 at Fontenay near la Rochelle, and died in Paris in 1603. He was brought

---

[1] The best account of Vieta's life and works is that by A. De Morgan in the *English Cyclopaedia*, London, vol. vi, 1858.

up as a lawyer and practised for some time at the Parisian bar; he then became a member of the provincial parliament in Brittany; and finally in 1580, through the influence of the Duke de Rohan, he was made master of requests, an office attached to the parliament at Paris; the rest of his life was spent in the public service. He was a firm believer in the right divine of kings, and probably a zealous catholic. After 1580 he gave up most of his leisure to mathematics, though his great work, *In Artem Analyticam Isagoge*, in which he explained how algebra could be applied to the solution of geometrical problems, was not published till 1591.

His mathematical reputation was already considerable, when one day the ambassador from the Low Countries remarked to Henry IV. that France did not possess any geometricians capable of solving a problem which had been propounded in 1593 by his countryman Adrian Romanus to all the mathematicians of the world, and which required the solution of an equation of the 45th degree. The king thereupon summoned Vieta, and informed him of the challenge. Vieta saw that the equation was satisfied by the chord of a circle (of unit radius) which subtends an angle $2\pi/45$ at the centre, and in a few minutes he gave back to the king two solutions of the problem written in pencil. In explanation of this feat I should add that Vieta had previously discovered how to form the equation connecting $\sin n\theta$ with $\sin \theta$ and $\cos \theta$. Vieta in his turn asked Romanus to give a geometrical construction to describe a circle which should touch three given circles. This was the problem which Apollonius had treated in his *De Tactionibus*, a lost book which Vieta at a later time conjecturally restored. Romanus solved the problem by the use of conic sections, but failed to do it by Euclidean geometry. Vieta gave a Euclidean solution which so impressed Romanus that he travelled to Fontenay, where the French court was then settled, to make Vieta's acquaintance—an acquaintanceship which rapidly ripened into warm friendship.

Henry was much struck with the ability shown by Vieta in this matter. The Spaniards had at that time a cipher containing nearly 600 characters, which was periodically changed, and which they believed it was impossible to decipher. A despatch having been intercepted, the king gave it to Vieta, and asked him to try to read it and find the key to the system. Vieta succeeded, and for two years the French used it, greatly to their profit, in the war which was then raging. So convinced was Philip II. that the cipher could not be discovered, that when he found his plans known he complained to the Pope that the French were using sorcery against him, "contrary to the practice of the Christian faith."

Vieta wrote numerous works on algebra and geometry. The most important are the *In Artem Analyticam Isagoge*, Tours, 1591; the *Supplementum Geometriae*, and a collection of geometrical problems, Tours, 1593; and the *De Numerosa Potestatum Resolutione*, Paris, 1600. All of these were printed for private circulation only, but they were collected by F. van Schooten and published in one volume at Leyden in 1646. Vieta also wrote the *De Aequationum Recognitione et Emendatione*, which was published after his death in 1615 by Alexander Anderson.

The *In Artem* is the earliest work on symbolic algebra. It also introduced the use of letters for both known and unknown (positive) quantities, a notation for the powers of quantities, and enforced the advantage of working with homogeneous equations. To this an appendix called *Logistice Speciosa* was added on addition and multiplication of algebraical quantities, and on the powers of a binomial up to the sixth. Vieta implies that he knew how to form the coefficients of these six expansions by means of the arithmetical triangle as Tartaglia had previously done, but Pascal gave the general rule for forming it for any order, and Stifel had already indicated the method in the expansion of $(1 + x)^n$ if those in the expansion of $(1 + x)^{n-1}$ were known; Newton was the first to give the general expression for the coefficient of x^p in the expansion of $(1 + x)^n$. Another appendix known as *Zetetica* on the solution of equations was subsequently added to the *In Artem*.

The *In Artem* is memorable for two improvements in algebraic notation which were introduced here, though it is probable that Vieta took the idea of both from other authors.

One of these improvements was that he denoted the known quantities by the consonants B, C, D, &c., and the unknown quantities by the vowels A, E, I, &c. Thus in any problem he was able to use a number of unknown quantities. In this particular point he seems to have been forestalled by Jordanus and by Stifel. The present custom of using the letters at the beginning of the alphabet a, b, c, &c., to represent known quantities and those towards the end, x, y, z, &c., to represent the unknown quantities was introduced by Descartes in 1637.

The other improvement was this. Till this time it had been generally the custom to introduce new symbols to represent the square, cube, &c., of quantities which had already occurred in the equations; thus, if R or N stood for x, Z or C or Q stood for x^2, and C or K for x^3, &c.

So long as this was the case the chief advantage of algebra was that it afforded a concise statement of results every statement of which was reasoned out. But when Vieta used A to denote the unknown quantity x, he sometimes employed A *quadratus*, A *cubus*, ... to represent $x^2, x^3,$ ..., which at once showed the connection between the different powers; and later the successive powers of A were commonly denoted by the abbreviations Aq, Ac, Aqq, &c. Thus Vieta would have written the equation

$$3BA^2 - DA + A^3 = Z,$$

as *B 3 in A quad. — D plano in A + A cubo aequatur Z solido.* It will be observed that the dimensions of the constants $(B, D,$ and $Z)$ are chosen so as to make the equation homogeneous: this is characteristic of all his work. It will be also noticed that he does not use a sign for equality; and in fact the particular sign = which we use to denote equality was employed by him to represent "the difference between." Vieta's notation is not so convenient as that previously used by Stifel, Bombelli, and Stevinus, but it was more generally adopted.

These two steps were almost essential to any further progress in algebra. In both of them Vieta had been forestalled, but it was his good luck in emphasising their importance to be the means of making them generally known at a time when opinion was ripe for such an advance.

The *De Aequationum Recognitione et Emendatione* is mostly on the theory of equations. It was not published till twelve years after Vieta's death, and it is possible that the editor made additions to it. Vieta here indicated how from a given equation another could be obtained whose roots were equal to those of the original increased by a given quantity, or multiplied by a given quantity; he used this method to get rid of the coefficient of x in a quadratic equation and of the coefficient of x^2 in a cubic equation, and was thus enabled to give the general algebraic solution of both. It would seem that he knew that the first member of an algebraical equation $\phi(x) = 0$ could be resolved into linear factors, and that the coefficients of x could be expressed as functions of the roots; perhaps the discovery of both these theorems should be attributed to him.

His solution of a cubic equation is as follows. First reduce the equation to the form $x^3 + 3a^2x = 2b^3$. Next let $x = a^2/y - y$, and we get $y^6 + 2b^3y^3 = a^6$, which is a quadratic in y^3. Hence y can be found, and therefore x can be determined.

His solution of a biquadratic is similar to that known as Ferrari's, and essentially as follows. He first got rid of the term involving x^3, thus reducing the equation to the form $x^4 + a^2 x^2 + b^3 x = c^4$. He then took the forms involving x^2 and x to the right-hand side of the equation and added $x^2 y^2 + \frac{1}{4} y^4$ to each side, so that the equation became

$$(x^2 + \frac{1}{2} y^2)^2 = x^2(y^2 - a^2) - b^3 x + \frac{1}{4} y^4 + c^4.$$

He then chose y so that the right-hand side of this equality is a perfect square. Substituting this value of y, he was able to take the square root of both sides, and thus obtain two quadratic equations for x, each of which can be solved.

The *De Numerosa Potestatum Resolutione* deals with numerical equations. In this a method for approximating to the values of positive roots is given, but it is prolix and of little use, though the principle (which is similar to that of Newton's rule) is correct. Negative roots are uniformly rejected. This work is hardly worthy of Vieta's reputation.

Vieta's trigonometrical researches are included in various tracts which are collected in Van Schooten's edition. Besides some trigonometrical tables he gave the general expression for the sine (or chord) of an angle in terms of the sine and cosine of its submultiples. Delambre considers this as the completion of the Arab system of trigonometry. We may take it then that from this time the results of elementary trigonometry were familiar to mathematicians. Vieta also elaborated the theory of right-angled spherical triangles.

Among Vieta's miscellaneous tracts will be found a proof that each of the famous geometrical problems of the trisection of an angle and the duplication of the cube depends on the solution of a cubic equation. There are also some papers connected with an angry controversy with Clavius, in 1594, on the subject of the reformed calendar, in which Vieta was not well advised.

Vieta's works on geometry are good, but they contain nothing which requires mention here. He applied algebra and trigonometry to help him in investigating the properties of figures. He also, as I have already said, laid great stress on the desirability of always working with homogeneous equations, so that if a square or a cube were given it should be denoted by expressions like a^2 or b^3, and not by terms like m or n which do not indicate the dimensions of the quantities they represent. He had a lively dispute with Scaliger on the latter publishing a solution of the quadrature of the circle, and Vieta succeeded in showing the mistake into which his rival had fallen. He gave a solution of his own which as far as it goes is correct, and stated that the area of a square is to that of the circumscribing circle as

$$\sqrt{\tfrac{1}{2}} \times \sqrt{(\tfrac{1}{2} + \sqrt{\tfrac{1}{2}})} \times \sqrt{\{\tfrac{1}{2} + \sqrt{(\tfrac{1}{2} + \sqrt{\tfrac{1}{2}})}\}} \ldots \textit{ad inf.} : 1.$$

This is one of the earliest attempts to find the value of π by means of an infinite series. He was well acquainted with the extant writings of the Greek geometricians, and introduced the curious custom, which during the seventeenth and eighteenth centuries became fashionable, of restoring lost classical works. He himself produced a conjectural restoration of the *De Tactionibus* of Apollonius.

Girard.   Vieta's results in trigonometry and the theory of equations were extended by *Albert Girard*, a Dutch mathematician, who was born in Lorraine in 1595, and died on December 9, 1632.

In 1626 Girard published at the Hague a short treatise on trigonometry, to which were appended tables of the values of the trigonometrical functions. This work contains the earliest use of the abbreviations *sin, tan, sec* for sine, tangent, and secant. The supplemental triangles in spherical trigonometry are also discussed; their properties seem to have been discovered by Girard and Snell at about the same time. Girard also gave the expression for the area of a spherical triangle in terms of the spherical excess—this was discovered independently by Cavalieri. In 1627 Girard brought out an edition of Marolois's Geometry with considerable additions.

Girard's algebraical investigations are contained in his *Invention nouvelle en l'algèbre*, published at Amsterdam in 1629.[1] This contains the earliest use of brackets; a geometrical interpretation of the negative sign; the statement that the number of roots of an algebraical question is equal to its degree; the distinct recognition of imaginary roots; the theorem, known as Newton's rule, for finding the sum of like powers of the roots of an equation; and (in the opinion of some writers) implies also a knowledge that the first member of an algebraical equation $\phi(x) = 0$ could be resolved into linear factors. Girard's investigations were unknown to most of his contemporaries, and exercised no appreciable influence on the development of mathematics.

The invention of logarithms by Napier of Merchiston in 1614, and their introduction into England by Briggs and others, have been already

---

[1] It was reissued by B. de Haan at Leyden in 1884.

mentioned in chapter XI. A few words on these mathematicians may be here added.

Napier.[1]   *John Napier* was born at Merchiston in 1550, and died on April 4, 1617. He spent most of his time on the family estate near Edinburgh, and took an active part in the political and religious controversies of the day; the business of his life was to show that the Pope was Antichrist, but his favourite amusement was the study of mathematics and science.

As soon as the use of exponents became common in algebra the introduction of logarithms would naturally follow, but Napier reasoned out the result without the use of any symbolic notation to assist him, and the invention of logarithms was the result of the efforts of many years with a view to abbreviate the processes of multiplication and division. It is likely that Napier's attention may have been partly directed to the desirability of facilitating computations by the stupendous arithmetical efforts of some of his contemporaries, who seem to have taken a keen pleasure in surpassing one another in the extent to which they carried multiplications and divisions. The trigonometrical tables by Rheticus, which were published in 1596 and 1613, were calculated in a most laborious way: Vieta himself delighted in arithmetical calculations which must have taken days of hard work, and of which the results often served no useful purpose: L. van Ceulen (1539–1610) practically devoted his life to finding a numerical approximation to the value of π, finally in 1610 obtaining it correct to 35 places of decimals: while, to cite one more instance, P. A. Cataldi (1548–1626), who is chiefly known for his invention in 1613 of the form of continued fractions, must have spent years in numerical calculations.

In regard to Napier's other work I may again mention that in his *Rabdologia*, published in 1617, he introduced an improved form of rod by the use of which the product of two numbers can be found in a mechanical way, or the quotient of one number by another. He also invented two other rods called "virgulae," by which square and cube roots can be extracted. I should add that in spherical trigonometry he discovered certain formulae known as Napier's analogies, and enunciated the "rule of circular parts" for the solution of right-angled spherical

---

[1] See the *Napier Tercentenary Memorial Volume*, Edinburgh, 1915. An edition of all his works was issued at Edinburgh in 1839: a bibliography of his writings is appended to a translation of the *Constructio* by W. R. Macdonald, Edinburgh, 1889.

triangles.

Briggs.   The name of Briggs is inseparably associated with the history of logarithms. *Henry Briggs*[1] was born near Halifax in 1561: he was educated at St. John's College, Cambridge, took his degree in 1581, and obtained a fellowship in 1588: he was elected to the Gresham professorship of geometry in 1596, and in 1619 or 1620 became Savilian professor at Oxford, a chair which he held until his death on January 26, 1631. It may be interesting to add that the chair of geometry founded by Sir Thomas Gresham was the earliest professorship of mathematics established in Great Britain. Some twenty years earlier Sir Henry Savile had given at Oxford open lectures on Greek geometry and geometricians, and in 1619 he endowed the chairs of geometry and astronomy in that university which are still associated with his name. Both in London and at Oxford Briggs was the first occupant of the chair of geometry. He began his lectures at Oxford with the ninth proposition of the first book of Euclid—that being the furthest point to which Savile had been able to carry his audiences. At Cambridge the Lucasian chair was established in 1663, the earliest occupants being Barrow and Newton.

The almost immediate adoption throughout Europe of logarithms for astronomical and other calculations was mainly the work of Briggs, who undertook the tedious work of calculating and preparing tables of logarithms. Amongst others he convinced Kepler of the advantages of Napier's discovery, and the spread of the use of logarithms was rendered more rapid by the zeal and reputation of Kepler, who by his tables of 1625 and 1629 brought them into vogue in Germany, while Cavalieri in 1624 and Edmund Wingate in 1626 did a similar service for Italian and French mathematicians respectively. Briggs also was instrumental in bringing into common use the method of long division now generally employed.

Harriot.   *Thomas Harriot*, who was born at Oxford in 1560, and died in London on July 2, 1621, did a great deal to extend and codify the theory of equations. The early part of his life was spent in America with Sir Walter Raleigh; while there he made the first survey of Virginia and North Carolina, the maps of these being subsequently presented to Queen Elizabeth. On his return to England he settled in London, and gave up most of his time to mathematical studies.

---

[1]See pp. 27–30 of my *History of the Study of Mathematics at Cambridge*, Cambridge, 1889.

The majority of the propositions I have assigned to Vieta are to be found in Harriot's writings, but it is uncertain whether they were discovered by him independently of Vieta or not. In any case it is probable that Vieta had not fully realised all that was contained in the propositions he had enunciated. Some of the consequences of these, with extensions and a systematic exposition of the theory of equations, were given by Harriot in his *Artis Analyticae Praxis*, which was first printed in 1631. The *Praxis* is more analytical than any algebra that preceded it, and marks an advance both in symbolism and notation, though negative and imaginary roots are rejected. It was widely read, and proved one of the most powerful instruments in bringing analytical methods into general use. Harriot was the first to use the signs > and < to represent greater than and less than. When he denoted the unknown quantity by a he represented a^2 by aa, a^3 by aaa, and so on. This is a distinct improvement on Vieta's notation. The same symbolism was used by Wallis as late as 1685, but concurrently with the modern index notation which was introduced by Descartes. I need not allude to the other investigations of Harriot, as they are comparatively of small importance; extracts from some of them were published by S. P. Rigaud in 1833.

Oughtred.   Among those who contributed to the general adoption in England of these various improvements and additions to algorism and algebra was *William Oughtred*,[1] who was born at Eton on March 5, 1574, and died at his vicarage of Albury in Surrey on June 30, 1660: it is sometimes said that the cause of his death was the excitement and delight which he experienced "at hearing the House of Commons [or Convention] had voted the King's return"; a recent critic adds that it should be remembered "by way of excuse that he [Oughtred] was then eighty-six years old," but perhaps the story is sufficiently discredited by the date of his death. Oughtred was educated at Eton and King's College, Cambridge, of the latter of which colleges he was a fellow and for some time mathematical lecturer.

His *Clavis Mathematicae* published in 1631 is a good systematic text-book on arithmetic, and it contains practically all that was then known on the subject. In this work he introduced the symbol × for multiplication. He also introduced the symbol : : in proportion: previously to his time a proportion such as $a : b = c : d$ was usually written

---

[1] See *William Oughtred*, by F. Cajori, Chicago, 1916. A complete edition of Oughtred's works was published at Oxford in 1677.

as $a - b - c - d$; he denoted it by $a \cdot b :: c \cdot d$. Wallis says that some found fault with the book on account of the style, but that they only displayed their own incompetence, for Oughtred's "words be always full but not redundant." Pell makes a somewhat similar remark.

Oughtred also wrote a treatise on trigonometry published in 1657, in which abbreviations for *sine, cosine,* &c., were employed. This was really an important advance, but the works of Girard and Oughtred, in which they were used, were neglected and soon forgotten, and it was not until Euler reintroduced contractions for the trigonometrical functions that they were generally adopted. In this work the colon (*i.e.* the symbol :) was used to denote a ratio.

We may say roughly that henceforth elementary arithmetic, algebra, and trigonometry were treated in a manner which is not substantially different from that now in use; and that the subsequent improvements introduced were additions to the subjects as then known, and not a rearrangement of them on new foundations.

The origin of the more common symbols in algebra.

It may be convenient if I collect here in one place the scattered remarks I have made on the introduction of the various symbols for the more common operations in algebra.[1]

The later Greeks, the Hindoos, and Jordanus indicated *addition* by mere juxtaposition. It will be observed that this is still the custom in arithmetic, where, for instance, $2\frac{1}{2}$ stands for $2 + \frac{1}{2}$. The Italian algebraists, when they gave up expressing every operation in words at full length and introduced syncopated algebra, usually denoted *plus* by its initial letter P or p, a line being sometimes drawn through the letter to show that it was a contraction, or a symbol of operation, and not a quantity. The practice, however, was not uniform; Pacioli, for example, sometimes denoted plus by $\bar{p}$, and sometimes by e, and Tartaglia commonly denoted it by ϕ. The German and English algebraists, on the other hand, introduced the sign + almost as soon as they used algorism, but they spoke of it as *signum additorum* and employed it only to indicate excess; they also used it with a special meaning in solutions by the method of false assumption. Widman used it as an abbreviation

---

[1]See also two articles by C. Henry in the June and July numbers of the *Revue Archéologique*, 1879, vol. xxxvii, pp. 324–333, vol. xxxviii, pp. 1–10.

for excess in 1489: by 1630 it was part of the recognised notation of algebra, and was used as a symbol of operation.

Subtraction was indicated by Diophantus by an inverted and truncated ψ. The Hindoos denoted it by a dot. The Italian algebraists when they introduced syncopated algebra generally denoted *minus* by M or m, a line being sometimes drawn through the letter; but the practice was not uniform—Pacioli, for example, denoting it sometimes by $\bar{m}$, and sometimes by *de* for *demptus*. The German and English algebraists introduced the present symbol which they described as *signum subtractorum*. It is most likely that the vertical bar in the symbol for plus was superimposed on the symbol for minus to distinguish the two. It may be noticed that Pacioli and Tartaglia found the sign − already used to denote a division, a ratio, or a proportion indifferently. The present sign for minus was in general use by about the year 1630, and was then employed as a symbol of operation.

Vieta, Schooten, and others among their contemporaries employed the sign = written between two quantities to denote the difference between them; thus $a = b$ means with them what we denote by $a \sim b$. On the other hand, Barrow wrote —: for the same purpose. I am not aware when or by whom the current symbol $\sim$ was first used with this signification.

Oughtred in 1631 used the sign × to indicate *multiplication*; Harriot in 1631 denoted the operation by a dot; Descartes in 1637 indicated it by juxtaposition. I am not aware of any symbols for it which were in previous use. Leibnitz in 1686 employed the sign $\frown$ to denote multiplication.

Division was ordinarily denoted by the Arab way of writing the quantities in the form of a fraction by means of a line drawn between them in any of the forms $a - b$, a/b, or $\dfrac{a}{b}$. Oughtred in 1631 employed a dot to denote either division or a ratio. Leibnitz in 1686 employed the sign $\smile$ to denote division. The colon (or symbol :), used to denote a *ratio*, occurs on the last two pages of Oughtred's *Canones Sinuum*, published in 1657. I believe that the current symbol for division ÷ is only a combination of the − and the symbol : for a ratio; it was used by Johann Heinrich Rahn at Zürich in 1659, and by John Pell in London in 1668. The symbol ∹ was used by Barrow and other writers of his time to indicate continued proportion.

The current symbol for *equality* was introduced by Record in 1557; Xylander in 1575 denoted it by two parallel vertical lines; but in general till the year 1600 the word was written at length; and from then until the time of Newton, say about 1680, it was more frequently represented by ∞ or by ∞ than by any other symbol. Either of these latter signs was used as a contraction for the first two letters of the word *aequalis*.

The symbol :: to denote *proportion*, or the equality of two ratios, was introduced by Oughtred in 1631, and was brought into common use by Wallis in 1686. There is no object in having a symbol to indicate the equality of two ratios which is different from that used to indicate the equality of other things, and it is better to replace it by the sign =.

The sign > for *is greater than* and the sign < for *is less than* were introduced by Harriot in 1631, but Oughtred simultaneously invented the symbols ‾⌋ and ⌊‾ for the same purpose; and these latter were frequently used till the beginning of the eighteenth century, *ex. gr.* by Barrow.

The symbols $\neq$ for *is not equal to*, $\not>$ for *is not greater than*, and $\not<$ for *is not less than*, are, I believe, now rarely used outside Great Britain; they were employed, if not invented, by Euler. The symbols $\geq$ and $\leq$ were introduced by P. Bouguer in 1734.

The vinculum was introduced by Vieta in 1591; and brackets were first used by Girard in 1629.

The symbol $\sqrt{}$ to denote the square root was introduced by Rudolff in 1526; a similar notation had been used by Bhaskara and by Chuquet.

The different methods of representing the power to which a magnitude was raised have been already briefly alluded to. The earliest known attempt to frame a symbolic notation was made by Bombelli in 1572, when he represented the unknown quantity by ⌣, its square by ⌣₂⌣, its cube by ⌣₃⌣, &c. In 1586 Stevinus used ①, ②, ③, &c., in a similar way; and suggested, though he did not use, a corresponding notation for fractional indices. In 1591 Vieta improved on this by denoting the different powers of A by A, A *quad.*, A *cub.*, &c., so that he could indicate the powers of different magnitudes; Harriot in 1631 further improved on Vieta's notation by writing aa for a^2, aaa for a^3, &c., and this remained in use for fifty years concurrently with the index notation. In 1634 P. Herigonus, in his *Cursus mathematicus*, published in five volumes at Paris in 1634–1637, wrote $a, a2, a3, \ldots$ for $a, a^2, a^3 \ldots$.

The idea of using exponents to mark the power to which a quantity was raised was due to Descartes, and was introduced by him in 1637; but he used only positive integral indices $a^1, a^2, a^3, \ldots$. Wallis in 1659 explained the meaning of negative and fractional indices in expressions such as a^{-1}, $ax^{1/2}$, &c.; the latter conception having been foreshadowed by Oresmus and perhaps by Stevinus. Finally the idea of an index unrestricted in magnitude, such as the n in the expression a^n, is, I believe, due to Newton, and was introduced by him in connection with the binomial theorem in the letters for Leibnitz written in 1676.

The symbol ∞ for infinity was first employed by Wallis in 1655 in his *Arithmetica Infinitorum*; but does not occur again until 1713, when it is used in James Bernoulli's *Ars Conjectandi*. This sign was sometimes employed by the Romans to denote the number 1000, and it has been conjectured that this led to its being applied to represent any very large number.

There are but few special symbols in trigonometry; I may, however, add here the following note which contains all that I have been able to learn on the subject. The current sexagesimal division of angles is derived from the Babylonians through the Greeks. The Babylonian unit angle was the angle of an equilateral triangle; following their usual practice this was divided into sixty equal parts or degrees, a degree was subdivided into sixty equal parts or minutes, and so on; it is said that 60 was assumed as the base of the system in order that the number of degrees corresponding to the circumference of a circle should be the same as the number of days in a year which it is alleged was taken (at any rate in practice) to be 360.

The word *sine* was used by Regiomontanus and was derived from the Arabs; the terms *secant* and *tangent* were introduced by Thomas Finck (born in Denmark in 1561 and died in 1646) in his *Geometriae Rotundi*, Bâle, 1583; the word *cosecant* was (I believe) first used by Rheticus in his *Opus Palatinum*, 1596; the terms *cosine* and *cotangent* were first employed by E. Gunter in his *Canon Triangulorum*, London, 1620. The abbreviations *sin, tan, sec* were used in 1626 by Girard, and those of *cos* and *cot* by Oughtred in 1657; but these contractions did not come into general use till Euler reintroduced them in 1748. The idea of trigonometrical *functions* originated with John Bernoulli, and this view of the subject was elaborated in 1748 by Euler in his *Introductio in Analysin Infinitorum*.

CHAPTER XIII.

THE CLOSE OF THE RENAISSANCE.[1]
CIRC. 1586–1637.

THE closing years of the renaissance were marked by a revival of interest in nearly all branches of mathematics and science. As far as pure mathematics is concerned we have already seen that during the last half of the sixteenth century there had been a great advance in algebra, theory of equations, and trigonometry; and we shall shortly see (in the second section of this chapter) that in the early part of the seventeenth century some new processes in geometry were invented. If, however, we turn to applied mathematics it is impossible not to be struck by the fact that even as late as the middle or end of the sixteenth century no marked progress in the theory had been made from the time of Archimedes. Statics (of solids) and hydrostatics remained in much the state in which he had left them, while dynamics as a science did not exist. It was Stevinus who gave the first impulse to the renewed study of statics, and Galileo who laid the foundation of dynamics; and to their works the first section of this chapter is devoted.

The development of mechanics and experimental methods.

Stevinus.[2] *Simon Stevinus* was born at Bruges in 1548, and died at the Hague in 1620. We know very little of his life save that he was

---

[1] See footnote to chapter xii.

[2] An analysis of his works is given in the *Histoire des sciences mathématiques et physiques chez les Belges*, by L. A. J. Quetelet, Brussels, 1866, pp. 144–168; see also *Notice historique sur la vie et les ouvrages de Stevinus*, by J. V. Göthals, Brussels, 1841; and *Les travaux de Stevinus*, by M. Steichen, Brussels, 1846. The works of Stevinus were collected by Snell, translated into Latin, and published at Leyden in 1608 under the title *Hypomnemata Mathematica*.

originally a merchant's clerk at Antwerp, and at a later period of his life was the friend of Prince Maurice of Orange, by whom he was made quartermaster-general of the Dutch army.

To his contemporaries he was best known for his works on fortifications and military engineering, and the principles he laid down are said to be in accordance with those which are now usually accepted. To the general populace he was also well known on account of his invention of a carriage which was propelled by sails; this ran on the sea-shore, carried twenty-eight people, and easily outstripped horses galloping by the side; his model of it was destroyed in 1802 by the French when they invaded Holland. It was chiefly owing to the influence of Stevinus that the Dutch and French began a proper system of book-keeping in the national accounts.

I have already alluded to the introduction in his *Arithmetic*, published in 1585, of exponents to mark the power to which quantities were raised; for instance, he wrote $3x^2 - 5x + 1$ as $3 \; ② - 5 \; ① + 1 \; ⓪$. His notation for decimal fractions was of a similar character. He further suggested the use of fractional (but not negative) exponents. In the same book he likewise suggested a decimal system of weights and measures.

He also published a geometry which is ingenious though it does not contain many results which were not previously known; in it some theorems on perspective are enunciated.

It is, however, on his *Statics and Hydrostatics*, published (in Flemish) at Leyden in 1586, that his fame rests. In this work he enunciates the triangle of forces—a theorem which some think was first propounded by Leonardo da Vinci. Stevinus regards this as the fundamental proposition of the subject. Previous to the publication of his work the science of statics had rested on the theory of the lever, but subsequently it became usual to commence by proving the possibility of representing forces by straight lines, and thus reducing many theorems to geometrical propositions, and in particular to obtaining in this way a proof of the parallelogram (which is equivalent to the triangle) of forces. Stevinus is not clear in his arrangement of the various propositions or in their logical sequence, and the new treatment of the subject was not definitely established before the appearance in 1687 of Varignon's work on mechanics. Stevinus also found the force which must be exerted along the line of greatest slope to support a given weight on an inclined plane—a problem the solution of which had been long in dispute. He further distinguishes between stable and unstable equilibrium. In hydrostatics he discusses the question of the pressure which a fluid can exercise, and explains the so-called hydrostatic paradox.

His method[1] of finding the resolved part of a force in a given direction, as illustrated by the case of a weight resting on an inclined plane, is a good specimen of his work and is worth quoting.

He takes a wedge ABC whose base AC is horizontal [and whose sides BA, BC are in the ratio of 2 to 1]. A thread connecting a number of small equal equidistant weights is placed over the wedge as indicated in the figure below (which I reproduce from his demonstration) so that the number of these weights on BA is to the number on BC in the same proportion as BA is to BC. This is always possible if the dimensions of the wedge be properly chosen, and he places four weights resting on BA and two on BC; but we may replace these weights by a heavy uniform chain $TSLVT$ without altering his argument. He says in effect, that experience shews that such a chain will remain at rest; if not, we could obtain perpetual motion. Thus the effect in the direction BA of the weight of the part TS of the chain must balance the effect in the direction BC of the weight of the part TV of the chain. Of course BC may be vertical, and if so the above statement is equivalent to saying that the effect in the direction BA of the weight of the chain on it is diminished in the proportion of BC to BA; in other words, if a weight W rests on an inclined plane of inclination α the component of W down the line of greatest slope is $W \sin \alpha$.

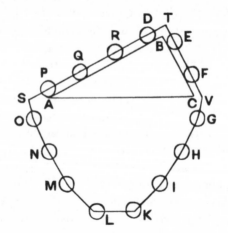

[1] *Hypomnemata Mathematica*, vol. iv, *de Statica*, prop. 19.

Stevinus was somewhat dogmatic in his statements, and allowed no one to differ from his conclusions, "and those," says he, in one place, "who cannot see this, may the Author of nature have pity upon their unfortunate eyes, for the fault is not in the thing, but in the sight which we are unable to give them."

Galileo.[1]   Just as the modern treatment of statics originates with Stevinus, so the foundation of the science of dynamics is due to Galileo. *Galileo Galilei* was born at Pisa on February 18, 1564, and died near Florence on January 8, 1642. His father, a poor descendant of an old and noble Florentine house, was himself a fair mathematician and a good musician. Galileo was educated at the monastery of Vallombrosa, where his literary ability and mechanical ingenuity attracted considerable attention. He was persuaded to become a novitiate of the order in 1579, but his father, who had other views, at once removed him, and sent him in 1581 to the university of Pisa to study medicine. It was there that he noticed that the great bronze lamp, hanging from the roof of the cathedral, performed its oscillations in equal times, and independently of whether the oscillations were large or small—a fact which he verified by counting his pulse. He had been hitherto kept in ignorance of mathematics, but one day, by chance hearing a lecture on geometry (by Ricci), he was so fascinated by the science that thenceforward he devoted his leisure to its study, and finally got leave to discontinue his medical studies. He left the university in 1585, and almost immediately commenced his original researches.

He published in 1586 an account of the hydrostatic balance, and in 1588 an essay on the centre of gravity in solids; these were not printed till later. The fame of these works secured for him in 1589 the appointment to the mathematical chair at Pisa—the stipend, as was then the case with most professorships, being very small. During the next three years he carried on, from the leaning tower, that series of experiments on falling bodies which established the first principles of dynamics. Unfortunately, the manner in which he promulgated his discoveries, and the ridicule he threw on those who opposed him, gave not unnatural offence, and in 1591 he was obliged to resign his position.

[1]See the biography of Galileo, by J. J. Fahie, London, 1903. An edition of Galileo's works was issued in 16 volumes by E. Albèri, Florence, 1842–1856. A good many of his letters on various mathematical subjects have been since discovered, and a new and complete edition is in process of issue by the Italian Government, Florence; vols. i to xix and a bibliography, 1890–1907.

At this time he seems to have been much hampered by want of money. Influence was, however, exerted on his behalf with the Venetian senate, and he was appointed professor at Padua, a chair which he held for eighteen years, 1592–1610. His lectures there seem to have been chiefly on mechanics and hydrostatics, and the substance of them is contained in his treatise on mechanics, which was published in 1612. In these lectures he repeated his Pisan experiments, and demonstrated that falling bodies did not (as was then commonly believed) descend with velocities proportional, amongst other things, to their weights. He further shewed that, if it were assumed that they descended with a uniformly accelerated motion, it was possible to deduce the relations connecting velocity, space, and time which did actually exist. At a later date, by observing the times of descent of bodies sliding down inclined planes, he shewed that this hypothesis was true. He also proved that the path of a projectile is a parabola, and in doing so implicitly used the principles laid down in the first two laws of motion as enunciated by Newton. He gave an accurate definition of momentum which some writers have thought may be taken to imply a recognition of the truth of the third law of motion. The laws of motion are, however, nowhere enunciated in a precise and definite form, and Galileo must be regarded rather as preparing the way for Newton than as being himself the creator of the science of dynamics.

In statics he laid down the principle that in machines what was gained in power was lost in speed, and in the same ratio. In the statics of solids he found the force which can support a given weight on an inclined plane; in hydrostatics he propounded the more elementary theorems on pressure and on floating bodies; while among hydrostatical instruments he used, and perhaps invented, the thermometer, though in a somewhat imperfect form.

It is, however, as an astronomer that most people regard Galileo, and though, strictly speaking, his astronomical researches lie outside the subject-matter of this book, it may be interesting to give the leading facts. It was in the spring of 1609 that Galileo heard that a tube containing lenses had been made by an optician, Hans Lippershey, of Middleburg, which served to magnify objects seen through it. This gave him the clue, and he constructed a telescope of that kind which still bears his name, and of which an ordinary opera-glass is an example. Within a few months he had produced instruments which were capable of magnifying thirty-two diameters, and within a year he had made and published observations on the solar spots, the lunar mountains,

Jupiter's satellites, the phases of Venus, and Saturn's ring. The discovery of the microscope followed naturally from that of the telescope. Honours and emoluments were showered on him, and he was enabled in 1610 to give up his professorship and retire to Florence. In 1611 he paid a temporary visit to Rome, and exhibited in the gardens of the Quirinal the new worlds revealed by the telescope.

It would seem that Galileo had always believed in the Copernican system, but was afraid of promulgating it on account of the ridicule it excited. The existence of Jupiter's satellites seemed, however, to make its truth almost certain, and he now boldly preached it. The orthodox party resented his action, and in February 1616 the Inquisition declared that to suppose the sun the centre of the solar system was false, and opposed to Holy Scripture. The edict of March 5, 1616, which carried this into effect, has never been repealed, though it has been long tacitly ignored. It is well known that towards the middle of the seventeenth century the Jesuits evaded it by treating the theory as an hypothesis from which, though false, certain results would follow.

In January 1632 Galileo published his dialogues on the system of the world, in which in clear and forcible language he expounded the Copernican theory. In these, apparently through jealousy of Kepler's fame, he does not so much as mention Kepler's laws (the first two of which had been published in 1609, and the third in 1619); he rejects Kepler's hypothesis that the tides are caused by the attraction of the moon, and tries to explain their existence (which he alleges is a confirmation of the Copernican hypothesis) by the statement that different parts of the earth rotate with different velocities. He was more successful in showing that mechanical principles would account for the fact that a stone thrown straight up falls again to the place from which it was thrown—a fact which previously had been one of the chief difficulties in the way of any theory which supposed the earth to be in motion.

The publication of this book was approved by the papal censor, but substantially was contrary to the edict of 1616. Galileo was summoned to Rome, forced to recant, do penance, and was released only on promise of obedience. The documents recently printed show that he was threatened with the torture, but probably there was no intention of carrying the threat into effect.

When released he again took up his work on mechanics, and by 1636 had finished a book which was published under the title *Discorsi intorno a due nuove scienze* at Leyden in 1638. In 1637 he lost his sight, but with the aid of pupils he continued his experiments on mechanics and hydrostatics, and in particular on the possibility of using a pendulum to regulate a clock, and on the theory of impact.

An anecdote of this time has been preserved which, though probably not authentic, is sufficiently interesting to bear repetition. According to one version of the story, Galileo was interviewed by some members of a Florentine guild who wanted their pumps altered so as to raise water to a height which was greater than thirty feet; and thereupon he remarked that it might be desirable to first find out why the water rose at all. A bystander intervened and said there was no difficulty about that, because nature abhorred a vacuum. Yes, said Galileo, but apparently it is only a vacuum which is less than thirty feet. His favourite pupil Torricelli was present, and thus had his attention directed to the question, which he subsequently elucidated.

Galileo's work may, I think, be fairly summed up by saying that his researches on mechanics are deserving of high praise, and that they are memorable for clearly enunciating the fact that science must be founded on laws obtained by experiment; his astronomical observations and his deductions therefrom were also excellent, and were expounded with a literary skill which leaves nothing to be desired; but though he produced some of the evidence which placed the Copernican theory on a satisfactory basis, he did not himself make any special advance in the theory of astronomy.

Francis Bacon.[1]   The necessity of an experimental foundation for science was also advocated with considerable effect by Galileo's contemporary *Francis Bacon* (Lord Verulam), who was born at London on Jan. 22, 1561, and died on April 9, 1626. He was educated at Trinity College, Cambridge. His career in politics and at the bar culminated in his becoming Lord Chancellor, with the title of Lord Verulam. The story of his subsequent degradation for accepting bribes is well known.

His chief work is the *Novum Organum*, published in 1620, in which he lays down the principles which should guide those who are making experiments on which they propose to found a theory of any branch of physics or applied mathematics. He gave rules by which the results of induction could be tested, hasty generalisations avoided, and experiments used to check one another. The influence of this treatise in the eighteenth century was great, but it is probable that during the preceding century it was little read, and the remark repeated by several

---

[1] See his life by J. Spedding, London, 1872–74. The best edition of his works is that by Ellis, Spedding, and Heath, in 7 volumes, London, second edition, 1870.

French writers that Bacon and Descartes are the creators of modern philosophy rests on a misapprehension of Bacon's influence on his contemporaries; any detailed account of this book belongs, however, to the history of scientific ideas rather than to that of mathematics.

Before leaving the subject of applied mathematics I may add a few words on the writings of Guldinus, Wright, and Snell.

Guldinus.   *Habakkuk Guldinus*, born at St. Gall on June 12, 1577, and died at Grätz on Nov. 3, 1643, was of Jewish descent, but was brought up as a Protestant; he was converted to Roman Catholicism, and became a Jesuit, when he took the Christian name of Paul, and it was to him that the Jesuit colleges at Rome and Grätz owed their mathematical reputation. The two theorems known by the name of Pappus (to which I have alluded above) were published by Guldinus in the fourth book of his *De Centro Gravitatis*, Vienna, 1635–1642. Not only were the rules in question taken without acknowledgment from Pappus, but (according to Montucla) the proof of them given by Guldinus was faulty, though he was successful in applying them to the determination of the volumes and surfaces of certain solids. The theorems were, however, previously unknown, and their enunciation excited considerable interest.

Wright.[1]   I may here also refer to *Edward Wright*, who is worthy of mention for having put the art of navigation on a scientific basis. Wright was born in Norfolk about 1560, and died in 1615. He was educated at Caius College, Cambridge, of which society he was subsequently a fellow. He seems to have been a good sailor, and he had a special talent for the construction of instruments. About 1600 he was elected lecturer on mathematics by the East India Company; he then settled in London, and shortly afterwards was appointed mathematical tutor to Henry, Prince of Wales, the son of James I. His mechanical ability may be illustrated by an orrery of his construction by which it was possible to predict eclipses; it was shewn in the Tower as a curiosity as late as 1675.

In the maps in use before the time of Gerard Mercator a degree, whether of latitude or longitude, had been represented in all cases by the same length, and the course to be pursued by a vessel was marked on the map by a straight line joining the ports of arrival and departure. Mercator had seen that this led to considerable errors, and had realised

---

[1]See pp. 25–27 of my *History of the Study of Mathematics at Cambridge*, Cambridge, 1889.

that to make this method of tracing the course of a ship at all accurate the space assigned on the map to a degree of latitude ought gradually to increase as the latitude increased. Using this principle, he had empirically constructed some charts, which were published about 1560 or 1570. Wright set himself the problem to determine the theory on which such maps should be drawn, and succeeded in discovering the law of the scale of the maps, though his rule is strictly correct for small arcs only. The result was published in the second edition of Blundeville's *Exercises*.

In 1599 Wright published his *Certain Errors in Navigation Detected and Corrected*, in which he explained the theory and inserted a table of meridional parts. The reasoning shews considerable geometrical power. In the course of the work he gives the declinations of thirty-two stars, explains the phenomena of the dip, parallax, and refraction, and adds a table of magnetic declinations; he assumes the earth to be stationary. In the following year he published some maps constructed on his principle. In these the northernmost point of Australia is shewn; the latitude of London is taken to be 51°32′.

Snell.   A contemporary of Guldinus and Wright was *Willebrod Snell*, whose name is still well known through his discovery in 1619 of the law of refraction in optics. Snell was born at Leyden in 1581, occupied a chair of mathematics at the university there, and died there on Oct. 30, 1626. He was one of those infant prodigies who occasionally appear, and at the age of twelve he is said to have been acquainted with the standard mathematical works. I will here only add that in geodesy he laid down the principles for determining the length of the arc of a meridian from the measurement of any base line, and in spherical trigonometry he discovered the properties of the polar or supplemental triangle.

Revival of interest in pure geometry.

The close of the sixteenth century was marked not only by the attempt to found a theory of dynamics based on laws derived from experiment, but also by a revived interest in geometry. This was largely due to the influence of Kepler.

Kepler.[1]   *Johann Kepler*, one of the founders of modern astron-

---

[1] See *Johann Kepplers Leben und Wirken*, by J. L. E. von Breitschwert, Stuttgart, 1831; and R. Wolf's *Geschichte der Astronomie*, Munich, 1877. A complete edition

omy, was born of humble parents near Stuttgart on Dec. 27, 1571, and died at Ratisbon on Nov. 15, 1630. He was educated under Mästlin at Tübingen. In 1593 he was appointed professor at Grätz, where he made the acquaintance of a wealthy widow, whom he married, but found too late that he had purchased his freedom from pecuniary troubles at the expense of domestic happiness. In 1599 he accepted an appointment as assistant to Tycho Brahe, and in 1601 succeeded his master as astronomer to the emperor Rudolph II. But his career was dogged by bad luck: first his stipend was not paid; next his wife went mad and then died, and a second marriage in 1611 did not prove fortunate; while, to complete his discomfort, he was expelled from his chair, and narrowly escaped condemnation for heterodoxy. During this time he depended for his income on telling fortunes and casting horoscopes, for, as he says, "nature which has conferred upon every animal the means of existence has designed astrology as an adjunct and ally to astronomy." He seems, however, to have had no scruple in charging heavily for his services, and to the surprise of his contemporaries was found at his death to possess a considerable hoard of money. He died while on a journey to try and recover for the benefit of his children some of the arrears of his stipend.

In describing Galileo's work I alluded briefly to the three laws in astronomy that Kepler had discovered, and in connection with which his name will be always associated. I may further add that he suggested that the planets might be retained in their orbits by magnetic vortices, but this was little more than a crude conjecture. I have also already mentioned the prominent part he took in bringing logarithms into general use on the continent. These are familiar facts; but it is not known so generally that Kepler was also a geometrician and algebraist of considerable power, and that he, Desargues, and perhaps Galileo, may be considered as forming a connecting link between the mathematicians of the renaissance and those of modern times.

Kepler's work in geometry consists rather in certain general principles enunciated, and illustrated by a few cases, than in any systematic exposition of the subject. In a short chapter on conics inserted in his *Paralipomena*, published in 1604, he lays down what has been called the principle of continuity, and gives as an example the statement that

of Kepler's works was published by C. Frisch at Frankfort, in 8 volumes, 1858–71; and an analysis of the mathematical part of his chief work, the *Harmonice Mundi*, is given by Chasles in his *Aperçu historique*. See also Cantor, vol. ii, part xv.

a parabola is at once the limiting case of an ellipse and of a hyperbola; he illustrates the same doctrine by reference to the foci of conics (the word *focus* was introduced by him); and he also explains that parallel lines should be regarded as meeting at infinity. He introduced the use of the eccentric angle in discussing properties of the ellipse.

In his *Stereometria*, which was published in 1615, he determines the volumes of certain vessels and the areas of certain surfaces, by means of infinitesimals instead of by the long and tedious method of exhaustions. These investigations as well as those of 1604 arose from a dispute with a wine merchant as to the proper way of gauging the contents of a cask. This use of infinitesimals was objected to by Guldinus and other writers as inaccurate, but though the methods of Kepler are not altogether free from objection he was substantially correct, and by applying the law of continuity to infinitesimals he prepared the way for Cavalieri's method of indivisibles, and the infinitesimal calculus of Newton and Leibnitz.

Kepler's work on astronomy lies outside the scope of this book. I will mention only that it was founded on the observations of Tycho Brahe,[1] whose assistant he was. His three laws of planetary motion were the result of many and laborious efforts to reduce the phenomena of the solar system to certain simple rules. The first two were published in 1609, and stated that the planets describe ellipses round the sun, the sun being in a focus; and that the line joining the sun to any planet sweeps over equal areas in equal times. The third was published in 1619, and stated that the squares of the periodic times of the planets are proportional to the cubes of the major axes of their orbits. The laws were deduced from observations on the motions of Mars and the earth, and were extended by analogy to the other planets. I ought to add that he attempted to explain why these motions took place by a hypothesis which is not very different from Descartes's theory of vortices. He suggested that the tides were caused by the attraction of the moon. Kepler also devoted considerable time to the elucidation of the theories of vision and refraction in optics.

While the conceptions of the geometry of the Greeks were being extended by Kepler, a Frenchman, whose works until recently were almost unknown, was inventing a new method of investigating the subject—a method which is now known as projective geometry. This was the discovery of Desargues, whom I put (with some hesitation) at the close of

---

[1] For an account of Tycho Brahe, born at Knudstrup in 1546 and died at Prague in 1601, see his life by J. L. E. Dreyer, Edinburgh, 1890.

this period, and not among the mathematicians of modern times.

Desargues.[1]   *Gérard Desargues*, born at Lyons in 1593, and died in 1662, was by profession an engineer and architect, but he gave some courses of gratuitous lectures in Paris from 1626 to about 1630 which made a great impression upon his contemporaries. Both Descartes and Pascal had a high opinion of his work and abilities, and both made considerable use of the theorems he had enunciated.

In 1636 Desargues issued a work on perspective; but most of his researches were embodied in his *Brouillon proiect* on conics, published in 1639, a copy of which was discovered by Chasles in 1845. I take the following summary of it from C. Taylor's work on conics. Desargues commences with a statement of the doctrine of continuity as laid down by Kepler: thus the points at the opposite ends of a straight line are regarded as coincident, parallel lines are treated as meeting at a point at infinity, and parallel planes on a line at infinity, while a straight line may be considered as a circle whose centre is at infinity. The theory of involution of six points, with its special cases, is laid down, and the projective property of pencils in involution is established. The theory of polar lines is expounded, and its analogue in space suggested. A tangent is defined as the limiting case of a secant, and an asymptote as a tangent at infinity. Desargues shows that the lines which join four points in a plane determine three pairs of lines in involution on any transversal, and from any conic through the four points another pair of lines can be obtained which are in involution with any two of the former. He proves that the points of intersection of the diagonals and the two pairs of opposite sides of any quadrilateral inscribed in a conic are a conjugate triad with respect to the conic, and when one of the three points is at infinity its polar is a diameter; but he fails to explain the case in which the quadrilateral is a parallelogram, although he had formed the conception of a straight line which was wholly at infinity. The book, therefore, may be fairly said to contain the fundamental theorems on involution, homology, poles and polars, and perspective.

The influence exerted by the lectures of Desargues on Descartes, Pascal, and the French geometricians of the seventeenth century was considerable; but the subject of projective geometry soon fell into oblivion, chiefly because the analytical geometry of Descartes was so much more powerful as a method of proof or discovery.

[1] See *Oeuvres de Desargues*, by M. Poudra, 2 vols., Paris, 1864; and a note in the *Bibliotheca Mathematica*, 1885, p. 90.

The researches of Kepler and Desargues will serve to remind us that as the geometry of the Greeks was not capable of much further extension, mathematicians were now beginning to seek for new methods of investigation, and were extending the conceptions of geometry. The invention of analytical geometry and of the infinitesimal calculus temporarily diverted attention from pure geometry, but at the beginning of the last century there was a revival of interest in it, and since then it has been a favourite subject of study with many mathematicians.

Mathematical knowledge at the close of the renaissance.

Thus by the beginning of the seventeenth century we may say that the fundamental principles of arithmetic, algebra, theory of equations, and trigonometry had been laid down, and the outlines of the subjects as we know them had been traced. It must be, however, remembered that there were no good elementary text-books on these subjects; and a knowledge of them was therefore confined to those who could extract it from the ponderous treatises in which it lay buried. Though much of the modern algebraical and trigonometrical notation had been introduced, it was not familiar to mathematicians, nor was it even universally accepted; and it was not until the end of the seventeenth century that the language of these subjects was definitely fixed. Considering the absence of good text-books, I am inclined rather to admire the rapidity with which it came into universal use, than to cavil at the hesitation to trust to it alone which many writers showed.

If we turn to applied mathematics, we find, on the other hand, that the science of statics had made but little advance in the eighteen centuries that had elapsed since the time of Archimedes, while the foundations of dynamics were laid by Galileo only at the close of the sixteenth century. In fact, as we shall see later, it was not until the time of Newton that the science of mechanics was placed on a satisfactory basis. The fundamental conceptions of mechanics are difficult, but the ignorance of the principles of the subject shown by the mathematicians of this time is greater than would have been anticipated from their knowledge of pure mathematics.

With this exception, we may say that the principles of analytical geometry and of the infinitesimal calculus were needed before there was likely to be much further progress. The former was employed by Descartes in 1637, the latter was invented by Newton some thirty or forty years later, and their introduction may be taken as marking the commencement of the period of modern mathematics.

THIRD PERIOD.

𝔐𝔬𝔡𝔢𝔯𝔫 𝔐𝔞𝔱𝔥𝔢𝔪𝔞𝔱𝔦𝔠𝔰.

The history of modern mathematics begins with the invention of analytical geometry and the infinitesimal calculus. The mathematics is far more complex than that produced in either of the preceding periods; but, during the seventeenth and eighteenth centuries, it may be generally described as characterized by the development of analysis, and its application to the phenomena of nature.

I continue the chronological arrangement of the subject. Chapter XV contains the history of the forty years from 1635 to 1675, and an account of the mathematical discoveries of Descartes, Cavalieri, Pascal, Wallis, Fermat, and Huygens. Chapter XVI is given up to a discussion of Newton's researches. Chapter XVII contains an account of the works of Leibnitz and his followers during the first half of the eighteenth century (including D'Alembert), and of the contemporary English school to the death of Maclaurin. The works of Euler, Lagrange, Laplace, and their contemporaries form the subject-matter of chapter XVIII.

Lastly, in chapter XIX I have added some notes on a few of the mathematicians of recent times; but I exclude all detailed reference to living writers, and partly because of this, partly for other reasons there given, the account of contemporary mathematics does not profess to cover the subject.

CHAPTER XIV.

THE HISTORY OF MODERN MATHEMATICS.

THE division between this period and that treated in the last six chapters is by no means so well defined as that which separates the history of Greek mathematics from the mathematics of the middle ages. The methods of analysis used in the seventeenth century and the kind of problems attacked changed but gradually; and the mathematicians at the beginning of this period were in immediate relations with those at the end of that last considered. For this reason some writers have divided the history of mathematics into two parts only, treating the schoolmen as the lineal successors of the Greek mathematicians, and dating the creation of modern mathematics from the introduction of the Arab text-books into Europe. The division I have given is, I think, more convenient, for the introduction of analytical geometry and of the infinitesimal calculus revolutionized the development of the subject, and therefore it seems preferable to take their invention as marking the commencement of modern mathematics.

The time that has elapsed since these methods were invented has been a period of incessant intellectual activity in all departments of knowledge, and the progress made in mathematics has been immense. The greatly extended range of knowledge, the mass of materials to be mastered, the absence of perspective, and even the echoes of old controversies, combine to increase the difficulties of an author. As, however, the leading facts are generally known, and the works published during this time are accessible to any student, I may deal more concisely with the lives and writings of modern mathematicians than with those of their predecessors, and confine myself more strictly than before to those who have materially affected the progress of the subject.

To give a sense of unity to a history of mathematics it is necessary to treat it chronologically, but it is possible to do this in two ways. We may discuss separately the development of different branches of mathematics during a certain period (not too long), and deal with the works of each mathematician under such heads as they may fall. Or we may describe in succession the lives and writings of the mathematicians of a certain period, and deal with the development of different subjects under the heads of those who studied them. Personally, I prefer the latter course; and not the least advantage of this, from my point of view, is that it adds a human interest to the narrative. No doubt as the subject becomes more complex this course becomes more difficult, and it may be that when the history of mathematics in the nineteenth century is written it will be necessary to deal separately with the separate branches of the subject, but, as far as I can, I continue to present the history biographically.

Roughly speaking, we may say that five distinct stages in the history of modern mathematics can be discerned.

First of all, there is the invention of analytical geometry by Descartes in 1637; and almost at the same time the introduction of the method of indivisibles, by the use of which areas, volumes, and the positions of centres of mass can be determined by summation in a manner analogous to that effected nowadays by the aid of the integral calculus. The method of indivisibles was soon superseded by the integral calculus. Analytical geometry, however, maintains its position as part of the necessary training of every mathematician, and for all purposes of research is incomparably more potent than the geometry of the ancients. The latter is still, no doubt, an admirable intellectual training, and it frequently affords an elegant demonstration of some proposition the truth of which is already known, but it requires a special procedure for every particular problem attacked. The former, on the other hand, lays down a few simple rules by which any property can be at once proved or disproved.

In the *second* place, we have the invention, some thirty years later, of the fluxional or differential calculus. Wherever a quantity changes according to some continuous law—and most things in nature do so change—the differential calculus enables us to measure its rate of increase or decrease; and, from its rate of increase or decrease, the integral calculus enables us to find the original quantity. Formerly every separate function of x such as $(1+x)^n$, $\log(1+x)$, $\sin x$, $\tan^{-1} x$, &c., could be expanded in ascending powers of x only by means of such special procedure as was suitable for that particular problem; but, by the aid of the calculus, the expansion of any function of x in ascending powers of x is in general reducible to one rule which covers all cases alike. So, again, the theory of maxima and minima, the determination of the lengths of curves and the areas enclosed by them, the determination of surfaces, of volumes, and of centres of mass, and many other problems, are each reducible to a single rule. The theories of differential equations, of the calculus of variations, of finite differences, &c., are the developments of the ideas of the calculus.

These two subjects—analytical geometry and the calculus—became the chief instruments of further progress in mathematics. In both of them a sort of machine was constructed: to solve a problem, it was only necessary to put in the particular function dealt with, or the equation of the particular curve or surface considered, and on performing certain simple operations the result came out. The validity of the process was proved once for all, and it was no longer requisite to invent some special method for every separate function, curve, or surface.

In the *third* place, Huygens, following Galileo, laid the foundation of a satisfactory treatment of dynamics, and Newton reduced it to an exact science. The latter mathematician proceeded to apply the new analytical methods not only to numerous problems in the mechanics of solids and fluids on the earth, but to the solar system; the whole of mechanics terrestrial and celestial was thus brought within the domain of mathematics. There is no doubt that Newton used the calculus to obtain many of his results, but he seems to have thought that, if his demonstrations were established by the aid of a new science which was at that time generally unknown, his critics (who would not understand the fluxional calculus) would fail to realise the truth and importance of his discoveries. He therefore determined to give geometrical proofs of all his results. He accordingly cast the *Principia* into a geometrical form, and thus presented it to the world in a language which all men could then understand. The theory of mechanics was extended, systematized, and put in its modern form by Lagrange and Laplace towards the end of the eighteenth century.

In the *fourth* place, we may say that during this period the chief branches of physics have been brought within the scope of mathematics. This extension of the domain of mathematics was commenced by Huygens and Newton when they propounded their theories of light; but it was not until the beginning of the last century that sufficiently accurate observations were made in most physical subjects to enable mathematical reasoning to be applied to them.

Numerous and far-reaching conclusions have been obtained in physics by the application of mathematics to the results of observations and experiments, but we now want some more simple hypotheses from which we can deduce those laws which at present form our starting-point. If, to take one example, we could say in what electricity consisted, we might get some simple laws or hypotheses from which by the aid of mathematics all the observed phenomena could be deduced, in the same way as Newton deduced all the results of physical astronomy from the law of gravitation. All lines of research seem, moreover, to indicate that there is an intimate connection between the different branches of physics, e.g. between light, heat, elasticity, electricity, and magnetism. The ultimate explanation of this and of the leading facts in physics seems to demand a study of molecular physics; a knowledge of molecular physics in its turn seems to require some theory as to the constitution of matter; it would further appear that the key to the constitution of matter is to be found in electricity or chemical physics. So the matter stands at present; the connection between the different branches of physics, and the fundamental laws of those branches (if there be any simple ones), are riddles which are yet unsolved. This history does not pretend to treat of problems which are now the subject of investigation; the fact also that mathematical physics is mainly the creation of the nineteenth century would exclude all detailed discussion of the subject.

Fifthly, this period has seen an immense extension of pure mathematics. Much of this is the creation of comparatively recent times, and I regard the details of it as outside the limits of this book, though in chapter XIX I have allowed myself to mention some of the subjects discussed. The most striking features of this extension are the critical discussion of fundamental principles, the developments of higher geometry, of higher arithmetic or the theory of numbers, of higher algebra (including the theory of forms), and of the theory of equations, also the discussion of functions of double and multiple periodicity, and the creation of a theory of functions.

This hasty summary will indicate the subjects treated and the limitations I have imposed on myself. The history of the origin and growth of analysis and its application to the material universe comes within my purview. The extensions in the latter half of the nineteenth century of pure mathematics and of the application of mathematics to physical problems open a new period which lies beyond the limits of this book; and I allude to these subjects only so far as they may indicate the directions in which the future history of mathematics appears to be developing.

CHAPTER XV.

HISTORY OF MATHEMATICS FROM DESCARTES TO HUYGENS.[1]
CIRC. 1635–1675.

I PROPOSE in this chapter to consider the history of mathematics during the forty years in the middle of the seventeenth century. I regard Descartes, Cavalieri, Pascal, Wallis, Fermat, and Huygens as the leading mathematicians of this time. I shall treat them in that order, and I shall conclude with a brief list of the more eminent remaining mathematicians of the same date.

I have already stated that the mathematicians of this period— and the remark applies more particularly to Descartes, Pascal, and Fermat—were largely influenced by the teaching of Kepler and Desargues, and I would repeat again that I regard these latter and Galileo as forming a connecting link between the writers of the renaissance and those of modern times. I should also add that the mathematicians considered in this chapter were contemporaries, and, although I have tried to place them roughly in such an order that their chief works shall come in a chronological arrangement, it is essential to remember that they were in relation one with the other, and in general were acquainted with one another's researches as soon as these were published.

Descartes.[2]   Subject to the above remarks, we may consider Descartes as the first of the modern school of mathematics. *René Des-*

---

[1]See Cantor, part xv, vol. ii, pp. 599–844; other authorities for the mathematicians of this period are mentioned in the footnotes.

[2]See *Descartes*, by E. S. Haldane, London, 1905; and *Descartes Savant*, by G. Milhaud, Paris, 1921. A complete edition of his works, edited by C. Adam and P. Tanner, is in process of issue by the French Government; vols. i–ix, 1897–1904. A tolerably complete account of Descartes's mathematical and physical investigations is given in Ersch and Gruber's *Encyclopädie*. The most complete edition of his works is that by Victor Cousin in 11 vols., Paris, 1824–26. Some minor papers subsequently discovered were printed by F. de Careil, Paris, 1859.

cartes was born near Tours on March 31, 1596, and died at Stockholm on February 11, 1650; thus he was a contemporary of Galileo and Desargues. His father, who, as the name implies, was of a good family, was accustomed to spend half the year at Rennes when the local parliament, in which he held a commission as councillor, was in session, and the rest of the time on his family estate of *Les Cartes* at La Haye. René, the second of a family of two sons and one daughter, was sent at the age of eight years to the Jesuit School at La Flêche, and of the admirable discipline and education there given he speaks most highly. On account of his delicate health he was permitted to lie in bed till late in the mornings; this was a custom which he always followed, and when he visited Pascal in 1647 he told him that the only way to do good work in mathematics and to preserve his health was never to allow any one to make him get up in the morning before he felt inclined to do so; an opinion which I chronicle for the benefit of any schoolboy into whose hands this work may fall.

On leaving school in 1612 Descartes went to Paris to be introduced to the world of fashion. Here, through the medium of the Jesuits, he made the acquaintance of Mydorge, and renewed his schoolboy friendship with Mersenne, and together with them he devoted the two years of 1615 and 1616 to the study of mathematics. At that time a man of position usually entered either the army or the church; Descartes chose the former profession, and in 1617 joined the army of Prince Maurice of Orange, then at Breda. Walking through the streets there he saw a placard in Dutch which excited his curiosity, and stopping the first passer, asked him to translate it into either French or Latin. The stranger, who happened to be Isaac Beeckman, the head of the Dutch College at Dort, offered to do so if Descartes would answer it; the placard being, in fact, a challenge to all the world to solve a certain geometrical problem.[1] Descartes worked it out within a few hours, and a warm friendship between him and Beeckman was the result. This unexpected test of his mathematical attainments made the uncongenial life of the army distasteful to him, and though, under family influence and tradition, he remained a soldier, he continued to occupy his leisure with mathematical studies. He was accustomed to date the first ideas of his new philosophy and of his analytical geometry from three dreams which he experienced on the night of November 10, 1619, at Neuberg,

---

[1]Some doubt has been recently expressed as to whether the story is well founded: see *L'Intermédiaire des Mathématiciens*, Paris, 1909, vol. xvi, pp. 12–13.

when campaigning on the Danube, and he regarded this as the critical day of his life, and one which determined his whole future.

He resigned his commission in the spring of 1621, and spent the next five years in travel, during most of which time he continued to study pure mathematics. In 1626 we find him settled at Paris, "a little well-built figure, modestly clad in green taffety, and only wearing sword and feather in token of his quality as a gentleman." During the first two years there he interested himself in general society, and spent his leisure in the construction of optical instruments; but these pursuits were merely the relaxations of one who failed to find in philosophy that theory of the universe which he was convinced finally awaited him.

In 1628 Cardinal de Berulle, the founder of the Oratorians, met Descartes, and was so much impressed by his conversation that he urged on him the duty of devoting his life to the examination of truth. Descartes agreed, and the better to secure himself from interruption moved to Holland, then at the height of its power. There for twenty years he lived, giving up all his time to philosophy and mathematics. Science, he says, may be compared to a tree; metaphysics is the root, physics is the trunk, and the three chief branches are mechanics, medicine, and morals, these forming the three applications of our knowledge, namely, to the external world, to the human body, and to the conduct of life.

He spent the first four years, 1629 to 1633, of his stay in Holland in writing *Le Monde*, which embodies an attempt to give a physical theory of the universe; but finding that its publication was likely to bring on him the hostility of the church, and having no desire to pose as a martyr, he abandoned it: the incomplete manuscript was published in 1664. He then devoted himself to composing a treatise on universal science; this was published at Leyden in 1637 under the title *Discours de la méthode pour bien conduire sa raison et chercher la vérité dans les sciences*, and was accompanied with three appendices (which possibly were not issued till 1638) entitled *La Dioptrique*, *Les Météores*, and *La Géométrie*; it is from the last of these that the invention of analytical geometry dates. In 1641 he published a work called *Meditationes*, in which he explained at some length his views of philosophy as sketched out in the *Discours*. In 1644 he issued the *Principia Philosophiae*, the greater part of which was devoted to physical science, especially the laws of motion and the theory of vortices. In 1647 he received a pension from the French court in honour of his discoveries. He went to Sweden on the invitation of the Queen in 1649, and died a few months later of inflammation of the lungs.

In appearance, Descartes was a small man with large head, projecting brow, prominent nose, and black hair coming down to his eyebrows. His voice was feeble. In disposition he was cold and selfish. Considering the range of his studies he was by no means widely read, and he despised both learning and art unless something tangible could be extracted therefrom. He never married, and left no descendants, though he had one illegitimate daughter, who died young.

As to his philosophical theories, it will be sufficient to say that he discussed the same problems which have been debated for the last two thousand years, and probably will be debated with equal zeal two thousand years hence. It is hardly necessary to say that the problems themselves are of importance and interest, but from the nature of the case no solution ever offered is capable either of rigid proof or of disproof; all that can be effected is to make one explanation more probable than another, and whenever a philosopher like Descartes believes that he has at last finally settled a question it has been possible for his successors to point out the fallacy in his assumptions. I have read somewhere that philosophy has always been chiefly engaged with the inter-relations of God, Nature, and Man. The earliest philosophers were Greeks who occupied themselves mainly with the relations between God and Nature, and dealt with Man separately. The Christian Church was so absorbed in the relation of God to Man as entirely to neglect Nature. Finally, modern philosophers concern themselves chiefly with the relations between Man and Nature. Whether this is a correct historical generalization of the views which have been successively prevalent I do not care to discuss here, but the statement as to the scope of modern philosophy marks the limitations of Descartes's writings.

Descartes's chief contributions to mathematics were his analytical geometry and his theory of vortices, and it is on his researches in connection with the former of these subjects that his mathematical reputation rests.

Analytical geometry does not consist merely (as is sometimes loosely said) in the application of algebra to geometry; that had been done by Archimedes and many others, and had become the usual method of procedure in the works of the mathematicians of the sixteenth century. The great advance made by Descartes was that he saw that a point in a plane could be completely determined if its distances, say x and y, from two fixed lines drawn at right angles in the plane were given, with the convention familiar to us as to the interpretation of positive and negative values; and that though an equation $f(x, y) = 0$ was indeterminate and could be satisfied by an infinite number of values of x and y, yet these values of x and y determined the co-ordinates of a number of points which form a curve, of which the equation $f(x, y) = 0$ expresses some geometrical property, that is, a property true of the curve at every point on it. Descartes asserted that a point in space could be similarly determined by three co-ordinates, but he confined his attention to plane curves.

It was at once seen that in order to investigate the properties of a curve it was sufficient to select, as a definition, any characteristic geometrical property, and to express it by means of an equation between the (current) co-ordinates of any point on the curve, that is, to translate the definition into the language of analytical geometry. The equation so obtained contains implicitly every property of the curve, and any particular property can be deduced from it by ordinary algebra without troubling about the geometry of the figure. This may have been dimly recognized or foreshadowed by earlier writers, but Descartes went further and pointed out the very important facts that two or more curves can be referred to one and the same system of co-ordinates, and that the points in which two curves intersect can be determined by finding the roots common to their two equations. I need not go further into details, for nearly everyone to whom the above is intelligible will have read analytical geometry, and is able to appreciate the value of its invention.

Descartes's *Géométrie* is divided into three books: the first two of these treat of analytical geometry, and the third includes an analysis of the algebra then current. It is somewhat difficult to follow the reasoning, but the obscurity was intentional. "Je n'ai rien omis," says he, "qu'à dessein ... j'avois prévu que certaines gens qui se vantent de sçavoir tout n'auroient pas manqué de dire que je n'avois rien écrit qu'ils n'eussent sçu auparavant, si je me fusse rendu assez intelligible pour eux."

The first book commences with an explanation of the principles of analytical geometry, and contains a discussion of a certain problem which had been propounded by Pappus in the seventh book of his Συναγωγή and of which some particular cases had been considered by Euclid and Apollonius. The general theorem had baffled previous geometricians, and it was in the attempt to solve it that Descartes was led to the invention of analytical geometry. The full enunciation of the problem is rather involved, but the most important case is to find the locus of a point such that the product of the perpendiculars on m

given straight lines shall be in a constant ratio to the product of the perpendiculars on n other given straight lines. The ancients had solved this geometrically for the case $m = 1$, $n = 1$, and the case $m = 1$, $n = 2$. Pappus had further stated that, if $m = n = 2$, the locus is a conic, but he gave no proof; Descartes also failed to prove this by pure geometry, but he shewed that the curve is represented by an equation of the second degree, that is, is a conic; subsequently Newton gave an elegant solution of the problem by pure geometry.

In the second book Descartes divides curves into two classes, namely, geometrical and mechanical curves. He defines geometrical curves as those which can be generated by the intersection of two lines each moving parallel to one co-ordinate axis with "commensurable" velocities; by which terms he means that dy/dx is an algebraical function, as, for example, is the case in the ellipse and the cissoid. He calls a curve mechanical when the ratio of the velocities of these lines is "incommensurable"; by which term he means that dy/dx is a transcendental function, as, for example, is the case in the cycloid and the quadratrix. Descartes confined his discussion to geometrical curves, and did not treat of the theory of mechanical curves. The classification into algebraical and transcendental curves now usual is due to Newton.[1]

Descartes also paid particular attention to the theory of the tangents to curves—as perhaps might be inferred from his system of classification just alluded to. The then current definition of a tangent at a point was a straight line through the point such that between it and the curve no other straight line could be drawn, that is, the straight line of closest contact. Descartes proposed to substitute for this a statement equivalent to the assertion that the tangent is the limiting position of the secant; Fermat, and at a later date Maclaurin and Lagrange, adopted this definition. Barrow, followed by Newton and Leibnitz, considered a curve as the limit of an inscribed polygon when the sides become indefinitely small, and stated that a side of the polygon when produced became in the limit a tangent to the curve. Roberval, on the other hand, defined a tangent at a point as the direction of motion at that instant of a point which was describing the curve. The results are the same whichever definition is selected, but the controversy as to which definition was the correct one was none the less lively. In his letters Descartes illustrated his theory by giving the general rule for drawing tangents and normals to a roulette.

[1] See below, page 279.

The method used by Descartes to find the tangent or normal at any point of a given curve was substantially as follows. He determined the centre and radius of a circle which should cut the curve in two consecutive points there. The tangent to the circle at that point will be the required tangent to the curve. In modern text-books it is usual to express the condition that two of the points in which a straight line (such as $y = mx + c$) cuts the curve shall coincide with the given point: this enables us to determine m and c, and thus the equation of the tangent there is determined. Descartes, however, did not venture to do this, but selecting a circle as the simplest curve and one to which he knew how to draw a tangent, he so fixed his circle as to make it touch the given curve at the point in question, and thus reduced the problem to drawing a tangent to a circle. I should note in passing that he only applied this method to curves which are symmetrical about an axis, and he took the centre of the circle on the axis.

The obscure style deliberately adopted by Descartes diminished the circulation and immediate appreciation of these books; but a Latin translation of them, with explanatory notes, was prepared by F. de Beaune, and an edition of this, with a commentary by F. van Schooten, issued in 1659, was widely read.

The third book of the *Géométrie* contains an analysis of the algebra then current, and it has affected the language of the subject by fixing the custom of employing the letters at the beginning of the alphabet to denote known quantities, and those at the end of the alphabet to denote unknown quantities.[1] Descartes further introduced the system of indices now in use; very likely it was original on his part, but I would here remind the reader that the suggestion had been made by previous writers, though it had not been generally adopted. It is doubtful whether or not Descartes recognised that his letters might represent any quantities, positive or negative, and that it was sufficient to prove a proposition for one general case. He was the earliest writer to realize the advantage to be obtained by taking all the terms of an equation to one side of it, though Stifel and Harriot had sometimes employed that form by choice. He realised the meaning of negative quantities and used them freely. In this book he made use of the rule for finding a limit to the number of positive and of negative roots of an algebraical equation, which is still known by his name; and introduced the method of inde-

---

[1] On the origin of the custom of using x to represent an unknown example, see a note by G. Eneström in the *Bibliotheca Mathematica*, 1885, p. 43.

terminate coefficients for the solution of equations. He believed that he had given a method by which algebraical equations of any order could be solved, but in this he was mistaken. It may be also mentioned that he enunciated the theorem, commonly attributed to Euler, on the relation between the numbers of faces, edges, and angles of a polyhedron: this is in one of the papers published by Careil.

Of the two other appendices to the *Discours* one was devoted to *optics*. The chief interest of this consists in the statement given of the law of refraction. This appears to have been taken from Snell's work, though, unfortunately, it is enunciated in a way which might lead a reader to suppose that it is due to the researches of Descartes. Descartes would seem to have repeated Snell's experiments when in Paris in 1626 or 1627, and it is possible that he subsequently forgot how much he owed to the earlier investigations of Snell. A large part of the optics is devoted to determining the best shape for the lenses of a telescope, but the mechanical difficulties in grinding a surface of glass to a required form are so great as to render these investigations of little practical use. Descartes seems to have been doubtful whether to regard the rays of light as proceeding from the eye and so to speak touching the object, as the Greeks had done, or as proceeding from the object, and so affecting the eye; but, since he considered the velocity of light to be infinite, he did not deem the point particularly important.

The other appendix, on *meteors*, contains an explanation of numerous atmospheric phenomena, including the rainbow; the explanation of the latter is necessarily incomplete, since Descartes was unacquainted with the fact that the refractive index of a substance is different for lights of different colours.

Descartes's physical theory of the universe, embodying most of the results contained in his earlier and unpublished *Le Monde*, is given in his *Principia*, 1644, and rests on a metaphysical basis. He commences with a discussion on motion; and then lays down ten laws of nature, of which the first two are almost identical with the first two laws of motion as given by Newton; the remaining eight laws are inaccurate. He next proceeds to discuss the nature of matter which he regards as uniform in kind though there are three forms of it. He assumes that the matter of the universe must be in motion, and that the motion must result in a number of vortices. He states that the sun is the centre of an immense whirlpool of this matter, in which the planets float and are swept round like straws in a whirlpool of water. Each planet is supposed to be the centre of a secondary whirlpool by which its satellites are carried: these secondary whirlpools are supposed to produce variations of density in the surrounding medium which constitute the primary whirlpool, and so cause the planets to move in ellipses and not in circles. All these assumptions are arbitrary and unsupported by any investigation. It is not difficult to prove that on his hypothesis the sun would be in the centre of these ellipses, and not at a focus (as Kepler had shewn was the case), and that the weight of a body at every place on the surface of the earth except the equator would act in a direction which was not vertical; but it will be sufficient here to say that Newton in the second book of his *Principia*, 1687, considered the theory in detail, and shewed that its consequences are not only inconsistent with each of Kepler's laws and with the fundamental laws of mechanics, but are also at variance with the laws of nature assumed by Descartes. Still, in spite of its crudeness and its inherent defects, the theory of vortices marks a fresh era in astronomy, for it was an attempt to explain the phenomena of the whole universe by the same mechanical laws which experiment shews to be true on the earth.

Cavalieri.[1]　　Almost contemporaneously with the publication in 1637 of Descartes's geometry, the principles of the integral calculus, so far as they are concerned with summation, were being worked out in Italy. This was effected by what was called the principle of indivisibles, and was the invention of Cavalieri. It was applied by him and his contemporaries to numerous problems connected with the quadrature of curves and surfaces, the determination of volumes, and the positions of centres of mass. It served the same purpose as the tedious method of exhaustions used by the Greeks; in principle the methods are the same, but the notation of indivisibles is more concise and convenient. It was, in its turn, superseded at the beginning of the eighteenth century by the integral calculus.

Bonaventura Cavalieri was born at Milan in 1598, and died at Bologna on November 27, 1647. He became a Jesuit at an early age; on the recommendation of the Order he was in 1629 made professor of mathematics at Bologna; and he continued to occupy the chair there until his death. I have already mentioned Cavalieri's name in connection with the introduction of the use of logarithms into Italy, and have alluded to his discovery of the expression for the area of a spherical

---

[1]Cavalieri's life has been written by P. Frisi, Milan, 1778; by F. Predari, Milan, 1843; by Gabrio Piola, Milan, 1844; and by A. Favaro, Bologna, 1888. An analysis of his works is given in Marie's *Histoire des Sciences*, Paris, 1885–8, vol. iv, pp. 69–90.

triangle in terms of the spherical excess. He was one of the most influential mathematicians of his time, but his subsequent reputation rests mainly on his invention of the principle of indivisibles.

The principle of indivisibles had been used by Kepler in 1604 and 1615 in a somewhat crude form. It was first stated by Cavalieri in 1629, but he did not publish his results till 1635. In his early enunciation of the principle in 1635 Cavalieri asserted that a line was made up of an infinite number of points (each without magnitude), a surface of an infinite number of lines (each without breadth), and a volume of an infinite number of surfaces (each without thickness). To meet the objections of Guldinus and others, the statement was recast, and in its final form as used by the mathematicians of the seventeenth century it was published in Cavalieri's *Exercitationes Geometricae* in 1647; the third exercise is devoted to a defence of the theory. This book contains the earliest demonstration of the properties of Pappus.[1] Cavalieri's works on indivisibles were reissued with his later corrections in 1653.

The method of indivisibles rests, in effect, on the assumption that any magnitude may be divided into an infinite number of small quantities which can be made to bear any required ratios (*ex. gr.* equality) one to the other. The analysis given by Cavalieri is hardly worth quoting except as being one of the first steps taken towards the formation of an infinitesimal calculus. One example will suffice. Suppose it be required to find the area of a right-angled triangle. Let the base be made up of, or contain n points (or indivisibles), and similarly let the other side contain na points, then the ordinates at the successive points of the base will contain $a, 2a \ldots, na$ points. Therefore the number of points in the area is $a+2a+\ldots+na$; the sum of which is $\frac{1}{2}n^2a+\frac{1}{2}na$. Since n is very large, we may neglect $\frac{1}{2}na$, for it is inconsiderable compared with $\frac{1}{2}n^2a$. Hence the area is equal to $\frac{1}{2}(na)n$, that is, $\frac{1}{2} \times$ altitude $\times$ base. There is no difficulty in criticizing such a proof, but, although the form in which it is presented is indefensible, the substance of it is correct.

It would be misleading to give the above as the only specimen of the method of indivisibles, and I therefore quote another example, taken from a later writer, which will fairly illustrate the use of the method when modified and corrected by the method of limits. Let it be required to find the area outside a parabola APC and bounded by the curve, the tangent at A, and a line DC parallel to AB the diameter at A. Complete the parallelogram $ABCD$. Divide AD into n equal parts, let

[1]See above, pp. 84, 209.

AM contain r of them, and let MN be the $(r + 1)$th part. Draw MP and NQ parallel to AB, and draw PR parallel to AD. Then when n becomes indefinitely large, the curvilinear area $APCD$ will be the limit of the sum of all parallelograms like PN. Now

$$\text{area } PN : \text{ area } BD = MP \cdot MN : DC \cdot AD.$$

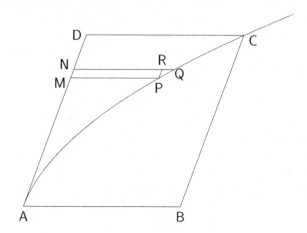

But by the properties of the parabola

$$MP : DC = AM^2 : AD^2 = r^2 : n^2,$$

and $\qquad\qquad\qquad MN : AD = 1 : n.$

Hence $\qquad\qquad MP \cdot MN : DC \cdot AD = r^2 : n^3.$

Therefore $\qquad\qquad \text{area } PN : \text{area } BD = r^2 : n^3.$

Therefore, ultimately,

$$\text{area } APCD : \text{area } BD = 1^2 + 2^2 + \ldots + (n-1)^2 : n^3$$
$$= \tfrac{1}{6}n(n-1)(2n-1) : n^3$$

which, in the limit, $\qquad\qquad\qquad = 1 : 3.$

It is perhaps worth noticing that Cavalieri and his successors always used the method to find the ratio of two areas, volumes, or magnitudes of the same kind and dimensions, that is, they never thought of an area as containing so many units of area. The idea of comparing a magnitude with a unit of the same kind seems to have been due to Wallis.

It is evident that in its direct form the method is applicable to only a few curves. Cavalieri proved that, if m be a positive integer, then the limit, when n is infinite, of $(1^m + 2^m + \cdots + n^m)/n^{m+1}$ is $1/(m + 1)$, which is equivalent to saying that he found the integral to x of x^m from $x = 0$ to $x = 1$; he also discussed the quadrature of the hyperbola.

Pascal.[1]   Among the contemporaries of Descartes none displayed greater natural genius than Pascal, but his mathematical reputation rests more on what he might have done than on what he actually effected, as during a considerable part of his life he deemed it his duty to devote his whole time to religious exercises.

Blaise Pascal was born at Clermont on June 19, 1623, and died at Paris on Aug. 19, 1662. His father, a local judge at Clermont, and himself of some scientific reputation, moved to Paris in 1631, partly to prosecute his own scientific studies, partly to carry on the education of his only son, who had already displayed exceptional ability. Pascal was kept at home in order to ensure his not being overworked, and with the same object it was directed that his education should be at first confined to the study of languages, and should not include any mathematics. This naturally excited the boy's curiosity, and one day, being then twelve years old, he asked in what geometry consisted. His tutor replied that it was the science of constructing exact figures and of determining the proportions between their different parts. Pascal, stimulated no doubt by the injunction against reading it, gave up his play-time to this new study, and in a few weeks had discovered for himself many properties of figures, and in particular the proposition that the sum of the angles of a triangle is equal to two right angles. I have read somewhere, but I cannot lay my hand on the authority, that his proof merely consisted in turning the angular points of a triangular piece of paper over so as to meet in the centre of the inscribed circle: a similar demonstration can be got by turning the angular points over so as to meet at the foot of the perpendicular drawn from the biggest angle to the opposite side. His father, struck by this display of ability, gave him a copy of Euclid's *Elements*, a book which Pascal read with avidity and soon mastered.

At the age of fourteen he was admitted to the weekly meetings of Roberval, Mersenne, Mydorge, and other French geometricians; from

---

[1]See *Pascal* by J. Bertrand, Paris, 1891; and *Pascal, sein Leben und seine Kämpfe*, by J. G. Dreydorff, Leipzig, 1870. Pascal's life, written by his sister Mme. Périer, was edited by A. P. Faugère, Paris, 1845, and has formed the basis for several works. An edition of his writings was published in five volumes at the Hague in 1779, second edition, Paris, 1819; some additional pamphlets and letters were published in three volumes at Paris in 1858.

which, ultimately, the French Academy sprung. At sixteen Pascal wrote an essay on conic sections; and in 1641, at the age of eighteen, he constructed the first arithmetical machine, an instrument which, eight years later, he further improved. His correspondence with Fermat about this time shews that he was then turning his attention to analytical geometry and physics. He repeated Torricelli's experiments, by which the pressure of the atmosphere could be estimated as a weight, and he confirmed his theory of the cause of barometrical variations by obtaining at the same instant readings at different altitudes on the hill of Puy-de-Dôme.

In 1650, when in the midst of these researches, Pascal suddenly abandoned his favourite pursuits to study religion, or, as he says in his *Pensées*, "to contemplate the greatness and the misery of man"; and about the same time he persuaded the younger of his two sisters to enter the Port Royal society.

In 1653 he had to administer his father's estate. He now took up his old life again, and made several experiments on the pressure exerted by gases and liquids; it was also about this period that he invented the arithmetical triangle, and together with Fermat created the calculus of probabilities. He was meditating marriage when an accident again turned the current of his thoughts to a religious life. He was driving a four-in-hand on November 23, 1654, when the horses ran away; the two leaders dashed over the parapet of the bridge at Neuilly, and Pascal was saved only by the traces breaking. Always somewhat of a mystic, he considered this a special summons to abandon the world. He wrote an account of the accident on a small piece of parchment, which for the rest of his life he wore next to his heart, to perpetually remind him of his covenant; and shortly moved to Port Royal, where he continued to live until his death in 1662. Constitutionally delicate, he had injured his health by his incessant study; from the age of seventeen or eighteen he suffered from insomnia and acute dyspepsia, and at the time of his death was physically worn out.

His famous *Provincial Letters* directed against the Jesuits, and his *Pensées*, were written towards the close of his life, and are the first example of that finished form which is characteristic of the best French literature. The only mathematical work that he produced after retiring to Port Royal was the essay on the cycloid in 1658. He was suffering from sleeplessness and toothache when the idea occurred to him, and to his surprise his teeth immediately ceased to ache. Regarding this as a divine intimation to proceed with the problem, he worked incessantly for eight days at it, and completed a tolerably full account of the geometry of the cycloid.

I now proceed to consider his mathematical works in rather greater detail.

His early essay on the *geometry of conics*, written in 1639, but not published till 1779, seems to have been founded on the teaching of Desargues. Two of the results are important as well as interesting. The first of these is the theorem known now as "Pascal's theorem," namely, that if a hexagon be inscribed in a conic, the points of intersection of the opposite sides will lie in a straight line. The second, which is really due to Desargues, is that if a quadrilateral be inscribed in a conic, and a straight line be drawn cutting the sides taken in order in the points $A, B, C,$ and D, and the conic in P and Q, then

$$PA \cdot PC : PB \cdot PD = QA \cdot QC : QB \cdot QD.$$

Pascal employed his *arithmetical triangle* in 1653, but no account of his method was printed till 1665. The triangle is constructed as in the figure below, each horizontal line being formed from the one above it by making every number in it equal to the sum of those above and to the left of it in the row immediately above it; *ex. gr.* the fourth number in the fourth line, namely, 20, is equal to $1 + 3 + 6 + 10$. The numbers in each line are what are now called *figurate* numbers. Those in the first line are called numbers of the first order; those in the second line, natural numbers or numbers of the second order; those in the third line, numbers of the third order, and so on. It is easily shewn that the mth number in the nth row is $(m + n - 2)!/(m - 1)!(n - 1)!$

Pascal's arithmetical triangle, to any required order, is got by drawing a diagonal downwards from right to left as in the figure. The numbers in any diagonal give the coefficients of the expansion of a binomial; for example, the figures in the fifth diagonal, namely, 1, 4, 6, 4, 1, are the coefficients in the expansion $(a + b)^4$. Pascal used the triangle partly for this purpose, and partly to find the numbers of combinations of m things taken n at a time, which he stated, correctly, to be $(n + 1)(n + 2)(n + 3)\ldots m/(m - n)!$

Perhaps as a mathematician Pascal is best known in connection with his correspondence with Fermat in 1654, in which he laid down the principles of the *theory of probabilities*. This correspondence arose from a problem proposed by a gamester, the Chevalier de Méré, to Pascal, who communicated it to Fermat. The problem was this. Two players of equal skill want to leave the table before finishing their game. Their scores and the number of points which constitute the game being given, it is desired to find in what proportion they should divide the stakes. Pascal and Fermat agreed on the answer, but gave different proofs. The following is a translation of Pascal's solution. That of Fermat is given later.

The following is my method for determining the share of each player when, for example, two players play a game of three points and each player has staked 32 pistoles.

Suppose that the first player has gained two points, and the second player one point; they have now to play for a point on this condition, that, if the first player gain, he takes all the money which is at stake, namely, 64 pistoles; while, if the second player gain, each player has two points, so that they are on terms of equality, and, if they leave off playing, each ought to take 32 pistoles. Thus, if the first player gain, then 64 pistoles belong to him, and, if he lose, then 32 pistoles belong to him. If therefore the players do not wish to play this game, but to separate without playing it, the first player would say to the second, "I am certain of 32 pistoles even if I lose this game, and as for the other 32 pistoles perhaps I shall have them and perhaps you will have them; the chances are equal. Let us then divide these 32 pistoles equally, and give me also the 32 pistoles of which I am certain." Thus the first player will have 48 pistoles and the second 16 pistoles.

Next, suppose that the first player has gained two points and the second player none, and that they are about to play for a point; the condition then is that, if the first player gain this point, he secures the game and takes the 64 pistoles, and, if the second player gain this point, then the players will be in the situation already examined, in which the first player is entitled to 48 pistoles and the second to 16 pistoles. Thus, if they do not wish to play, the first player would say to the second, "If I gain the point I gain 64 pistoles; if I lose it, I am entitled to 48 pistoles. Give me then the 48 pistoles of which I am certain, and divide the other 16 equally, since our chances of gaining the point are equal." Thus the first player will have 56 pistoles and the second player 8 pistoles.

Finally, suppose that the first player has gained one point and the second player none. If they proceed to play for a point, the condition is that, if the first player gain it, the players will be in the situation first examined, in which the first player is entitled to 56 pistoles; if the first player lose the point, each player has then a point, and each is entitled to 32 pistoles. Thus, if they do not wish to play, the first player would say to the second, "Give me the 32 pistoles of which I am certain, and divide the remainder of the 56 pistoles equally, that is, divide 24 pistoles equally." Thus the first player will have the sum of 32 and 12 pistoles, that is, 44 pistoles, and consequently the second will have 20 pistoles.

Pascal proceeds next to consider the similar problems when the game is won by whoever first obtains $m + n$ points, and one player has m while the other has n points. The answer is obtained by using the arithmetical triangle. The general solution (in which the skill of the players is unequal) is given in many modern text-books on algebra, and agrees with Pascal's result, though of course the notation of the latter is different and less convenient.

Pascal made an illegitimate use of the new theory in the seventh chapter of his *Pensées*. In effect, he puts his argument that, as the value of eternal happiness must be infinite, then, even if the probability of a religious life ensuring eternal happiness be very small, still the expectation (which is measured by product of the two) must be of sufficient magnitude to make it worth while to be religious. The argument, if worth anything, would apply equally to any religion which promised eternal happiness to those who accepted its doctrines. If any conclusion may be drawn from the statement, it is the undesirability of applying mathematics to questions of morality of which some of the data are necessarily outside the range of an exact science. It is only fair to add that no one had more contempt than Pascal for those who changed their opinions according to the prospect of material benefit, and this isolated passage is at variance with the spirit of his writings.

The last mathematical work of Pascal was that on the *cycloid* in 1658. The cycloid is the curve traced out by a point on the circumference of a circular hoop which rolls along a straight line. Galileo, in 1630, had called attention to this curve, the shape of which is particularly graceful, and had suggested that the arches of bridges should be built in this form.[1] Four years later, in 1634, Roberval found the area of the cycloid; Descartes thought little of this solution and defied him to find its tangents, the same challenge being also sent to Fermat who at once solved the problem. Several questions connected with the curve, and with the surface and volume generated by its revolution about its axis, base, or the tangent at its vertex, were then proposed by various mathematicians. These and some analogous questions, as well as the positions of the centres of the mass of the solids formed, were solved by Pascal in 1658, and the results were issued as a challenge to the world. Wallis succeeded in solving all the questions except those connected with the centre of mass. Pascal's own solutions were effected by the method of indivisibles, and are similar to those which a modern mathematician would give by the aid of the integral calculus. He obtained by summation what are equivalent to the integrals of $\sin \phi$, $\sin^2 \phi$, and $\phi \sin \phi$, one limit being either 0 or $\frac{1}{2}\pi$. He also investigated the geometry of the Archimedean spiral. These researches, according to D'Alembert, form a connecting link between the geometry of Archimedes and the infinitesimal calculus of Newton.

Wallis.[2]   *John Wallis* was born at Ashford on November 22, 1616, and died at Oxford on October 28, 1703. He was educated at Felstead school, and one day in his holidays, when fifteen years old, he happened to see a book of arithmetic in the hands of his brother; struck with curiosity at the odd signs and symbols in it he borrowed the book, and in a fortnight, with his brother's help, had mastered the subject. As it was intended that he should be a doctor, he was sent to Emmanuel College, Cambridge, while there he kept an "act" on the doctrine of the circulation of the blood; that is said to have been the first occasion in Europe on which this theory was publicly maintained in a disputation. His interests, however, centred on mathematics.

He was elected to a fellowship at Queens' College, Cambridge, and subsequently took orders, but on the whole adhered to the Puritan party, to whom he rendered great assistance in deciphering the royalist despatches. He, however, joined the moderate Presbyterians in sign-

---

[1] The bridge, by Essex, across the Cam in the grounds of Trinity College, Cambridge, has cycloidal arches. On the history of the cycloid before Galileo, see S. Günther, *Bibliotheca Mathematica*, 1887, vol. i, pp. 7–14.

[2] See my *History of the Study of Mathematics at Cambridge*, pp. 41–46. An edition of Wallis's mathematical works was published in three volumes at Oxford, 1693–98.

ing the remonstrance against the execution of Charles I., by which he incurred the lasting hostility of the Independents. In spite of their opposition he was appointed in 1649 to the Savilian chair of geometry at Oxford, where he lived until his death on October 28, 1703. Besides his mathematical works he wrote on theology, logic, and philosophy, and was the first to devise a system for teaching deaf-mutes. I confine myself to a few notes on his more important mathematical writings. They are notable partly for the introduction of the use of infinite series as an ordinary part of analysis, and partly for the fact that they revealed and explained to all students the principles of the new methods of analysis introduced by his contemporaries and immediate predecessors.

In 1655 Wallis published a treatise on *conic sections* in which they were defined analytically. I have already mentioned that the *Géométrie* of Descartes is both difficult and obscure, and to many of his contemporaries, to whom the method was new, it must have been incomprehensible. This work did something to make the method intelligible to all mathematicians: it is the earliest book in which these curves are considered and defined as curves of the second degree.

The most important of Wallis's works was his *Arithmetica Infinitorum*, which was published in 1656. In this treatise the methods of analysis of Descartes and Cavalieri were systematised and greatly extended, but their logical exposition is open to criticism. It at once became the standard book on the subject, and is constantly referred to by subsequent writers. It is prefaced by a short tract on conic sections. He commences by proving the law of indices; shews that $x^0, x^{-1}, x^{-2} \ldots$ represents $1, 1/x, 1/x^2 \ldots$; that $x^{1/2}$ represents the square root of x, that $x^{2/3}$ represents the cube root of x^2, and generally that x^{-n} represents the reciprocal of x^n, and that $x^{p/q}$ represents the qth root of x^p.

Leaving the numerous algebraical applications of this discovery he next proceeds to find, by the method of indivisibles, the area enclosed between the curve $y = x^m$, the axis of x, and any ordinate $x = h$; and he proves that the ratio of this area to that of the parallelogram on the same base and of the same altitude is equal to the ratio $1 : m + 1$. He apparently assumed that the same result would be true also for the curve $y = ax^m$, where a is any constant, and m any number positive or negative; but he only discusses the case of the parabola in which $m = 2$, and that of the hyperbola in which $m = -1$: in the latter case his interpretation of the result is incorrect. He then shews that similar results might be written down for any curve of the form $y = \Sigma ax^m$; and hence that, if the ordinate y of a curve can be expanded in powers of the abscissa x, its quadrature can be determined: thus he says that if the equation of a curve were $y = x^0 + x^1 + x^2 + \ldots$, its area would be $x + \frac{1}{2}x^2 + \frac{1}{3}x^3 + \ldots$. He then applies this to the quadrature of the curves $y = (x - x^2)^0$, $y = (x - x^2)^1$, $y = (x - x^2)^2$, $y = (x - x^2)^3$, etc. taken between the limits $x = 0$ and $x = 1$; and shews that the areas are respectively 1, $\frac{1}{6}$, $\frac{1}{30}$, $\frac{1}{140}$, etc. He next considers curves of the form $y = x^{-m}$ and establishes the theorem that the area bounded by the curve, the axis of x, and the ordinate $x = 1$, is to the area of the rectangle on the same base and of the same altitude as $m : m + 1$. This is equivalent to finding the value of $\int_0^1 x^{1/m}\, dx$. He illustrates this by the parabola in which $m = 2$. He states, but does not prove, the corresponding result for a curve of the form $y = x^{p/q}$.

Wallis shewed considerable ingenuity in reducing the equations of curves to the forms given above, but, as he was unacquainted with the binomial theorem, he could not effect the quadrature of the circle, whose equation is $y = (x - x^2)^{1/2}$, since he was unable to expand this in powers of x. He laid down, however, the principle of interpolation. Thus, as the ordinate of the circle $y = (x - x^2)^{1/2}$ is the geometrical mean between the ordinates of the curves $y = (x - x^2)^0$ and $y = (x - x^2)^1$, it might be supposed that, as an approximation, the area of the semicircle $\int_0^1 (x - x^2)^{1/2}\, dx$, which is $\frac{1}{8}\pi$, might be taken as the geometrical mean between the values of

$$\int_0^1 (x - x^2)^0\, dx \quad \text{and} \quad \int_0^1 (x - x^2)^1\, dx,$$

that is, 1 and $\frac{1}{6}$; this is equivalent to taking $4\sqrt{\frac{2}{3}}$ or $3 \cdot 26 \ldots$ as the value of π. But, Wallis argued, we have in fact a series 1, $\frac{1}{6}$, $\frac{1}{30}$, $\frac{1}{140}$, $\ldots$, and therefore the term interpolated between 1 and $\frac{1}{6}$ ought to be so chosen as to obey the law of this series. This, by an elaborate method, which I need not describe in detail, leads to a value for the interpolated term which is equivalent to taking

$$\pi = 2\frac{2 \cdot 2 \cdot 4 \cdot 4 \cdot 6 \cdot 6 \cdot 8 \cdot 8 \ldots}{1 \cdot 3 \cdot 3 \cdot 5 \cdot 5 \cdot 7 \cdot 7 \cdot 9 \ldots}$$

The mathematicians of the seventeenth century constantly used interpolation to obtain results which we should attempt to obtain by direct analysis.

In this work also the formation and properties of continued fractions are discussed, the subject having been brought into prominence by Brouncker's use of these fractions.

A few years later, in 1659, Wallis published a tract containing the solution of the problems on the cycloid which had been proposed by Pascal. In this he incidentally explained how the principles laid down in his *Arithmetica Infinitorum* could be used for the rectification of algebraic curves; and gave a solution of the problem to rectify the semi-cubical parabola $x^3 = ay^2$, which had been discovered in 1657 by his pupil William Neil. Since all attempts to rectify the ellipse and hyperbola had been (necessarily) ineffectual, it had been supposed that no curves could be rectified, as indeed Descartes had definitely asserted to be the case. The logarithmic spiral had been rectified by Torricelli, and was the first curved line (other than the circle) whose length was determined by mathematics, but the extension by Neil and Wallace to an algebraical curve was novel. The cycloid was the next curve rectified; this was done by Wren in 1658.

Early in 1658 a similar discovery, independent of that of Neil, was made by van Heuraët,[1] and this was published by van Schooten in his edition of Descartes's *Geometria* in 1659. Van Heuraët's method is as follows. He supposes the curve to be referred to rectangular axes; if this be so, and if (x, y) be the co-ordinates of any point on it, and n the length of the normal, and if another point whose co-ordinates are (x, η) be taken such that $\eta : h = n : y$ where h is a constant; then, if ds be the element of the length of the required curve, we have by similar triangles $ds : dx = n : y$. Therefore $hds = \eta dx$. Hence, if the area of the locus of the point (x, η) can be found, the first curve can be rectified. In this way van Heuraët effected the rectification of the curve $y^3 = ax^2$; but added that the rectification of the parabola $y^2 = ax$ is impossible since it requires the quadrature of the hyperbola. The solutions given by Neil and Wallis are somewhat similar to that given by van Heuraët, though no general rule is enunciated, and the analysis is clumsy. A third method was suggested by Fermat in 1660, but it is inelegant and laborious.

The theory of the collision of bodies was propounded by the Royal Society in 1668 for the consideration of mathematicians. Wallis, Wren, and Huygens sent correct and similar solutions, all depending on what is now called the conservation of momentum; but, while Wren and Huygens confined their theory to perfectly elastic bodies, Wallis considered also imperfectly elastic bodies. This was followed in 1669 by a work on statics (centres of gravity), and in 1670 by one on dynamics: these

---

[1]On van Heuraët, see the *Bibliotheca Mathematica*, 1887, vol. i, pp. 76–80.

provide a convenient synopsis of what was then known on the subject.

In 1685 Wallis published an *Algebra*, preceded by a historical account of the development of the subject, which contains a great deal of valuable information. The second edition, issued in 1693 and forming the second volume of his *Opera*, was considerably enlarged. This algebra is noteworthy as containing the first systematic use of formulae. A given magnitude is here represented by the numerical ratio which it bears to the unit of the same kind of magnitude: thus, when Wallis wants to compare two lengths he regards each as containing so many units of length. This perhaps will be made clearer if I say that the relation between the space described in any time by a particle moving with a uniform velocity would be denoted by Wallis by the formula $s = vt$, where s is the number representing the ratio of the space described to the unit of length; while previous writers would have denoted the same relation by stating what is equivalent[1] to the proposition $s_1 : s_2 = v_1 t_1 : v_2 t_2$. It is curious to note that Wallis rejected as absurd the now usual idea of a negative number as being less than nothing, but accepted the view that it is something greater than infinity. The latter opinion may be tenable and not inconsistent with the former, but it is hardly a more simple one.

Fermat.[2]   While Descartes was laying the foundations of analytical geometry, the same subject was occupying the attention of another and not less distinguished Frenchman. This was Fermat. *Pierre de Fermat*, who was born near Montauban in 1601, and died at Castres on January 12, 1665, was the son of a leather-merchant; he was educated at home; in 1631 he obtained the post of councillor for the local parliament at Toulouse, and he discharged the duties of the office with scrupulous accuracy and fidelity. There, devoting most of his leisure to mathematics, he spent the remainder of his life—a life which, but for a somewhat acrimonious dispute with Descartes on the validity of certain analysis used by the latter, was unruffled by any event which calls for special notice. The dispute was chiefly due to the obscurity of Descartes, but the tact and courtesy of Fermat brought it to a friendly conclusion. Fermat was a good scholar, and amused himself by conjec-

---

[1]See *ex. gr.* Newton's *Principia*, bk. i, sect. i, lemma 10 or 11.

[2]The best edition of Fermat's works is that in three volumes, edited by S. P. Tannery and C. Henry, and published by the French government, 1891–6. Of earlier editions, I may mention one of his papers and correspondence, printed at Toulouse in two volumes, 1670 and 1679: of which a summary, with notes, was published by E. Brassinne at Toulouse in 1853, and a reprint was issued at Berlin in 1861.

turally restoring the work of Apollonius on plane loci.

Except a few isolated papers, Fermat published nothing in his life-time, and gave no systematic exposition of his methods. Some of the most striking of his results were found after his death on loose sheets of paper or written in the margins of works which he had read and annotated, and are unaccompanied by any proof. It is thus somewhat difficult to estimate the character of his investigations. He was con-stitutionally modest and retiring, and does not seem to have intended his papers to be published. It is probable that he revised his notes as occasion required, and that his published works represent the final form of his researches, and therefore cannot be dated much earlier than 1660. I shall consider separately (i) his investigations in the theory of numbers; (ii) his use in geometry of analysis and of infinitesimals; and (iii) his method of treating questions of probability.

(i) *The theory of numbers* appears to have been the favourite study of Fermat. He prepared an edition of Diophantus, and the notes and comments thereon contain numerous theorems of considerable elegance. Most of the proofs of Fermat are lost, and it is possible that some of them were not rigorous—an induction by analogy and the intuition of genius sufficing to lead him to correct results. The following examples will illustrate these investigations.

(a) If p be a prime and a be prime to p, then $a^{p-1} - 1$ is divisible by p, that is, $a^{p-1} - 1 \equiv 0$ (mod. p). A proof of this, first given by Euler, is well known. A more general theorem is that $a^{\phi(n)} - 1 \equiv 0$ (mod. n) where a is prime to n, and $\phi(n)$ is the number of integers less than n and prime to it.

(b) An odd prime can be expressed as the difference of two square integers in one and only one way. Fermat's proof is as follows. Let n be the prime, and suppose it equal to $x^2 - y^2$, that is, to $(x+y)(x-y)$. Now, by hypothesis, the only integral factors of n are n and unity, hence $x + y = n$ and $x - y = 1$. Solving these equations we get $x = \frac{1}{2}(n+1)$ and $y = \frac{1}{2}(n-1)$.

(c) He gave a proof of the statement made by Diophantus that the sum of the squares of two integers cannot be of the form $4n - 1$; and he added a corollary which I take to mean that it is impossible that the product of a square and a prime of the form $4n - 1$ [even if multiplied by a number prime to the latter], can be either a square or the sum of two squares. For example, 44 is a multiple of 11 (which is of the form $4 \times 3 - 1$) by 4, hence it cannot be expressed as the sum of two squares. He also stated that a number of the form $a^2 + b^2$, where a is prime to

b, cannot be divided by a prime of the form $4n - 1$.

(d) Every prime of the form $4n + 1$ is expressible, and that in one way only, as the sum of two squares. This problem was first solved by Euler, who shewed that a number of the form $2^m(4n+1)$ can be always expressed as the sum of two squares.

(e) If a, b, c, be integers, such that $a^2 + b^2 = c^2$, then ab cannot be a square. Lagrange gave a solution of this.

(f) The determination of a number x such that $x^2n + 1$ may be a square, where n is a given integer which is not a square. Lagrange gave a solution of this.

(g) There is only one integral solution of the equation $x^2 + 2 = y^3$; and there are only two integral solutions of the equation $x^2 + 4 = y^3$. The required solutions are evidently for the first equation $x = 5$, and for the second equation $x = 2$ and $x = 11$. This question was issued as a challenge to the English mathematicians Wallis and Digby.

(h) No integral values of x, y, z can be found to satisfy the equation $x^n + y^n = z^n$, if n be an integer greater than 2. This proposition[1] has acquired extraordinary celebrity from the fact that no general demonstration of it has been given, but there is no reason to doubt that it is true.

Probably Fermat discovered its truth first for the case $n = 3$, and then for the case $n = 4$. His proof for the former of these cases is lost, but that for the latter is extant, and a similar proof for the case of $n = 3$ can be given. These proofs depend upon shewing that, if three integral values of x, y, z can be found which satisfy the equation, then it will be possible to find three other and smaller integers which also satisfy it: in this way, finally, we shew that the equation must be satisfied by three values which obviously do not satisfy it. Thus no integral solution is possible. This method is inapplicable to the general case.

Fermat's discovery of the general theorem was made later. A proof can be given on the assumption that a number can be resolved into the product of powers of primes in one and only one way. It is possible that Fermat's argument rested on some such supposition, but this is an unsupported conjecture. The assumption is true of real integers, but is not necessarily true for algebraic integers—an algebraic integer being defined as a root of an algebraic equation $x^n + a_1 x^{n-1} + \ldots +$

---

[1] On this curious proposition, see L. J. Mordell, *Fermat's Last Theorem*, Cambridge, 1921; L. E. Dickson, *History of the Theory of Numbers*, vol. ii chap. 26, Washington, 1920.

$a_n = 0$, whose coefficients, a, are arithmetical integers; for instance, $a + b\sqrt{-m}$, where a, b, and m are arithmetical integers, is an algebraic integer. Thus, admitting the use of these generalised integers, 21 can be expressed in three ways as the product of primes, namely, of 3 and 7, or of $4 + \sqrt{-5}$ and $4 - \sqrt{-5}$, or of $1 + 2\sqrt{-5}$ and $1 - 2\sqrt{-5}$; and similarly there are values of n for which Fermat's equation leads to expressions which can be factorised in more than one way.

In 1823 Legendre obtained a proof for the case of $n = 5$; in 1832 Dirichlet gave one for $n = 14$, and in 1840 Lamé and Lebesgue gave proofs for $n = 7$. In 1849 Kummer proved the truth of the theorem for all numbers which satisfy certain Bernoullian conditions. The only numbers less than 100 which do not do so are $37, 59, 67$, and the theorem can, by other arguments, be proved for these three cases. It has now been extended far beyond 100, and no exception to it has yet been found. The following extracts, from a letter now in the university library at Leyden, will give an idea of Fermat's methods; the letter is undated, but it would appear that, at the time Fermat wrote it, he had proved the proposition (h) above only for the case when $n = 3$.

Je ne m'en servis au commencement que pour demontrer les propositions negatives, comme par exemple, qu'il n'y a aucū nombre moindre de l'unité qu'un multiple de 3 qui soit composé d'un quarré et du triple d'un autre quarré. Qu'il n'y a aucun triangle rectangle de nombres dont l'aire soit un nombre quarré. La preuve se fait par ἀπαγωγὴν τὴν εἰς ἀδύνατον en cette manière. S'il y auoit aucun triangle rectangle en nombres entiers, qui eust son aire esgale à un quarré, il y auroit un autre triangle moindre que celuy la qui auroit la mesme proprieté. S'il y en auoit un second moindre que le premier qui eust la mesme proprieté il y en auroit par un pareil raisonnement un troisieme moindre que ce second qui auroit la mesme proprieté et enfin un quatrieme, un cinquieme etc. a l'infini en descendant. Or est il qu'estant donné un nombre il n'y en a point infinis en descendant moindres que celuy la, j'entens parler tousjours des nombres entiers. D'ou on conclud qu'il est donc impossible qu'il y ait aucun triangle rectangle dont l'aire soit quarré. Vide foliū post sequens....

Je fus longtemps sans pouuoir appliquer ma methode aux questions affirmatiues, parce que le tour et le biais pour y venir est beaucoup plus malaisé que celuy dont je me sers aux negatives. De sorte que lors qu'il me falut demonstrer que tout nombre premier qui surpasse de l'unité un multiple de 4, est composé de deux quarrez je me treuuay en belle peine. Mais enfin une meditation diverses fois reiterée me donna les lumieres qui me manquoient. Et les questions affirmatiues passerent par ma methode a l'ayde de quelques nouueaux principes qu'il y fallust joindre par necessité. Ce progres de mon raisonnement en ces questions affirmatives estoit tel. Si un nombre premier pris a discretion qui surpasse de l'unité un multiple de 4 n'est point composé de deux quarrez il y aura un nombre premier de mesme nature moindre que le donné; et ensuite un troisieme encore moindre, etc. en descendant a l'infini jusques a ce que uous arriviez au nombre 5, qui est le moindre de tous ceux de cette nature, lequel il s'en suivroit n'estre pas composé de deux quarrez, ce qu'il est pourtant d'ou on doit inferer par la deduction a l'impossible que tous ceux de cette nature sont par consequent composez de 2 quarrez.

Il y a infinies questions de cette espece. Mais il y en a quelques autres qui demandent de nouveaux principes pour y appliquer la descente, et la recherche en est quelques fois si mal aisée, qu'on n'y peut venir qu'auec une peine extreme. Telle est la question suiuante que Bachet sur Diophante avoüe n'avoir jamais peu demonstrer, sur le suject de laquelle M^r. Descartes fait dans une de ses lettres la mesme declaration, jusques la qu'il confesse qu'il la juge si difficile, qu'il ne voit point de voye pour la resoudre. Tout nombre est quarré, ou composé de deux, de trois, ou de quatre quarréz. Je l'ay enfin rangée sous ma methode et je demonstre que si un nombre donné n'estoit point de cette nature il y en auroit un moindre qui ne le seroit pas non plus, puis un troisieme moindre que le second &c. a l'infini, d'ou l'on infere que tous les nombres sont de cette nature....

J'ay ensuite consideré certaines questions qui bien que negatives ne restent pas de receuoir tres-grande difficulté, la methode pour y pratiquer la descente estant tout a fait diuerse des precedentes comme il sera aisé d'esprouuer. Telles sont les suiuantes. Il n'y a aucun cube diuisible en deux cubes. Il n'y a qu'un seul quarré en entiers qui augmenté du binaire fasse un cube, ledit quarré est 25. Il n'y a que deux quarrez en entiers lesquels augmentés de 4 fassent cube, lesdits quarrez sont 4 et 121....

Apres auoir couru toutes ces questions la plupart de diuerses (*sic*) nature et de differente façon de demonstrer, j'ay passé a l'inuention des regles generales pour resoudre les equations simples et doubles de Diophante. On propose par exemple 2 quarr. $+7957$ esgaux a un quarré (hoc est $2xx + 7967 \propto$ quadr.) J'ay une regle generale pour resoudre cette equation si elle est possible, ou decouvrir son impossibilité. Et ainsi en tons les cas et en tous nombres tant des quarrez que des unitez. On propose cette equation double $2x + 3$ et $3x + 5$ esgaux chaucon a un quarré. Bachet se glorifie en ses commentaires sur Diophante d'auoir trouvé une regle en deux cas particuliers. Je la donne generale en toute sorte de cas. Et determine par regle si elle est possible ou non....

Voila sommairement le conte de mes recherches sur le suject des nombres. Je ne l'ay escrit que parce que j'apprehende que le loisir d'estendre et de mettre au long toutes ces demonstrations et ces méthodes me manquera. En tout cas cette indication seruira aux sçauants pour trouver d'eux mesmes ce que je n'estens point, principalement si M^r. de Carcaui et Frénicle leur font part de quelques demonstrations par la descente que je leur ay enuoyees sur le suject de quelques propositions negatiues. Et peut estre la posterité me scaura gré de luy avoir fait connoistre que les anciens n'ont pas tout sceu, et cette relation pourra passer dans l'esprit de ceux qui viendront apres moy pour traditio lampadis ad filios, comme parle le grand Chancelier d'Angleterre, suiuant le sentiment et la deuise duquel j'adjousteray, multi pertransibunt et augebitur scientia.

(ii) I next proceed to mention Fermat's *use in geometry of analysis and of infinitesimals*. It would seem from his correspondence that he had thought out the principles of analytical geometry for himself before reading Descartes's *Géométrie*, and had realised that from the equation, or, as he calls it, the "specific property," of a curve all its properties could be deduced. His extant papers on geometry deal, however, mainly with the application of infinitesimals to the determination of the tangents to curves, to the quadrature of curves, and to questions of maxima and minima; probably these papers are a revision of his original manuscripts (which he destroyed), and were written about 1663, but there is no doubt that he was in possession of the general idea of his method for finding maxima and minima as early as 1628 or 1629.

He obtained the subtangent to the ellipse, cycloid, cissoid, conchoid, and quadratrix by making the ordinates of the curve and a straight line the same for two points whose abscissae were x and $x - e$; but there is nothing to indicate that he was aware that the process was general, and, though in the course of his work he used the principle, it is probable that he never separated it, so to speak, from the symbols of the particular problem he was considering. The first definite statement of the method was due to Barrow,[1] and was published in 1669.

Fermat also obtained the areas of parabolas and hyperbolas of any order, and determined the centres of mass of a few simple laminae and of a paraboloid of revolution. As an example of his method of solving these questions I will quote his solution of the problem to find the area between the parabola $y^3 = px^2$, the axis of x, and the line $x = a$. He says that, if the several ordinates at the points for which x is equal to a, $a(1 - e)$, $a(1 - e)^2$, ... be drawn, then the area will be split into a number of little rectangles whose areas are respectively

$$ae(pa^2)^{1/3}, \ ae(1 - e)\{pa^2(1 - e)^2\}^{1/3}, \ \ldots.$$

[1] See below, pp. 256–257.

The sum of these is $p^{1/3}a^{5/3}e/\{1 - (1 - e)^{5/3}\}$; and by a subsidiary proposition (for he was not acquainted with the binomial theorem) he finds the limit of this, when e vanishes, to be $\frac{3}{5}p^{1/3}a^{5/3}$. The theorems last mentioned were published only after his death; and probably they were not written till after he had read the works of Cavalieri and Wallis.

Kepler had remarked that the values of a function immediately adjacent to and on either side of a maximum (or minimum) value must be equal. Fermat applied this principle to a few examples. Thus, to find the maximum value of $x(a - x)$, his method is essentially equivalent to taking a consecutive value of x, namely $x - e$ where e is very small, and putting $x(a - x) = (x - e)(a - x + e)$. Simplifying, and ultimately putting $e = 0$, we get $x = \frac{1}{2}a$. This value of x makes the given expression a maximum.

(iii) Fermat must share with Pascal the honour of having founded *the theory of probabilities*. I have already mentioned the problem proposed to Pascal, and which he communicated to Fermat, and have there given Pascal's solution. Fermat's solution depends on the theory of combinations, and will be sufficiently illustrated by the following example, the substance of which is taken from a letter dated August 24, 1654, which occurs in the correspondence with Pascal. Fermat discusses the case of two players, A and B, where A wants two points to win and B three points. Then the game will be certainly decided in the course of four trials. Take the letters a and b, and write down all the combinations that can be formed of four letters. These combinations are 16 in number, namely, *aaaa, aaab, aaba, aabb*; *abaa, abab, abba, abbb*; *baaa, baab, baba, babb*; *bbaa, bbab, bbba, bbbb*. Now every combination in which a occurs twice or oftener represents a case favourable to A, and every combination in which b occurs three times or oftener represents a case favourable to B. Thus, on counting them, it will be found that there are 11 cases favourable to A, and 5 cases favourable to B; and, since these cases are all equally likely, A's chance of winning the game is to B's chance as 11 is to 5.

The only other problem on this subject which, as far as I know, attracted the attention of Fermat was also proposed to him by Pascal, and was as follows. A person undertakes to throw a six with a die in eight throws; supposing him to have made three throws without success, what portion of the stake should he be allowed to take on condition of giving up his fourth throw? Fermat's reasoning is as follows. The chance of success is $1/6$, so that he should be allowed to take $1/6$ of the stake on condition of giving up his throw. But, if we wish to estimate the value of the fourth throw before any throw is made, then the first throw is worth 1/6 of the stake; the second is worth 1/6 of what remains, that is, 5/36 of the stake; the third throw is worth 1/6 of what now remains, that is, 25/216 of the stake; the fourth throw is worth 1/6 of what now remains, that is, 125/1296 of the stake.

Fermat does not seem to have carried the matter much further, but his correspondence with Pascal shows that his views on the fundamental principles of the subject were accurate: those of Pascal were not altogether correct.

Fermat's reputation is quite unique in the history of science. The problems on numbers which he had proposed long defied all efforts to solve them, and many of them yielded only to the skill of Euler. One still remains unsolved. This extraordinary achievement has overshadowed his other work, but in fact it is all of the highest order of excellence, and we can only regret that he thought fit to write so little.

Huygens.[1] *Christian Huygens* was born at the Hague on April 14, 1629, and died in the same town on June 8, 1695. He generally wrote his name as Hugens, but I follow the usual custom in spelling it as above: it is also sometimes written as Huyghens. His life was uneventful, and there is little more to record in it than a statement of his various memoirs and researches.

In 1651 he published an essay in which he shewed the fallacy in a system of quadratures proposed by Grégoire de Saint-Vincent, who was well versed in the geometry of the Greeks, but had not grasped the essential points in the more modern methods. This essay was followed by tracts on the quadrature of the conics and the approximate rectification of the circle.

In 1654 his attention was directed to the improvement of the telescope. In conjunction with his brother he devised a new and better way of grinding and polishing lenses. As a result of these improvements he was able during the following two years, 1655 and 1656, to resolve numerous astronomical questions; as, for example, the nature of Saturn's appendage. His astronomical observations required some exact means of measuring time, and he was thus led in 1656 to invent the pendu-

---

[1]A new edition of all Huygens's works and correspondence was issued at the Hague in ten volumes, 1888–1905. An earlier edition of his works was published in six volumes, four at Leyden in 1724, and two at Amsterdam in 1728 (a life by s'Gravesande is prefixed to the first volume): his scientific correspondence was published at the Hague in 1833.

lum clock, as described in his tract *Horologium*, 1658. The time-pieces previously in use had been balance clocks.

In the year 1657 Huygens wrote a small work on the calculus of probabilities founded on the correspondence of Pascal and Fermat. He spent a couple of years in England about this time. His reputation was now so great that in 1665 Louis XIV. offered him a pension if he would live in Paris, which accordingly then became his place of residence.

In 1668 he sent to the Royal Society of London, in answer to a problem they had proposed, a memoir in which (simultaneously with Wallis and Wren) he proved by experiment that the momentum in a certain direction before the collision of two bodies is equal to the momentum in that direction after the collision. This was one of the points in mechanics on which Descartes had been mistaken.

The most important of Huygens's work was his *Horologium Oscillatorium* published at Paris in 1673. The first chapter is devoted to pendulum clocks. The second chapter contains a complete account of the descent of heavy bodies under their own weights in a vacuum, either vertically down or on smooth curves. Amongst other propositions he shews that the cycloid is tautochronous. In the third chapter he defines evolutes and involutes, proves some of their more elementary properties, and illustrates his methods by finding the evolutes of the cycloid and the parabola. These are the earliest instances in which the envelope of a moving line was determined. In the fourth chapter he solves the problem of the compound pendulum, and shews that the centres of oscillation and suspension are interchangeable. In the fifth and last chapter he discusses again the theory of clocks, points out that if the bob of the pendulum were, by means of cycloidal checks, made to oscillate in a cycloid the oscillations would be isochronous; and finishes by shewing that the centrifugal force on a body which moves round a circle of radius r with a uniform velocity v varies directly as v^2 and inversely as r. This work contains the first attempt to apply dynamics to bodies of finite size and not merely to particles.

In 1675 Huygens proposed to regulate the motion of watches by the use of the balance spring, in the theory of which he had been perhaps anticipated in a somewhat ambiguous and incomplete statement made by Hooke in 1658. Watches or portable clocks had been invented early in the sixteenth century, and by the end of that century were not very uncommon, but they were clumsy and unreliable, being driven by a main spring and regulated by a conical pulley and verge escapement; moreover, until 1687 they had only one hand. The first watch whose motion was regulated by a balance spring was made at Paris under Huygens's directions, and presented by him to Louis XIV.

The increasing intolerance of the Catholics led to his return to Holland in 1681, and after the revocation of the edict of Nantes he refused to hold any further communication with France. He now devoted himself to the construction of lenses of enormous focal length: of these three of focal lengths 123 feet, 180 feet, and 210 feet, were subsequently given by him to the Royal Society of London, in whose possession they still remain. It was about this time that he discovered the achromatic eyepiece (for a telescope) which is known by his name. In 1689 he came from Holland to England in order to make the acquaintance of Newton, whose *Principia* had been published in 1687. Huygens fully recognized the intellectual merits of the work, but seems to have deemed any theory incomplete which did not explain gravitation by mechanical means.

On his return in 1690 Huygens published his treatise on *light* in which the undulatory theory was expounded and explained. Most of this had been written as early as 1678. The general idea of the theory had been suggested by Robert Hooke in 1664, but he had not investigated its consequences in any detail. Only three ways have been suggested in which light can be produced mechanically. Either the eye may be supposed to send out something which, so to speak, feels the object (as the Greeks believed); or the object perceived may send out something which hits or affects the eye (as assumed in the emission theory); or there may be some medium between the eye and the object, and the object may cause some change in the form or condition of this intervening medium and thus affect the eye (as Hooke and Huygens supposed in the wave or undulatory theory). According to this last theory space is filled with an extremely rare ether, and light is caused by a series of waves or vibrations in this ether which are set in motion by the pulsations of the luminous body. From this hypothesis Huygens deduced the laws of reflexion and refraction, explained the phenomena of double refraction, and gave a construction for the extraordinary ray in biaxal crystals; while he found by experiment the chief phenomena of polarization.

The immense reputation and unrivalled powers of Newton led to disbelief in a theory which he rejected, and to the general adoption of Newton's emission theory. Within the present century crucial experiments have been devised which give different results according as one or the other theory is adopted; all these experiments agree with the results of the undulatory theory and differ from the results of the Newtonian theory; the latter is therefore untenable. Until, however, the theory of interference, suggested by Young, was worked out by Fresnel, the hypothesis of Huygens failed to account for all the facts, and even now the properties which, under it, have to be attributed to the intervening medium or ether involve difficulties of which we still seek a solution. Hence the problem as to how the effects of light are really produced cannot be said to be finally solved.

Besides these works Huygens took part in most of the controversies and challenges which then played so large a part in the mathematical world, and wrote several minor tracts. In one of these he investigated the form and properties of the catenary. In another he stated in general terms the rule for finding maxima and minima of which Fermat had made use, and shewed that the subtangent of an algebraical curve $f(x, y) = 0$ was equal to yf_y/f_x where f_y is the derived function of $f(x, y)$ regarded as a function of y. In some posthumous works, issued at Leyden in 1703, he further shewed how from the focal lengths of the component lenses the magnifying power of a telescope could be determined; and explained some of the phenomena connected with haloes and parhelia.

I should add that almost all his demonstrations, like those of Newton, are rigidly geometrical, and he would seem to have made no use of the differential or fluxional calculus, though he admitted the validity of the methods used therein. Thus, even when first written, his works were expressed in an archaic language, and perhaps received less attention than their intrinsic merits deserved.

I have now traced the development of mathematics for a period which we may take roughly as dating from 1635 to 1675, under the influence of Descartes, Cavalieri, Pascal, Wallis, Fermat, and Huygens. The life of Newton partly overlaps this period; his works and influence are considered in the next chapter.

I may dismiss **the remaining mathematicians of this time**[1] with comparatively slight notice. The most eminent of them are *Bachet, Barrow, Brouncker, Collins, De la Hire, de Laloubère, Frénicle, James Gregory, Hooke, Hudde, Nicholas Mercator, Mersenne, Pell, Roberval, Roemer, Rolle, Saint-Vincent, Sluze, Torricelli, Tschirnhausen, van Schooten, Viviani,* and *Wren.* In the following notes I have arranged the above-mentioned mathematicians so that as far as possible their

[1]Notes on several of these mathematicians will be found in C. Hutton's *Mathematical Dictionary and Tracts*, 5 volumes, London, 1812–1815.

chief contributions shall come in chronological order.

Bachet.   *Claude Gaspard Bachet de Méziriac* was born at Bourg in 1581, and died in 1638. He wrote the *Problèmes plaisants*, of which the first edition was issued in 1612, a second and enlarged edition was brought out in 1624; this contains an interesting collection of arithmetical tricks and questions, many of which are quoted in my *Mathematical Recreations and Essays*. He also wrote *Les éléments arithmétiques*, which exists in manuscript; and a translation of the *Arithmetic* of Diophantus. Bachet was the earliest writer who discussed the solution of indeterminate equations by means of continued fractions.

Mersenne.   *Marin Mersenne*, born in 1588 and died at Paris in 1648, was a Franciscan friar, who made it his business to be acquainted and correspond with the French mathematicians of that date and many of their foreign contemporaries. In 1634 he published a translation of Galileo's mechanics; in 1644 he issued his *Cogitata Physico-Mathematica*, by which he is best known, containing an account of some experiments in physics; he also wrote a synopsis of mathematics, which was printed in 1664.

In the preface to the *Cogitata* a statement is made about perfect numbers, which implies that the only values of p not greater than 257 which make N prime, where $N = 2^p - 1$, are 1, 2, 3, 5, 7, 13, 17, 19, 31, 67, 127, and 257: all prime values of N are known as Mersenne's Numbers. Some years ago I gave reasons for thinking that 67 was a misprint for 61. Until 1911, no error in this corrected statement was established, and it was gradually verified for all except sixteen values of p. In 1911, however, it was proved that N was prime when $p = 89$, and three years later that it was prime when $p = 107$: two facts at variance with Mersenne's statement. The prime or composite character of N now remains unknown for only ten values of p, namely, 139, 149, 157, 167, 193, 199, 227, 229, 241, and 257. We may safely say that the methods used to-day in establishing the known results for many of the higher values of p could not have been employed by Mersenne. It would be interesting to discover how he reached his conclusions, which are true if p does not exceed 88. Some recent writers conjecture that his statement was the result of a guess, intelligent though erroneous, as to the possible forms of p: I find it difficult to accept this opinion, but further discussion of the problem would be out of place here.

The theory of perfect numbers depends directly on that of Mersenne's Numbers. It is probable that all perfect numbers are included in the formula $2^{p-1}(2^p - 1)$, where $2^p - 1$ is a prime. Euclid proved that any number of this form is perfect; Euler shewed that the formula includes all even perfect numbers; and there is reason to believe—though a rigid demonstration is wanting—that an odd number cannot be perfect. If we assume that the last of these statements is true, then every perfect number is of the above form. Thus, if p = 2, 3, 5, 7, 13, 17, 19, 31, 61, then, by Mersenne's rule, the corresponding values of $2^p - 1$ are prime; they are 3, 7, 31, 127, 8191, 131071, 524287, 2147483647, 2305843009213693951; and the corresponding perfect numbers are 6, 28, 496, 8128, 33550336, 8589869056, 137438691328, 2305843008139952128, and 2658455991569831744654692615953842176.

Roberval.[1] *Gilles Personier (de) Roberval*, born at Roberval in 1602 and died at Paris in 1675, described himself from the place of his birth as de Roberval, a seigniorial title to which he had no right. He discussed the nature of the tangents to curves, solved some of the easier questions connected with the cycloid, generalized Archimedes's theorems on the spiral, wrote on mechanics, and on the method of indivisibles, which he rendered more precise and logical. He was a professor in the university of Paris, and in correspondence with nearly all the leading mathematicians of his time.

Van Schooten. *Frans van Schooten*, to whom we owe an edition of Vieta's works, succeeded his father (who had taught mathematics to Huygens, Hudde, and Sluze) as professor at Leyden in 1646. He brought out in 1659 a Latin translation of Descartes's *Géométrie*, and in 1657 a collection of mathematical exercises in which he recommended the use of co-ordinates in space of three dimensions. He died in 1661.

Saint-Vincent.[2] *Grégoire de Saint-Vincent*, a Jesuit, born at Bruges in 1584 and died at Ghent in 1667, discovered the expansion of $\log(1 + x)$ in ascending powers of x. Although a circle-squarer he is worthy of mention for the numerous theorems of interest which he discovered in his search after the impossible, and Montucla ingeniously remarks that "no one ever squared the circle with so much ability or (except for his principal object) with so much success." He wrote two books on the subject, one published in 1647 and the other in 1668, which cover some two or three thousand closely printed pages; the fallacy in the quadrature was pointed out by Huygens. In the former

---

[1] His chief works are included in the *Divers Ouvrages* by Academicians, Paris, 1693; these were reprinted in the sixth volume of the old mémoires of the Academy of Sciences, Paris, 1730.

[2] See L. A. J. Quetelet's *Histoire des sciences chez les Belges*, Brussels, 1866.

work he used indivisibles. An earlier work entitled *Theoremata Mathematica*, published in 1624, contains a clear account of the method of exhaustions, which is applied to several quadratures, notably that of the hyperbola.

Torricelli.[1]   *Evangelista Torricelli*, born at Faenza on Oct. 15, 1608, and died at Florence in 1647, wrote on the quadrature of the cycloid and conics; the rectification of the logarithmic spiral; the theory of the barometer; the value of gravity found by observing the motion of two weights connected by a string passing over a fixed pulley; the theory of projectiles; and the motion of fluids.

Hudde.   *Johann Hudde*, burgomaster of Amsterdam, was born there in 1633, and died in the same town in 1704. He wrote two tracts in 1659: one was on the reduction of equations which have equal roots; in the other he stated what is equivalent to the proposition that if $f(x, y) = 0$ be the algebraical equation of a curve, then the subtangent is $-y \dfrac{\partial f}{\partial y} \Big/ \dfrac{\partial f}{\partial x}$; but being ignorant of the notation of the calculus his enunciation is involved.

Frénicle.[2]   *Bernard Frénicle de Bessy*, born in Paris circ. 1605 and died in 1670, wrote numerous papers on combinations and on the theory of numbers, also on magic squares. It may be interesting to add that he challenged Huygens to solve the following system of equations in integers, $x^2 + y^2 = z^2$, $x^2 = u^2 + v^2$, $x - y = u - v$. A solution was given by M. Pépin in 1880.

De Laloubère.   *Antoine de Laloubère*, a Jesuit, born in Languedoc in 1600 and died at Toulouse in 1664, is chiefly celebrated for an incorrect solution of Pascal's problems on the cycloid, which he gave in 1660, but he has a better claim to distinction in having been the first mathematician to study the properties of the helix.

N. Mercator.   *Nicholas Mercator* (sometimes known as *Kauffmann*) was born in Holstein about 1620, but resided most of his life in England. He went to France in 1683, where he designed and constructed the fountains at Versailles, but the payment agreed on was refused unless he would turn Catholic; he died of vexation and poverty in Paris in 1687. He wrote a treatise on logarithms entitled *Logarithmo-technica*,

---

[1] Torricelli's mathematical writings were published at Florence in 1644, under the title *Opera Geometrica*; see also a memoir by G. Loria, *Bibliotheca mathematica*, series 3, vol. i, pp. 75–89, Leipzig, 1900.

[2] Frénicle's miscellaneous works, edited by De la Hire, were published in the *Mémoires de l'Académie*, vol. v, 1691.

published in 1668, and discovered the series

$$\log(1 + x) = x - \frac{1}{2}x^2 + \frac{1}{3}x^3 - \frac{1}{4}x^4 + \ldots;$$

he proved this by writing the equation of a hyperbola in the form

$$y = \frac{1}{1 + x} = 1 - x + x^2 - x^3 + \ldots,$$

to which Wallis's method of quadrature could be applied. The same series had been independently discovered by Saint-Vincent.

Barrow.[1]   *Isaac Barrow* was born in London in 1630, and died at Cambridge in 1677. He went to school first at Charterhouse (where he was so troublesome that his father was heard to pray that if it pleased God to take any of his children he could best spare Isaac), and subsequently to Felstead. He completed his education at Trinity College, Cambridge; after taking his degree in 1648, he was elected to a fellowship in 1649; he then resided for a few years in college, but in 1655 he was driven out by the persecution of the Independents. He spent the next four years in the East of Europe, and after many adventures returned to England in 1659. He was ordained the next year, and appointed to the professorship of Greek at Cambridge. In 1662 he was made professor of geometry at Gresham College, and in 1663 was selected as the first occupier of the Lucasian chair at Cambridge. He resigned the latter to his pupil Newton in 1669, whose superior abilities he recognized and frankly acknowledged. For the remainder of his life he devoted himself to the study of divinity. He was appointed master of Trinity College in 1672, and held the post until his death.

He is described as "low in stature, lean, and of a pale complexion," slovenly in his dress, and an inveterate smoker. He was noted for his strength and courage, and once when travelling in the East he saved the ship by his own prowess from capture by pirates. A ready and caustic wit made him a favourite of Charles II., and induced the courtiers to respect even if they did not appreciate him. He wrote with a sustained and somewhat stately eloquence, and with his blameless life and scrupulous conscientiousness was an impressive personage of the time.

His earliest work was a complete edition of the *Elements* of Euclid, which he issued in Latin in 1655, and in English in 1660; in 1657 he published an edition of the *Data*. His lectures, delivered in 1664, 1665, and

---

[1] Barrow's mathematical works, edited by W. Whewell, were issued at Cambridge in 1860. On *Barrow's Geometry*, see J. M. Child, Chicago, 1916.

1666, were published in 1683 under the title *Lectiones Mathematicae*; these are mostly on the metaphysical basis for mathematical truths. His lectures for 1667 were published in the same year, and suggest the analysis by which Archimedes was led to his chief results. In 1669 he issued his *Lectiones Opticae et Geometricae*. It is said in the preface that Newton revised and corrected these lectures, adding matter of his own, but it seems probable from Newton's remarks in the fluxional controversy that the additions were confined to the parts which dealt with optics. This, which is his most important work in mathematics, was republished with a few minor alterations in 1674. In 1675 he published an edition with numerous comments of the first four books of the *Conics* of Apollonius, and of the extant works of Archimedes and Theodosius.

In the optical lectures many problems connected with the reflexion and refraction of light are treated with ingenuity. The geometrical focus of a point seen by reflexion or refraction is defined; and it is explained that the image of an object is the locus of the geometrical foci of every point on it. Barrow also worked out a few of the easier properties of thin lenses, and considerably simplified the Cartesian explanation of the rainbow.

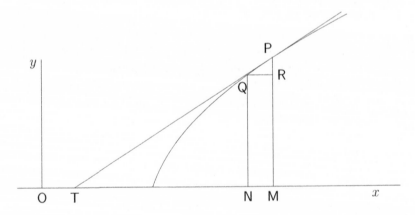

The geometrical lectures contain some new ways of determining the areas and tangents of curves. The most celebrated of these is the method given for the determination of tangents to curves, and this is sufficiently important to require a detailed notice, because it illustrates the way in which Barrow, Hudde, and Sluze were working on the lines suggested by Fermat towards the methods of the differential calculus. Fermat had observed that the tangent at a point P on a curve was determined if one other point besides P on it were known; hence, if the length of the subtangent MT could be found (thus determining the point T), then the line TP would be the required tangent. Now Barrow remarked that if the abscissa and ordinate at a point Q adjacent to P were drawn, he got a small triangle PQR (which he called the differential triangle, because its sides PR and PQ were the differences of the abscissae and ordinates of P and Q), so that

$$TM : MP = QR : RP.$$

To find $QR : RP$ he supposed that x, y were the co-ordinates of P, and $x - e, y - a$ those of Q (Barrow actually used p for x and m for y, but I alter these to agree with the modern practice). Substituting the co-ordinates of Q in the equation of the curve, and neglecting the squares and higher powers of e and a as compared with their first powers, he obtained $e : a$. The ratio a/e was subsequently (in accordance with a suggestion made by Sluze) termed the angular coefficient of the tangent at the point.

Barrow applied this method to the curves (i) $x^2(x^2 + y^2) = r^2y^2$; (ii) $x^3 + y^3 = r^3$; (iii) $x^3 + y^3 = rxy$, called *la galande*; (iv) $y = (r - x)\tan \pi x/2r$, the *quadratrix*; and (v) $y = r\tan \pi x/2r$. It will be sufficient here if I take as an illustration the simpler case of the parabola $y^2 = px$. Using the notation given above, we have for the point P, $y^2 = px$; and for the point Q, $(y - a)^2 = p(x - e)$. Subtracting we get $2ay - a^2 = pe$. But, if a be an infinitesimal quantity, a^2 must be infinitely smaller and therefore may be neglected when compared with the quantities $2ay$ and pe. Hence $2ay = pe$, that is, $e : a = 2y : p$. Therefore $TM : y = e : a = 2y : p$. Hence $TM = 2y^2/p = 2x$. This is exactly the procedure of the differential calculus, except that there we have a rule by which we can get the ratio a/e or dy/dx directly without the labour of going through a calculation similar to the above for every separate case.

Brouncker.   *William, Viscount Brouncker*, one of the founders of the Royal Society of London, born about 1620, and died on April 5, 1684, was among the most brilliant mathematicians of this time, and was in intimate relations with Wallis, Fermat, and other leading mathematicians. I mentioned above his curious reproduction of Brahmagupta's solution of a certain indeterminate equation. Brouncker proved that the area enclosed between the equilateral hyperbola $xy = 1$, the axis of x, and the ordinates $x = 1$ and $x = 2$, is equal either to

$$\frac{1}{1.2} + \frac{1}{3.4} + \frac{1}{5.6} + \dots, \text{ or to } 1 - \frac{1}{2} + \frac{1}{3} - \frac{1}{4} + \dots$$

He also worked out other similar expressions for different areas bounded by the hyberbola and straight lines. He wrote on the rectification of the parabola and of the cycloid.[1] It is noticeable that he used infinite series to express quantities whose values he could not otherwise determine. In answer to a request of Wallis to attempt the quadrature of the circle he shewed that the ratio of the area of a circle to the area of the circumscribed square, that is, the ratio of 4 to π, is equal to

$$\frac{1}{1} + \frac{1^2}{2} + \frac{3^2}{2} + \frac{5^2}{2} + \frac{7^2}{2} + \dots$$

Continued fractions[2] had been employed by Bombelli in 1572, and had been systematically used by Cataldi in his treatise on finding the square roots of numbers, published at Bologna in 1613. Their properties and theory were given by Huygens, 1703, and Euler, 1744.

James Gregory.   *James Gregory*, born at Drumoak near Aberdeen in 1638, and died at Edinburgh in October 1675, was successively professor at St. Andrews and Edinburgh. In 1660 he published his *Optica Promota*, in which the reflecting telescope known by his name is described. In 1667 he issued his *Vera Circuli et Hyperbolae Quadratura* in which he shewed how the areas of the circle and hyperbola could be obtained in the form of infinite convergent series, and here (I believe for the first time) we find a distinction drawn between convergent and divergent series. This work contains a remarkable geometrical proposition to the effect that the ratio of the area of any arbitrary sector of a circle to that of the inscribed or circumscribed regular polygons is not expressible by a finite number of algebraical terms. Hence he inferred that the quadrature of a circle was impossible; this was accepted by Montucla, but it is not conclusive, for it is conceivable that some particular sector might be squared, and this particular sector might be the whole circle. This book contains also the earliest enunciation of

---

[1] On these investigations, see his papers in the *Philosophical Transactions*, London, 1668, 1672, 1673, and 1678.

[2] On the history of continued fractions, see papers by S. Günther and A. Favaro in Boncompagni's *Bulletino di bibliografia*, Rome, 1874, vol. vii, pp. 213, 451, 533; and Cantor, vol. ii, pp. 622, 762, 766. Bombelli used them in 1572; but Cataldi introduced the usual notation for them.

the expansions in series of $\sin x$, $\cos x$, $\sin^{-1} x$ or arc $\sin x$, and $\cos^{-1} x$
or arc $\cos x$. It was reprinted in 1668 with an appendix, *Geometriae
Pars*, in which Gregory explained how the volumes of solids of revolution could be determined. In 1671, or perhaps earlier, he established
the theorem that

$$\theta = \tan \theta - \frac{1}{3} \tan^3 \theta + \frac{1}{5} \tan^5 \theta - \ldots,$$

the result being true only if θ lie between $-\frac{1}{4}\pi$ and $\frac{1}{4}\pi$. This is the theorem on which many of the subsequent calculations of approximations
to the numeral value of π have been based.

Wren.   *Sir Christopher Wren* was born at Knoyle, Wiltshire, on
October 20, 1632, and died in London on February 25, 1723. Wren's
reputation as a mathematician has been overshadowed by his fame as
an architect, but he was Savilian professor of astronomy at Oxford
from 1661 to 1673, and for some time president of the Royal Society.
Together with Wallis and Huygens he investigated the laws of collision
of bodies; he also discovered the two systems of generating lines on
the hyperboloid of one sheet, though it is probable that he confined
his attention to a hyperboloid of revolution.[1] Besides these he wrote
papers on the resistance of fluids, and the motion of the pendulum. He
was a friend of Newton and (like Huygens, Hooke, Halley, and others)
had made attempts to shew that the force under which the planets
move varies inversely as the square of the distance from the sun.

Wallis, Brouncker, Wren, and Boyle (the last-named being a chemist
and physicist rather than a mathematician) were the leading philosophers who founded the Royal Society of London. The society arose
from the self-styled "indivisible college" in London in 1645; most of its
members moved to Oxford during the civil war, where Hooke, who was
then an assistant in Boyle's laboratory, joined in their meetings; the
society was formally constituted in London in 1660, and was incorporated on July 15, 1662. The French Academy was founded in 1666, and
the Berlin Academy in 1700. The Accademia dei Lincei was founded
in 1603, but was dissolved in 1630.

Hooke.   *Robert Hooke*, born at Freshwater on July 18, 1635, and
died in London on March 3, 1703, was educated at Westminster, and
Christ Church, Oxford, and in 1665 became professor of geometry at
Gresham College, a post which he occupied till his death. He is still

---

[1]See the *Philosophical Transactions* London, 1669.

known by the law which he discovered, that the tension exerted by a stretched string is (within certain limits) proportional to the extension, or, in other words, that the stress is proportional to the strain. He invented and discussed the conical pendulum, and was the first to state explicitly that the motions of the heavenly bodies were merely dynamical problems. He was as jealous as he was vain and irritable, and accused both Newton and Huygens of unfairly appropriating his results. Like Huygens, Wren, and Halley, he made efforts to find the law of force under which the planets move about the sun, and he believed the law to be that of the inverse square of the distance. He, like Huygens, discovered that the small oscillations of a coiled spiral spring were practically isochronous, and was thus led to recommend (possibly in 1658) the use of the balance spring in watches. He had a watch of this kind made in London in 1675; it was finished just three months later than a similar one made in Paris under the directions of Huygens.

Collins.　*John Collins*, born near Oxford on March 5, 1625, and died in London on November 10, 1683, was a man of great natural ability, but of slight education. Being devoted to mathematics, he spent his spare time in correspondence with the leading mathematicians of the time, for whom he was always ready to do anything in his power, and he has been described—not inaptly—as the English Mersenne. To him we are indebted for much information on the details of the discoveries of the period.[1]

Pell.　Another mathematician who devoted a considerable part of his time to making known the discoveries of others, and to correspondence with leading mathematicians, was *John Pell*. Pell was born in Sussex on March 1, 1610, and died in London on December 10, 1685. He was educated at Trinity College, Cambridge; he occupied in succession the mathematical chairs at Amsterdam and Breda; he then entered the English diplomatic service; but finally settled in 1661 in London, where he spent the last twenty years of his life. His chief works were an edition, with considerable new matter, of the *Algebra* by Branker and Rhonius, London, 1668; and a table of square numbers, London, 1672.

Sluze.　*René François Walther de Sluze (Slusius)*, canon of Liége, born on July 7, 1622, and died on March 19, 1685, found for the subtangent of a curve $f(x, y) = 0$ an expression which is equivalent to

---

[1] See the *Commercium Epistolicum*, and S. P. Rigaud's *Correspondence of Scientific Men of the Seventeenth Century*, Oxford, 1841.

$-y\dfrac{\partial f}{\partial y}\Big/\dfrac{\partial f}{\partial x}$; he wrote numerous tracts,[1] and in particular discussed at some length spirals and points of inflexion.

Viviani. *Vincenzo Viviani*, a pupil of Galileo and Torricelli, born at Florence on April 5, 1622, and died there on September 22, 1703, brought out in 1659 a restoration of the lost book of Apollonius on conic sections, and in 1701 a restoration of the work of Aristaeus. He explained in 1677 how an angle could be trisected by the aid of the equilateral hyperbola or the conchoid. In 1692 he proposed the problem to construct four windows in a hemispherical vault so that the remainder of the surface can be accurately determined; a celebrated problem, of which analytical solutions were given by Wallis, Leibnitz, David Gregory, and James Bernoulli.

Tschirnhausen. *Ehrenfried Walther von Tschirnhausen* was born at Kislingswalde on April 10, 1631, and died at Dresden on October 11, 1708. In 1682 he worked out the theory of caustics by reflexion, or, as they were usually called, catacaustics, and shewed that they were rectifiable. This was the second case in which the envelope of a moving line was determined. He constructed burning mirrors of great power. The transformation by which he removed certain intermediate terms from a given algebraical equation is well known; it was published in the *Acta Eruditorum* for 1683.

De la Hire. *Philippe De la Hire* (or *Lahire*), born in Paris on March 18, 1640, and died there on April 21, 1719, wrote on graphical methods, 1673; on the conic sections, 1685; a treatise on epicycloids, 1694; one on roulettes, 1702; and, lastly, another on conchoids, 1708. His works on conic sections and epicycloids were founded on the teaching of Desargues, whose favourite pupil he was. He also translated the essay of Moschopulus on magic squares, and collected many of the theorems on them which were previously known; this was published in 1705.

Roemer. *Olof Roemer*, born at Aarhuus on September 25, 1644, and died at Copenhagen on September 19, 1710, was the first to measure the velocity of light; this was done in 1675 by means of the eclipses of Jupiter's satellites. He brought the transit and mural circle into common use, the altazimuth having been previously generally employed, and it was on his recommendation that astronomical observations of

---

[1] Some of his papers were published by Le Paige in vol. xvii of Boncompagni's *Bulletino di bibliografia*, Rome, 1884.

stars were subsequently made in general on the meridian. He was also the first to introduce micrometers and reading microscopes into an observatory. He also deduced from the properties of epicycloids the form of the teeth in toothed-wheels best fitted to secure a uniform motion.

Rolle.   *Michel Rolle*, born at Ambert on April 21, 1652, and died in Paris on November 8, 1719, wrote an algebra in 1689, which contains the theorem on the position of the roots of an equation which is known by his name. He published in 1696 a treatise on the solutions of equations, whether determinate or indeterminate, and he produced several other minor works. He taught that the differential calculus, which, as we shall see later, had been introduced towards the close of the seventeenth century, was nothing but a collection of ingenious fallacies.

CHAPTER XVI.

THE LIFE AND WORKS OF NEWTON.[1]

THE mathematicians considered in the last chapter commenced the creation of those processes which distinguish modern mathematics. The extraordinary abilities of Newton enabled him within a few years to perfect the more elementary of those processes, and to distinctly advance every branch of mathematical science then studied, as well as to create some new subjects. Newton was the contemporary and friend of Wallis, Huygens, and others of those mentioned in the last chapter, but though most of his mathematical work was done between the years 1665 and 1686, the bulk of it was not printed—at any rate in book-form—till some years later.

I propose to discuss the works of Newton more fully than those of other mathematicians, partly because of the intrinsic importance of his discoveries, and partly because this book is mainly intended for English readers, and the development of mathematics in Great Britain was for a century entirely in the hands of the Newtonian school.

Isaac Newton was born in Lincolnshire, near Grantham, on December 25, 1642, and died at Kensington, London, on March 20, 1727. He was educated at Trinity College, Cambridge, and lived there from 1661 till 1696, during which time he produced the bulk of his work in mathematics; in 1696 he was appointed to a valuable Government office, and moved to London, where he resided till his death.

[1]Newton's life and works are discussed in *The Memoirs of Newton*, by D. Brewster, 2 volumes, Edinburgh, second edition, 1860. An edition of most of Newton's works was published by S. Horsley in 5 volumes, London, 1779–1785; and a bibliography of them was issued by G. J. Gray, Cambridge, second edition, 1907; see also the catalogue of the Portsmouth Collection of Newton's papers, Cambridge, 1888. My *Essay on the Genesis, Contents, and History of Newton's Principia*, London, 1893, may be also consulted.

His father, who had died shortly before Newton was born, was a yeoman farmer, and it was intended that Newton should carry on the paternal farm. He was sent to school at Grantham, where his learning and mechanical proficiency excited some attention. In 1656 he returned home to learn the business of a farmer, but spent most of his time solving problems, making experiments, or devising mechanical models; his mother noticing this, sensibly resolved to find some more congenial occupation for him, and his uncle, having been himself educated at Trinity College, Cambridge, recommended that he should be sent there.

In 1661 Newton accordingly entered as a student at Cambridge, where for the first time he found himself among surroundings which were likely to develop his powers. He seems, however, to have had but little interest for general society or for any pursuits save science and mathematics. Luckily he kept a diary, and we can thus form a fair idea of the course of education of the most advanced students at an English university at that time. He had not read any mathematics before coming into residence, but was acquainted with Sanderson's *Logic*, which was then frequently read as preliminary to mathematics. At the beginning of his first October term he happened to stroll down to Stourbridge Fair, and there picked up a book on astrology, but could not understand it on account of the geometry and trigonometry. He therefore bought a Euclid, and was surprised to find how obvious the propositions seemed. He thereupon read Oughtred's *Clavis* and Descartes's *Géométrie*, the latter of which he managed to master by himself, though with some difficulty. The interest he felt in the subject led him to take up mathematics rather than chemistry as a serious study. His subsequent mathematical reading as an undergraduate was founded on Kepler's *Optics*, the works of Vieta, van Schooten's *Miscellanies*, Descartes's *Géométrie*, and Wallis's *Arithmetica Infinitorum*: he also attended Barrow's lectures. At a later time, on reading Euclid more carefully, he formed a high opinion of it as an instrument of education, and he used to express his regret that he had not applied himself to geometry before proceeding to algebraic analysis.

There is a manuscript of his, dated May 28, 1665, written in the same year as that in which he took his B.A. degree, which is the earliest documentary proof of his invention of fluxions. It was about the same time that he discovered the binomial theorem.[1]

On account of the plague the college was sent down during parts

---

[1]See below, pp. 269, 281.

of the year 1665 and 1666, and for several months at this time Newton lived at home. This period was crowded with brilliant discoveries.
He thought out the fundamental principles of his theory of gravitation, namely, that every particle of matter attracts every other particle, and he suspected that the attraction varied as the product of their
masses and inversely as the square of the distance between them. He
also worked out the fluxional calculus tolerably completely: thus in a
manuscript dated November 13, 1665, he used fluxions to find the tangent and the radius of curvature at any point on a curve, and in October
1666 he applied them to several problems in the theory of equations.
Newton communicated these results to his friends and pupils from and
after 1669, but they were not published in print till many years later.
It was also whilst staying at home at this time that he devised some
instruments for grinding lenses to particular forms other than spherical,
and perhaps he decomposed solar light into different colours.

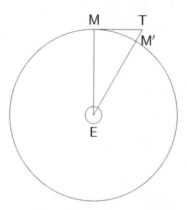

Leaving out details and taking round numbers only, his reasoning at
this time on the theory of gravitation seems to have been as follows. He
suspected that the force which retained the moon in its orbit about the
earth was the same as terrestrial gravity, and to verify this hypothesis
he proceeded thus. He knew that, if a stone were allowed to fall near
the surface of the earth, the attraction of the earth (that is, the weight
of the stone) caused it to move through 16 feet in one second. The
moon's orbit relative to the earth is nearly a circle; and as a rough
approximation, taking it to be so, he knew the distance of the moon,
and therefore the length of its path; he also knew the time the moon
took to go once round it, namely, a month. Hence he could easily
find its velocity at any point such as M. He could therefore find the

distance MT through which it would move in the next second if it were not pulled by the earth's attraction. At the end of that second it was however at M', and therefore the earth E must have pulled it through the distance TM' in one second (assuming the direction of the earth's pull to be constant). Now he and several physicists of the time had conjectured from Kepler's third law that the attraction of the earth on a body would be found to decrease as the body was removed farther away from the earth inversely as the square of the distance from the centre of the earth;[1] if this were the actual law and if gravity were the sole force which retained the moon in its orbit, then TM' should be to 16 feet inversely as the square of the distance of the moon from the centre of the earth to the square of the radius of the earth. In 1679, when he repeated the investigation, TM' was found to have the value which was required by the hypothesis, and the verification was complete; but in 1666 his estimate of the distance of the moon was inaccurate, and when he made the calculation he found that TM' was about one-eighth less than it ought to have been on his hypothesis.

This discrepancy does not seem to have shaken his faith in the belief that gravity extended as far as the moon and varied inversely as the square of the distance; but, from Whiston's notes of a conversation with Newton, it would seem that Newton inferred that some other force—probably Descartes's vortices—acted on the moon as well as gravity. This statement is confirmed by Pemberton's account of the investigation. It seems, moreover, that Newton already believed firmly in the principle of universal gravitation, that is, that every particle of matter attracts every other particle, and suspected that the attraction varied as the product of their masses and inversely as the square of the distance between them; but it is certain that he did not then know what the attraction of a spherical mass on any external point would be, and did not think it likely that a particle would be attracted by the earth as if the latter were concentrated into a single particle at its centre.

On his return to Cambridge in 1667 Newton was elected to a fellowship at his college, and permanently took up his residence there. In the early part of 1669, or perhaps in 1668, he revised Barrow's lectures for him. The end of the fourteenth lecture is known to have been written by Newton, but how much of the rest is due to his suggestions cannot now be determined. As soon as this was finished he was asked by Bar-

---

[1]An argument leading to this result is given below on page 274.

row and Collins to edit and add notes to a translation of Kinckhuysen's *Algebra*; he consented to do this, but on condition that his name should not appear in the matter. In 1670 he also began a systematic exposition of his analysis by infinite series, the object of which was to express the ordinate of a curve in an infinite algebraical series every term of which can be integrated by Wallis's rule; his results on this subject had been communicated to Barrow, Collins, and others in 1669. This was never finished: the fragment was published in 1711, but the substance of it had been printed as an appendix to the *Optics* in 1704. These works were only the fruit of Newton's leisure, most of his time during these two years being given up to optical researches.

In October, 1669, Barrow resigned the Lucasian chair in favour of Newton. During his tenure of the professorship, it was Newton's practice to lecture publicly once a week, for from half-an-hour to an hour at a time, in one term of each year, probably dictating his lectures as rapidly as they could be taken down; and in the week following the lecture to devote four hours to appointments which he gave to students who wished to come to his rooms to discuss the results of the previous lecture. He never repeated a course, which usually consisted of nine or ten lectures, and generally the lectures of one course began from the point at which the preceding course had ended. The manuscripts of his lectures for seventeen out of the first eighteen years of his tenure are extant.

When first appointed Newton chose optics for the subject of his lectures and researches, and before the end of 1669 he had worked out the details of his discovery of the decomposition of a ray of white light into rays of different colours by means of a prism. The complete explanation of the theory of the rainbow followed from this discovery. These discoveries formed the subject-matter of the lectures which he delivered as Lucasian professor in the years 1669, 1670, and 1671. The chief new results were embodied in a paper communicated to the Royal Society in February, 1672, and subsequently published in the *Philosophical Transactions*. The manuscript of his original lectures was printed in 1729 under the title *Lectiones Opticae*. This work is divided into two books, the first of which contains four sections and the second five. The first section of the first book deals with the decomposition of solar light by a prism in consequence of the unequal refrangibility of the rays that compose it, and a description of his experiments is added. The second section contains an account of the method which Newton invented for the determining the coefficients of refraction of different bodies. This is done by making a ray pass through a prism of the material so that the deviation is a minimum; and he proves that, if the angle of the prism be i and the deviation of the ray be δ, the refractive index will be $\sin \frac{1}{2}(i + \delta) \operatorname{cosec} \frac{1}{2}i$. The third section is on refractions at plane surfaces; he here shews that if a ray pass through a prism with minimum deviation, the angle of incidence is equal to the angle of emergence; most of this section is devoted to geometrical solutions of different problems. The fourth section contains a discussion of refractions at curved surfaces. The second book treats of his theory of colours and of the rainbow.

By a curious chapter of accidents Newton failed to correct the chromatic aberration of two colours by means of a couple of prisms. He therefore abandoned the hope of making a refracting telescope which should be achromatic, and instead designed a reflecting telescope, probably on the model of a small one which he had made in 1668. The form he used is that still known by his name; the idea of it was naturally suggested by Gregory's telescope. In 1672 he invented a reflecting microscope, and some years later he invented the sextant which was rediscovered by J. Hadley in 1731.

His professorial lectures from 1673 to 1683 were on algebra and the theory of equations, and are described below; but much of his time during these years was occupied with other investigations, and I may remark that throughout his life Newton must have devoted at least as much attention to chemistry and theology as to mathematics, though his conclusions are not of sufficient interest to require mention here. His theory of colours and his deductions from his optical experiments were at first attacked with considerable vehemence. The correspondence which this entailed on Newton occupied nearly all his leisure in the years 1672 to 1675, and proved extremely distasteful to him. Writing on December 9, 1675, he says, "I was so persecuted with discussions arising out of my theory of light, that I blamed my own imprudence for parting with so substantial a blessing as my quiet to run after a shadow." Again, on November 18, 1676, he observes, "I see I have made myself a slave to philosophy; but, if I get rid of Mr. Linus's business, I will resolutely bid adieu to it eternally, excepting what I do for my private satisfaction, or leave to come out after me; for I see a man must either resolve to put out nothing new, or to become a slave to defend it." The unreasonable dislike to have his conclusions doubted or to be involved in any correspondence about them was a prominent trait in Newton's character.

Newton was deeply interested in the question as to how the effects of light were really produced, and by the end of 1675 he had worked out the corpuscular or emission theory, and had shewn how it would account for all the various phenomena of geometrical optics, such as reflexion, refraction, colours, diffraction, &c. To do this, however, he was obliged to add a somewhat artificial rider, that the corpuscles had alternating fits of easy reflexion and easy refraction communicated to them by an ether which filled space. The theory is now known to be untenable, but it should be noted that Newton enunciated it as a hypothesis from which certain results would follow: it would seem that he believed the wave theory to be intrinsically more probable, but it was the difficulty of explaining diffraction on that theory that led him to suggest another hypothesis.

Newton's corpuscular theory was expounded in memoirs communicated to the Royal Society in December 1675, which are substantially reproduced in his *Optics*, published in 1704. In the latter work he dealt in detail with his theory of fits of easy reflexion and transmission, and the colours of thin plates, to which he added an explanation of the colours of thick plates [bk. II, part 4] and observations on the inflexion of light [bk. III].

Two letters written by Newton in the year 1676 are sufficiently interesting to justify an allusion to them.[1] Leibnitz, in 1674, in a correspondence with Oldenburg, wrote saying that he possessed "general analytical methods depending on infinite series." Oldenburg, in reply, told him that Newton and Gregory had used such series in their work. In answer to a request for information, Newton wrote on June 13, 1676, giving a brief account of his method. He here enunciated the binomial theorem, which he stated, in effect, in the form that if A, B, C, D, ... denote the successive terms in the expansion of $(P + PQ)^{m/n}$, then

$$(P+PQ)^{m/n} = A + \frac{m}{n}AQ + \frac{m-n}{2n}BQ + \frac{m-2n}{3n}CQ + \frac{m-3n}{4n}DQ + \ldots$$

where $A = P^{m/n}$. He gave examples of its use. He also gave the expansion of $\sin^{-1} x$, from which he deduced that of $\sin x$: this seems to be the earliest known instance of a reversion of series. He also inserted an expression for the rectification of an elliptic arc in an infinite series.

Leibnitz wrote on August 27 asking for fuller details; and Newton, on October 24, 1676, sent, through Oldenburg, an account of the way in

---

[1] See J. Wallis, *Opera*, vol. iii, Oxford, 1699, p. 622 *et seq.*

which he had been led to some of his results. The main results may be
briefly summarized. He begins by saying that altogether he had used
three methods for expansion in series. His first was arrived at from the
study of the method of interpolation. Thus, by considering the series
of expressions for $(1 - x^2)^{0/2}$, $(1 - x^2)^{2/2}$, $(1 - x^2)^{4/2}$, ... , he deduced
by interpolations a rule connecting the successive coefficients in the
expansions of $(1 - x^2)^{1/2}$, $(1 - x^2)^{3/2}$, ...; and then by analogy obtained
the expression for the general term in the expansion of a binomial.
He then tested his result in various ways; for instance in the case of
$(1 - x^2)^{1/2}$, by extracting the square root of $1 - x^2$, *more arithmetico*,
and by forming the square of the expansion of $(1-x^2)^{1/2}$, which reduced
to $1 - x^2$. He also used the series to determine the areas of the circle
and the hyperbola in infinite series, and found that the results were the
same as those he had arrived at by other means.

Having established this result, he then discarded the method of
interpolation, and employed his binomial theorem to express (when
possible) the ordinate of a curve in an infinite series in ascending powers
of the abscissa, and thus by Wallis's method he obtained expressions in
infinite series for the areas and arcs of curves in the manner described
in the appendix to his *Optics* and in his *De Analysi per Equationes
Numero Terminorum Infinitas*. He states that he had employed this
second method before the plague in 1665–66, and goes on to say that
he was then obliged to leave Cambridge, and subsequently (presumably
on his return to Cambridge) he ceased to pursue these ideas, as he found
that Nicholas Mercator had employed some of them in his *Logarithmo-
technica*, published in 1668; and he supposed that the remainder had
been or would be found out before he himself was likely to publish his
discoveries.

Newton next explains that he had also a third method, of which
(he says) he had about 1669 sent an account to Barrow and Collins,
illustrated by applications to areas, rectification, cubature, &c. This
was the method of fluxions; but Newton gives no description of it here,
though he adds some illustrations of its use. The first illustration is on
the quadrature of the curve represented by the equation

$$y = ax^m(b + cx^n)^p,$$

which he says can be effected as a sum of $(m + 1)/n$ terms if $(m + 1)/n$
be a positive integer, and which he thinks cannot otherwise be effected

except by an infinite series.[1] He also gives a list of other forms which are immediately integrable, of which the chief are

$$\frac{x^{mn-1}}{a + bx^n + cx^{2n}}, \quad \frac{x^{(m+1/2)n-1}}{a + bx^n + cx^{2n}}, \quad x^{mn-1}(a + bx^n + cx^{2n})^{\pm 1/2},$$

$$x^{mn-1}(a + bx^n)^{\pm 1/2}(c + dx^n)^{-1}, \quad x^{mn-n-1}(a + bx^n)(c + dx^n)^{-1/2};$$

where m is a positive integer and n is any number whatever. Lastly, he points out that the area of any curve can be easily determined approximately by the method of interpolation described below in discussing his *Methodus Differentialis*.

At the end of his letter Newton alludes to the solution of the "inverse problem of tangents," a subject on which Leibnitz had asked for information. He gives formulae for reversing any series, but says that besides these formulae he has two methods for solving such questions, which for the present he will not describe except by an anagram which, being read, is as follows, "Una methodus consistit in extractione fluentis quantitatis ex aequatione simul involvente fluxionem ejus: altera tantum in assumptione seriei pro quantitate qualibet incognita ex qua caetera commode derivari possunt, et in collatione terminorum homologorum aequationis resultantis, ad eruendos terminos assumptae seriei."

He implies in this letter that he is worried by the questions he is asked and the controversies raised about every new matter which he produces, which shew his rashness in publishing "quod umbram captando eatenus perdideram quietem meam, rem prorsus substantialem."

Leibnitz, in his answer, dated June 21, 1677, explains his method of drawing tangents to curves, which he says proceeds "not by fluxions of lines, but by the differences of numbers"; and he introduces his notation of dx and dy for the infinitesimal differences between the co-ordinates of two consecutive points on a curve. He also gives a solution of the problem to find a curve whose subtangent is constant, which shews that he could integrate.

In 1679 Hooke, at the request of the Royal Society, wrote to Newton expressing a hope that he would make further communications to the Society, and informing him of various facts then recently discovered. Newton replied saying that he had abandoned the study of philosophy, but he added that the earth's diurnal motion might be proved by the experiment of observing the deviation from the perpendicular of a

---

[1] This is not so, the integration is possible if $p + (m + 1)/n$ be an integer.

stone dropped from a height to the ground—an experiment which was subsequently made by the Society and succeeded. Hooke in his letter mentioned Picard's geodetical researches; in these Picard used a value of the radius of the earth which is substantially correct. This led Newton to repeat, with Picard's data, his calculations of 1666 on the lunar orbit, and he thus verified his supposition that gravity extended as far as the moon and varied inversely as the square of the distance. He then proceeded to consider the general theory of motion of a particle under a centripetal force, that is, one directed to a fixed point, and showed that the vector would sweep over equal areas in equal times. He also proved that, if a particle describe an ellipse under a centripetal force to a focus, the law must be that of the inverse square of the distance from the focus, and conversely, that the orbit of a particle projected under the influence of such a force would be a conic (or, it may be, he thought only an ellipse). Obeying his rule to publish nothing which could land him in a scientific controversy these results were locked up in his note-books, and it was only a specific question addressed to him five years later that led to their publication.

The *Universal Arithmetic*, which is on algebra, theory of equations, and miscellaneous problems, contains the substance of Newton's lectures during the years 1673 to 1683. His manuscript of it is still extant; Whiston[1] extracted a somewhat reluctant permission from Newton to print it, and it was published in 1707. Amongst several new theorems on various points in algebra and the theory of equations Newton here enunciates the following important results. He explains that the equation whose roots are the solution of a given problem will have as many roots as there are different possible cases; and he considers how it happens that the equation to which a problem leads may contain roots which do not satisfy the original question. He extends Descartes's rule of signs to give limits to the number of imaginary roots. He uses the principle of continuity to explain how two real and unequal roots may become imaginary in passing through equality, and illustrates this by geometrical considerations; thence he shews that imaginary roots must

---

[1] *William Whiston*, born in Leicestershire on December 9, 1667, educated at Clare College, Cambridge, of which society he was a fellow, and died in London on August 22, 1752, wrote several works on astronomy. He acted as Newton's deputy in the Lucasian chair from 1699, and in 1703 succeeded him as professor, but he was expelled in 1711, mainly for theological reasons. He was succeeded by Nicholas Saunderson, the blind mathematician, who was born in Yorkshire in 1682, and died at Christ's College, Cambridge, on April 19, 1739.

occur in pairs. Newton also here gives rules to find a superior limit to the positive roots of a numerical equation, and to determine the approximate values of the numerical roots. He further enunciates the theorem known by his name for finding the sum of the nth powers of the roots of an equation, and laid the foundation of the theory of symmetrical functions of the roots of an equation.

The most interesting theorem contained in the work is his attempt to find a rule (analogous to that of Descartes for real roots) by which the number of imaginary roots of an equation can be determined. He knew that the result which he obtained was not universally true, but he gave no proof and did not explain what were the exceptions to the rule. His theorem is as follows. Suppose the equation to be of the nth degree arranged in descending powers of x (the coefficient of x^n being positive), and suppose the $n + 1$ fractions

$$1, \quad \frac{n}{n-1} \frac{2}{1}, \quad \frac{n-1}{n-2} \frac{3}{2}, \quad \ldots, \quad \frac{n-p+1}{n-p} \frac{p+1}{p}, \quad \ldots, \quad \frac{2}{1} \frac{n}{n-1}, \quad 1$$

to be formed and written below the corresponding terms of the equation, then, if the square of any term when multiplied by the corresponding fraction be greater than the product of the terms on each side of it, put a plus sign above it: otherwise put a minus sign above it, and put a plus sign above the first and last terms. Now consider any two consecutive terms in the original equation, and the two symbols written above them. Then we may have any one of the four following cases: (α) the terms of the same sign and the symbols of the same sign; (β) the terms of the same sign and the symbols of opposite signs; (γ) the terms of opposite signs and the symbols of the same sign; (δ) the terms of opposite signs and the symbols of opposite signs. Then it has been shewn that the number of negative roots will not exceed the number of cases (α), and the number of positive roots will not exceed the number of cases (γ); and therefore the number of imaginary roots is not less than the number of cases (β) and (δ). In other words the number of changes of signs in the row of symbols written above the equation is an inferior limit to the number of imaginary roots. Newton, however, asserted that "you may almost know how many roots are impossible" by counting the changes of sign in the series of symbols formed as above. That is to say, he thought that in general the actual number of positive, negative, and imaginary roots could be got by the rule and not merely superior or inferior limits to these numbers. But though he knew that the rule was not universal he could not find (or at any rate did not state) what were the exceptions to it: this problem was subsequently discussed by Campbell, Maclaurin, Euler, and other writers; at last in 1865 Sylvester succeeded in proving the general result.[1]

In August, 1684, Halley came to Cambridge in order to consult Newton about the law of gravitation. Hooke, Huygens, Halley, and Wren had all conjectured that the force of the attraction of the sun or earth on an external particle varied inversely as the square of the distance. These writers seem independently to have shewn that, if Kepler's conclusions were rigorously true, as to which they were not quite certain, the law of attraction must be that of the inverse square. Probably their argument was as follows. If v be the velocity of a planet, r the radius of its orbit taken as a circle, and T its periodic time, $v = 2\pi r/T$. But, if f be the acceleration to the centre of the circle, we have $f = v^2/r$. Therefore, substituting the above value of v, $f = 4\pi^2 r/T^2$. Now, by Kepler's third law, T^2 varies as r^3; hence f varies inversely as r^2. They could not, however, deduce from the law the orbits of the planets. Halley explained that their investigations were stopped by their inability to solve this problem, and asked Newton if he could find out what the orbit of a planet would be if the law of attraction were that of the inverse square. Newton immediately replied that it was an ellipse, and promised to send or write out afresh the demonstration of it which he had found in 1679. This was sent in November, 1684.

Instigated by Halley, Newton now returned to the problem of gravitation; and before the autumn of 1684, he had worked out the substance of propositions 1–19, 21, 30, 32–35 in the first book of the *Principia*. These, together with notes on the laws of motion and various lemmas, were read for his lectures in the Michaelmas Term, 1684.

In November Halley received Newton's promised communication, which probably consisted of the substance of propositions 1, 11, and either proposition 17 or the first corollary of proposition 13; thereupon Halley again went to Cambridge, where he saw "a curious treatise, *De Motu*, drawn up since August." Most likely this contained Newton's manuscript notes of the lectures above alluded to: these notes are now in the university library, and are headed "*De Motu Corporum*." Halley begged that the results might be published, and finally secured a promise that they should be sent to the Royal Society: they were accordingly communicated to the Society not later than February, 1685, in the paper *De Motu*, which contains the substance of the following

[1]See the *Proceedings of the London Mathematical Society*, 1865, vol. i no. 2.

propositions in the *Principia*, book I, props. 1, 4, 6, 7, 10, 11, 15, 17, 32; book II, props. 2, 3, 4.

It seems also to have been due to the influence and tact of Halley at this visit in November, 1684, that Newton undertook to attack the whole problem of gravitation, and practically pledged himself to publish his results: these are contained in the *Principia*. As yet Newton had not determined the attraction of a spherical body on an external point, nor had he calculated the details of the planetary motions even if the members of the solar system could be regarded as points. The first problem was solved in 1685, probably either in January or February. "No sooner," to quote from Dr. Glaisher's address on the bicentenary of the publication of the *Principia*, "had Newton proved this superb theorem—and we know from his own words that he had no expectation of so beautiful a result till it emerged from his mathematical investigation—than all the mechanism of the universe at once lay spread before him. When he discovered the theorems that form the first three sections of book I, when he gave them in his lectures of 1684, he was unaware that the sun and earth exerted their attractions as if they were but points. How different must these propositions have seemed to Newton's eyes when he realized that these results, which he had believed to be only approximately true when applied to the solar system, were really exact! Hitherto they had been true only in so far as he could regard the sun as a point compared to the distance of the planets, or the earth as a point compared to the distance of the moon— a distance amounting to only about sixty times the earth's radius—but now they were mathematically true, excepting only for the slight deviation from a perfectly spherical form of the sun, earth, and planets. We can imagine the effect of this sudden transition from approximation to exactitude in stimulating Newton's mind to still greater efforts. It was now in his power to apply mathematical analysis with absolute precision to the actual problems of astronomy."

Of the three fundamental principles applied in the *Principia* we may say that the idea that every particle attracts every other particle in the universe was formed at least as early as 1666; the law of equable description of areas, its consequences, and the fact that if the law of attraction were that of the inverse square the orbit of a particle about a centre of force would be a conic were proved in 1679; and, lastly, the discovery that a sphere, whose density at any point depends only on the distance from the centre, attracts an external point as if the whole mass were collected at its centre was made in 1685. It was this last discovery that enabled him to apply the first two principles to the phenomena of bodies of finite size.

The draft of the first book of the *Principia* was finished before the summer of 1685, but the corrections and additions took some time, and the book was not presented to the Royal Society until April 28, 1686. This book is given up to the consideration of the motion of particles or bodies in free space either in known orbits, or under the action of known forces, or under their mutual attraction; and in particular to indicating how the effects of disturbing forces may be calculated. In it also Newton generalizes the law of attraction into a statement that every particle of matter in the universe attracts every other particle with a force which varies directly as the product of their masses, and inversely as the square of the distance between them; and he thence deduces the law of attraction for spherical shells of constant density. The book is prefaced by an introduction on the science of dynamics, which defines the limits of mathematical investigation. His object, he says, is to apply mathematics to the phenomena of nature; among these phenomena motion is one of the most important; now motion is the effect of force, and, though he does not know what is the nature or origin of force, still many of its effects can be measured; and it is these that form the subject-matter of the work.

The second book of the *Principia* was completed by the summer of 1686. This book treats of motion in a resisting medium, and of hydrostatics and hydrodynamics, with special applications to waves, tides, and acoustics. He concludes it by shewing that the Cartesian theory of vortices was inconsistent both with the known facts and with the laws of motion.

The next nine or ten months were devoted to the third book. Probably for this originally he had no materials ready. He commences by discussing when and how far it is justifiable to construct hypotheses or theories to account for known phenomena. He proceeds to apply the theorems obtained in the first book to the chief phenomena of the solar system, and to determine the masses and distances of the planets and (whenever sufficient data existed) of their satellites. In particular the motion of the moon, the various inequalities therein, and the theory of the tides are worked out in detail. He also investigates the theory of comets, shews that they belong to the solar system, explains how from three observations the orbit can be determined, and illustrates his results by considering certain special comets. The third book as we have it is but little more than a sketch of what Newton had finally proposed to himself to accomplish; his original scheme is among the "Portsmouth papers," and his notes shew that he continued to work at it for some years after the publication of the first edition of the *Principia*: the most interesting of his memoranda are those in which by means of fluxions he has carried his results beyond the point at which he was able to translate them into geometry.[1]

The demonstrations throughout the work are geometrical, but to readers of ordinary ability are rendered unnecessarily difficult by the absence of illustrations and explanations, and by the fact that no clue is given to the method by which Newton arrived at his results. The reason why it was presented in a geometrical form appears to have been that the infinitesimal calculus was then unknown, and, had Newton used it to demonstrate results which were in themselves opposed to the prevalent philosophy of the time, the controversy as to the truth of his results would have been hampered by a dispute concerning the validity of the methods used in proving them. He therefore cast the whole reasoning into a geometrical shape which, if somewhat longer, can at any rate be made intelligible to all mathematical students, So closely did he follow the lines of Greek geometry that he constantly used graphical methods, and represented forces, velocities, and other magnitudes in the Euclidean way by straight lines (*ex. gr.* book I, lemma 10), and not by a certain number of units. The latter and modern method had been introduced by Wallis, and must have been familiar to Newton. The effect of his confining himself rigorously to classical geometry is that the *Principia* is written in a language which is archaic, even if not unfamiliar.

The adoption of geometrical methods in the *Principia* for purposes of demonstration does not indicate a preference on Newton's part for geometry over analysis as an instrument of research, for it is known now that Newton used the fluxional calculus in the first instance in finding some of the theorems, especially those towards the end of book I and in book II; and in fact one of the most important uses of that calculus is stated in book II, lemma 2. But it is only just to remark that, at the time of its publication and for nearly a century afterwards, the differential and fluxional calculus were not fully developed, and did not possess the same superiority over the method he adopted which they do now; and it is a matter for astonishment that when Newton did employ

---

[1] For a fuller account of the *Principia* see my *Essay on the Genesis, Contents, and History of Newton's Principia*, London, 1893.

the calculus he was able to use it to so good an effect.

The printing of the work was slow, and it was not finally published till the summer of 1687. The cost was borne by Halley, who also corrected the proofs, and even put his own researches on one side to press the printing forward. The conciseness, absence of illustrations, and synthetical character of the book restricted the numbers of those who were able to appreciate its value; and, though nearly all competent critics admitted the validity of the conclusions, some little time elapsed before it affected the current beliefs of educated men. I should be inclined to say (but on this point opinions differ widely) that within ten years of its publication it was generally accepted in Britain as giving a correct account of the laws of the universe; it was similarly accepted within about twenty years on the continent, except in France, where the Cartesian hypothesis held its ground until Voltaire in 1733 took up the advocacy of the Newtonian theory.

The manuscript of the *Principia* was finished by 1686. Newton devoted the remainder of that year to his paper on physical optics, the greater part of which is given up to the subject of diffraction.

In 1687 James II. having tried to force the university to admit as a master of arts a Roman Catholic priest who refused to take the oaths of supremacy and allegiance, Newton took a prominent part in resisting the illegal interference of the king, and was one of the deputation sent to London to protect the rights of the university. The active part taken by Newton in this affair led to his being in 1689 elected member for the university. This parliament only lasted thirteen months, and on its dissolution he gave up his seat. He was subsequently returned in 1701, but he never took any prominent part in politics.

On his coming back to Cambridge in 1690 he resumed his mathematical studies and correspondence, but probably did not lecture. The two letters to Wallis, in which he explained his method of fluxions and fluents, were written in 1692 and published in 1693. Towards the close of 1692 and throughout the two following years, Newton had a long illness, suffering from insomnia and general nervous irritability. Perhaps he never quite regained his elasticity of mind, and, though after his recovery he shewed the same power in solving any question propounded to him, he ceased thenceforward to do original work on his own initiative, and it was somewhat difficult to stir him to activity in new subjects.

In 1694 Newton began to collect data connected with the irregularities of the moon's motion with the view of revising the part of the

Principia which dealt with that subject. To render the observations more accurate, he forwarded to Flamsteed[1] a table of corrections for refraction which he had previously made. This was not published till 1721, when Halley communicated it to the Royal Society. The original calculations of Newton and the papers connected with them are in the Portsmouth collection, and shew that Newton obtained it by finding the path of a ray, by means of quadratures, in a manner equivalent to the solution of a differential equation. As an illustration of Newton's genius, I may mention that even as late as 1754 Euler failed to solve the same problem. In 1782 Laplace gave a rule for constructing such a table, and his results agree substantially with those of Newton.

I do not suppose that Newton would in any case have produced much more original work after his illness; but his appointment in 1696 as warden, and his promotion in 1699 to the mastership of the Mint, at a salary of £1500 a year, brought his scientific investigations to an end, though it was only after this that many of his previous investigations were published in the form of books. In 1696 he moved to London, in 1701 he resigned the Lucasian chair, and in 1703 he was elected president of the Royal Society.

In 1704 Newton published his *Optics*, which contains the results of the papers already mentioned. To the first edition of this book were appended two minor works which have no special connection with optics; one being on cubic curves, the other on the quadrature of curves and on fluxions. Both of them were manuscripts with which his friends and pupils were familiar, but they were here published *urbi et orbi* for the first time.

The first of these appendices is entitled *Enumeratio Linearum Tertii Ordinis*;[2] the object seems to be to illustrate the use of analytical geometry, and as the application to conics was well known, Newton selected the theory of cubics.

---

[1] *John Flamsteed*, born at Derby in 1646 and died at Greenwich in 1719, was one of the most distinguished astronomers of this age, and the first astronomer-royal. Besides much valuable work in astronomy, he invented the system (published in 1680) of drawing maps by projecting the surface of the sphere on an enveloping cone, which can then be unwrapped. His life by R. F. Baily was published in London in 1835, but various statements in it should be read side by side with those in Brewster's life of Newton. Flamsteed was succeeded as astronomer-royal by Edmund Halley (see below, pp. 312–313).

[2] On this work and its bibliography, see my memoir in the *Transactions of the London Mathematical Society*, 1891, vol. xxii, pp. 104–143.

He begins with some general theorems, and classifies curves according as their equations are algebraical or transcendental; the former being cut by a straight line in a number of points (real or imaginary) equal to the degree of the curve, the latter being cut by a straight line in an infinite number of points. Newton then shews that many of the most important properties of conics have their analogues in the theory of cubics, and he discusses the theory of asymptotes and curvilinear diameters.

After these general theorems, he commences his detailed examination of cubics by pointing out that a cubic must have at least one real point at infinity. If the asymptote or tangent at this point be at a finite distance, it may be taken for the axis of y. This asymptote will cut the curve in three points altogether, of which at least two are at infinity. If the third point be at a finite distance, then (by one of his general theorems on asymptotes) the equation can be written in the form

$$xy^2 + hy = ax^3 + bx^2 + cx + d,$$

where the axes of x and y are the asymptotes of the hyperbola which is the locus of the middle points of all chords drawn parallel to the axis of y; while, if the third point in which this asymptote cuts the curve be also at infinity, the equation can be written in the form

$$xy = ax^3 + bx^2 + cx + d.$$

Next he takes the case where the tangent at the real point at infinity is not at a finite distance. A line parallel to the direction in which the curve goes to infinity may be taken as the axis of y. Any such line will cut the curve in three points altogether, of which one is by hypothesis at infinity, and one is necessarily at a finite distance. He then shews that if the remaining point in which this line cuts the curve be at a finite distance, the equation can be written in the form

$$y^2 = ax^3 + bx^2 + cx + d;$$

while if it be at an infinite distance, the equation can be written in the form

$$y = ax^3 + bx^2 + cx + d.$$

Any cubic is therefore reducible to one of four characteristic forms. Each of these forms is then discussed in detail, and the possibility of the existence of double points, isolated ovals, &c., is worked out. The final result is that in all there are seventy-eight possible forms which a cubic may take. Of these Newton enumerated only seventy-two; four of the remainder were mentioned by Stirling in 1717, one by Nicole in 1731, and one by Nicholas Bernoulli about the same time.

In the course of the work Newton states the remarkable theorem that, just as the shadow of a circle (cast by a luminous point on a plane) gives rise to all the conics, so the shadows of the curves represented by the equation $y^2 = ax^3 + bx^2 + cx + d$ give rise to all the cubics. This remained an unsolved puzzle until 1731, when Nicole and Clairaut gave demonstrations of it; a better proof is that given by Murdoch in 1740, which depends on the classification of these curves into five species according as to whether their points of intersection with the axis of x are real and unequal, real and two of them equal (two cases), real and all equal, or two imaginary and one real.

In this tract Newton also discusses double points in the plane and at infinity, the description of curves satisfying given conditions, and the graphical solution of problems by the use of curves.

The second appendix to the *Optics* is entitled *De Quadratura Curvarum*. Most of it had been communicated to Barrow in 1668 or 1669, and probably was familiar to Newton's pupils and friends from that time onwards. It consists of two parts.

The bulk of the first part is a statement of Newton's method of effecting the quadrature and rectification of curves by means of infinite series; it is noticeable as containing the earliest use in print of literal indices, and a printed statement of the binomial theorem, but these novelties are introduced only incidentally. The main object is to give rules for developing a function of x in a series in ascending powers of x, so as to enable mathematicians to effect the quadrature of any curve in which the ordinate y can be expressed as an explicit algebraical function of the abscissa x. Wallis had shewn how this quadrature could be found when y was given as a sum of a number of multiples of powers of x, and Newton's rules of expansion here established rendered possible the similar quadrature of any curve whose ordinate can be expressed as the sum of an infinite number of such terms. In this way he effects the quadrature of the curves

$$y = \frac{a^2}{b+x}, \quad y = (a^2 \pm x^2)^{\frac{1}{2}}, \quad y = (x - x^2)^{\frac{1}{2}}, \quad y = \left(\frac{1 + ax^2}{1 - bx^2}\right)^{\frac{1}{2}},$$

but naturally the results are expressed as infinite series. He then proceeds to curves whose ordinate is given as an implicit function of the abscissa; and he gives a method by which y can be expressed as an infinite series in ascending powers of x, but the application of the rule to any curve demands in general such complicated numerical calculations as to render it of little value. He concludes this part by shewing that the rectification of a curve can be effected in a somewhat similar way. His process is equivalent to finding the integral with regard to x of $(1 + \dot{y}^2)^{\frac{1}{2}}$ in the form of an infinite series. I should add that Newton indicates the importance of determining whether the series are convergent—an observation far in advance of his time—but he knew of no general test for the purpose; and in fact it was not until Gauss and Cauchy took up the question that the necessity of such limitations was commonly recognized.

The part of the appendix which I have just described is practically the same as Newton's manuscript *De Analysi per Equationes Numero Terminorum Infinitas*, which was subsequently printed in 1711. It is said that this was originally intended to form an appendix to Kinckhuysen's *Algebra*, which, as I have already said, he at one time intended to edit. The substance of it was communicated to Barrow, and by him to Collins, in letters of July 31 and August 12, 1669; and a summary of part of it was included in the letter of October 24, 1676, sent to Leibnitz.

It should be read in connection with Newton's *Methodus Differentialis*, also published in 1711. Some additional theorems are there given, and he discusses his method of interpolation, which had been briefly described in the letter of October 24, 1676. The principle is this. If $y = \phi(x)$ be a function of x, and if, when x is successively put equal to $a_1, a_2, \dots$, the values of y be known and be $b_1, b_2, \dots$, then a parabola whose equation is $y = p + qx + rx^2 + \dots$ can be drawn through the points $(a_1, b_1), (a_2, b_2), \dots$, and the ordinate of this parabola may be taken as an approximation to the ordinate of the curve. The degree of the parabola will of course be one less than the number of given points. Newton points out that in this way the areas of any curves can be approximately determined.

The second part of this appendix to the *Optics* contains a description of Newton's method of fluxions. This is best considered in connection with Newton's manuscript on the same subject which was published by John Colson in 1736, and of which it is a summary.

The invention of the infinitesimal calculus was one of the great intellectual achievements of the seventeenth century. This method of analysis, expressed in the notation of fluxions and fluents, was used by Newton in or before 1666, but no account of it was published until 1693, though its general outline was known by his friends and pupils long anterior to that year, and no complete exposition of his methods was given before 1736.

The idea of a fluxion or differential coefficient, as treated at this time, is simple. When two quantities—e.g. the radius of a sphere and its volume—are so related that a change in one causes a change in the other, the one is said to be a function of the other. The ratio of the rates at which they change is termed the differential coefficient or fluxion of the one with regard to the other, and the process by which this ratio is determined is known as differentiation. Knowing the differential coefficient and one set of corresponding values of the two quantities, it is possible by summation to determine the relation between them, as Cavalieri and others had shewn; but often the process is difficult. If, however, we can reverse the process of differentiation we can obtain this result directly. This process of reversal is termed integration. It was at once seen that problems connected with the quadrature of curves, and the determination of volumes (which were soluble by summation, as had been shewn by the employment of indivisibles), were reducible to integration. In mechanics also, by integration, velocities could be deduced from known accelerations, and distances traversed from known velocities. In short, wherever things change according to known laws, here was a possible method of finding the relation between them. It is true that, when we try to express observed phenomena in the language of the calculus, we usually obtain an equation involving the variables, and their differential coefficients—and possibly the solution may be beyond our powers. Even so, the method is often fruitful, and its use marked a real advance in thought and power.

I proceed to describe somewhat fully Newton's methods as described by Colson. Newton assumed that all geometrical magnitudes might be conceived as generated by continuous motion; thus a line may be considered as generated by the motion of a point, a surface by that of a line, a solid by that of a surface, a plane angle by the rotation of a line, and so on. The quantity thus generated was defined by him as the fluent or flowing quantity. The velocity of the moving magnitude was defined as the fluxion of the fluent. This seems to be the earliest definite recognition of the idea of a continuous function, though it had been foreshadowed in some of Napier's papers.

Newton's treatment of the subject is as follows. There are two kinds of problems. The object of the first is to find the fluxion of a given quantity, or more generally "the relation of the fluents being given, to find the relation of their fluxions." This is equivalent to differentiation. The object of the second or inverse method of fluxions is from the fluxion or some relations involving it to determine the fluent, or more generally "an equation being proposed exhibiting the relation of the fluxions of quantities, to find the relations of those quantities, or fluents, to one another."[1] This is equivalent either to integration which Newton termed the method of quadrature, or to the solution of a differential equation which was called by Newton the inverse method of tangents. The methods for solving these problems are discussed at considerable length.

Newton then went on to apply these results to questions connected with the maxima and minima of quantities, the method of drawing tangents to curves, and the curvature of curves (namely, the determination of the centre of curvature, the radius of curvature, and the rate at which the radius of curvature increases). He next considered the quadrature of curves, and the rectification of curves.[2] In finding the maximum and minimum of functions of one variable we regard the change of sign of the difference between two consecutive values of the function as the true criterion; but his argument is that when a quantity increasing has attained its maximum it can have no further increment, or when decreasing it has attained its minimum it can have no further decrement; consequently the fluxion must be equal to nothing.

It has been remarked that neither Newton nor Leibnitz produced a calculus, that is, a classified collection of rules; and that the problems they discussed were treated from first principles. That, no doubt, is the usual sequence in the history of such discoveries, though the fact is frequently forgotten by subsequent writers. In this case I think the statement, so far as Newton's treatment of the differential or fluxional part of the calculus is concerned, is incorrect, as the foregoing account sufficiently shews.

If a flowing quantity or fluent were represented by x, Newton denoted its fluxion by $\dot{x}$, the fluxion of $\dot{x}$ or second fluxion of x by $\ddot{x}$, and so on. Similarly the fluent of x was denoted by $\boxed{x}$, or sometimes by x' or $[x]$. The infinitely small part by which a fluent such as x increased in a small interval of time measured by o was called the moment of

---

[1] Colson's edition of Newton's manuscript, pp. 21, 22.
[2] *Ibid.* pp. 22, 23.

the fluent; and its value was shewn[1] to be $\dot{x}o$. Newton adds the important remark that thus we may in any problem neglect the terms multiplied by the second and higher powers of o, and we can always find an equation between the co-ordinates x, y of a point on a curve and their fluxions $\dot{x}$, $\dot{y}$. It is an application of this principle which constitutes one of the chief values of the calculus; for if we desire to find the effect produced by several causes on a system, then, if we can find the effect produced by each cause when acting alone in a very small time, the total effect produced in that time will be equal to the sum of the separate effects. I should here note the fact that Vince and other English writers in the eighteenth century used $\dot{x}$ to denote the increment of x and not the velocity with which it increased; that is, $\dot{x}$ in their writings stands for what Newton would have expressed by $\dot{x}o$ and what Leibnitz would have written as dx.

I need not discuss in detail the manner in which Newton treated the problems above mentioned. I will only add that, in spite of the form of his definition, the introduction into geometry of the idea of time was evaded by supposing that some quantity (ex. gr. the abscissa of a point on a curve) increased equably; and the required results then depend on the rate at which other quantities (ex. gr. the ordinate or radius of curvature) increase relatively to the one so chosen.[2] The fluent so chosen is what we now call the independent variable; its fluxion was termed the "principal fluxion"; and, of course, if it were denoted by x, then $\dot{x}$ was constant, and consequently $\ddot{x} = 0$.

There is no question that Newton used a method of fluxions in 1666, and it is practically certain that accounts of it were communicated in manuscript to friends and pupils from and after 1669. The manuscript, from which most of the above summary has been taken, is believed to have been written between 1671 and 1677, and to have been in circulation at Cambridge from that time onwards, though it is probable that parts were rewritten from time to time. It was unfortunate that it was not published at once. Strangers at a distance naturally judged of the method by the letter to Wallis in 1692, or by the *Tractatus de Quadratura Curvarum*, and were not aware that it had been so completely developed at an earlier date. This was the cause of numerous misunderstandings. At the same time it must be added that all mathematical analysis was leading up to the ideas and meth-

---

[1] Colson's edition of Newton's manuscript, p. 24.
[2] *Ibid.* p. 20

ods of the infinitesimal calculus. Foreshadowings of the principles and even of the language of that calculus can be found in the writings of Napier, Kepler, Cavalieri, Pascal, Fermat, Wallis, and Barrow. It was Newton's good luck to come at a time when everything was ripe for the discovery, and his ability enabled him to construct almost at once a complete calculus.

The infinitesimal calculus can also be expressed in the notation of the differential calculus: a notation which was invented by Leibnitz probably in 1675, certainly by 1677, and was published in 1684, some nine years before the earliest printed account of Newton's method of fluxions. But the question whether the general idea of the calculus expressed in that notation was obtained by Leibnitz from Newton, or whether it was discovered independently, gave rise to a long and bitter controversy. The leading facts are given in the next chapter.

The remaining events of Newton's life require little or no comment. In 1705 he was knighted. From this time onwards he devoted much of his leisure to theology, and wrote at great length on prophecies and predictions, subjects which had always been of interest to him. His *Universal Arithmetic* was published by Whiston in 1707, and his *Analysis by Infinite Series* in 1711; but Newton had nothing to do with the preparation of either of these for the press. His evidence before the House of Commons in 1714 on the determination of longitude at sea marks an important epoch in the history of navigation.

The dispute with Leibnitz as to whether he had derived the ideas of the differential calculus from Newton or invented it independently originated about 1708, and occupied much of Newton's time, especially between the years 1709 and 1716.

In 1709 Newton was persuaded to allow Cotes to prepare the long-talked-of second edition of the *Principia*; it was issued in March 1713. A third edition was published in 1726 under the direction of Henry Pemberton. In 1725 Newton's health began to fail. He died on March 20, 1727, and eight days later was buried in Westminster Abbey.

His chief works, taking them in their order of publication, are the *Principia*, published in 1687; the *Optics* (with appendices on *cubic curves*, the *quadrature and rectification of curves by the use of infinite series*, and the *method of fluxions*), published in 1704; the *Universal Arithmetic*, published in 1707; the *Analysis per Series, Fluxiones*, &c., and the *Methodus Differentialis*, published in 1711; the *Lectiones Opticae*, published in 1729; the *Method of Fluxions*, &c. (that is, *Newton's manuscript on fluxions*), translated by J. Colson and published in

1736; and the *Geometria Analytica,* printed in 1779 in the first volume of Horsley's edition of Newton's works.

In appearance Newton was short, and towards the close of his life rather stout, but well set, with a square lower jaw, brown eyes, a broad forehead, and rather sharp features. His hair turned grey before he was thirty, and remained thick and white as silver till his death.

As to his manners, he dressed slovenly, was rather languid, and was often so absorbed in his own thoughts as to be anything but a lively companion. Many anecdotes of his extreme absence of mind when engaged in any investigation have been preserved. Thus once when riding home from Grantham he dismounted to lead his horse up a steep hill; when he turned at the top to remount, he found that he had the bridle in his hand, while his horse had slipped it and gone away. Again, on the few occasions when he sacrificed his time to entertain his friends, if he left them to get more wine or for any similar reason, he would as often as not be found after the lapse of some time working out a problem, oblivious alike of his expectant guests and of his errand. He took no exercise, indulged in no amusements, and worked incessantly, often spending eighteen or nineteen hours out of the twenty-four in writing.

In character he was religious and conscientious, with an exceptionally high standard of morality, having, as Bishop Burnet said, "the whitest soul" he ever knew. Newton was always perfectly straightforward and honest; but in his controversies with Leibnitz, Hooke, and others, though scrupulously just, he was not generous; and it would seem that he frequently took offence at a chance expression when none was intended. He modestly attributed his discoveries largely to the admirable work done by his predecessors; and once explained that, if he had seen farther than other men, it was only because he had stood on the shoulders of giants. He summed up his own estimate of his work in the sentence, "I do not know what I may appear to the world; but to myself I seem to have been only like a boy, playing on the sea-shore, and diverting myself, in now and then finding a smoother pebble, or a prettier shell than ordinary, whilst the great ocean of truth lay all undiscovered before me." He was morbidly sensitive to being involved in any discussions. I believe that, with the exception of his papers on optics, every one of his works was published only under pressure from his friends and against his own wishes. There are several instances of his communicating papers and results on condition that his name should not be published: thus when in 1669 he had, at Collins's request, solved some problems on harmonic series and on annuities which had previously baffled investigation, he only gave permission that his results should be published "so it be," as he says, "without my name to it; for I see not what there is desirable in public esteem, were I able to acquire and maintain it: it would perhaps increase my acquaintance, the thing which I chiefly study to decline."

Perhaps the most wonderful single illustration of his powers was the composition in seven months of the first book of the *Principia*, and the expression of the numerous and complex results in classical geometrical form. As other illustrations of his ability I may mention his solutions of the problem of Pappus, of John Bernoulli's challenge, and of the question of orthogonal trajectories. The problem of Pappus, here alluded to, is to find the locus of a point such that the rectangle under its distances from two given straight lines shall be in a given ratio to the rectangle under its distances from two other given straight lines. Many geometricians from the time of Apollonius had tried to find a geometrical solution and had failed, but what had proved insuperable to his predecessors seems to have presented little difficulty to Newton who gave an elegant demonstration that the locus was a conic. Geometry, said Lagrange when recommending the study of analysis to his pupils, is a strong bow, but it is one which only a Newton can fully utilize. As another example I may mention that in 1696 John Bernoulli challenged mathematicians (i) to determine the brachistochrone, and (ii) to find a curve such that if any line drawn from a fixed point O cut it in P and Q then $OP^n + OQ^n$ would be constant. Leibnitz solved the first of these questions after an interval of rather more than six months, and then suggested they should be sent as a challenge to Newton and others. Newton received the problems on Jan. 29, 1697, and the next day gave the complete solutions of both, at the same time generalising the second question. An almost exactly similar case occurred in 1716 when Newton was asked to find the orthogonal trajectory of a family of curves. In five hours Newton solved the problem in the form in which it was propounded to him, and laid down the principles for finding trajectories.

It is almost impossible to describe the effect of Newton's writings without being suspected of exaggeration. But, if the state of mathematical knowledge in 1669 or at the death of Pascal or Fermat be compared with what was known in 1700 it will be seen how immense was the advance. In fact we may say that it took mathematicians half a century or more before they were able to assimilate the work produced in those years.

In pure geometry Newton did not establish any new methods, but no modern writer has shewn the same power in using those of classical geometry. In algebra and the theory of equations he introduced the system of literal indices, established the binomial theorem, and created no inconsiderable part of the theory of equations: one rule which he enunciated in this subject remained till a few years ago an unsolved riddle which had overtaxed the resources of succeeding mathematicians. In analytical geometry, he introduced the modern classification of curves into algebraical and transcendental; and established many of the fundamental properties of asymptotes, multiple points, and isolated loops, illustrated by a discussion of cubic curves. The fluxional or infinitesimal calculus was invented by Newton in or before the year 1666, and circulated in manuscript amongst his friends in and after the year 1669, though no account of the method was printed till 1693. The fact that the results are nowadays expressed in a different notation has led to Newton's investigations on this subject being somewhat overlooked.

Newton, further, was the first to place dynamics on a satisfactory basis, and from dynamics he deduced the theory of statics: this was in the introduction to the *Principia* published in 1687. The theory of attractions, the application of the principles of mechanics to the solar system, the creation of physical astronomy, and the establishment of the law of universal gravitation are due to him, and were first published in the same work, but of the nature of gravity he confessed his ignorance, though he found inconceivable the idea of action at a distance. The particular questions connected with the motion of the earth and moon were worked out as fully as was then possible. The theory of hydrodynamics was created in the second book of the *Principia*, and he added considerably to the theory of hydrostatics which may be said to have been first discussed in modern times by Pascal. The theory of the propagation of waves, and in particular the application to determine the velocity of sound, is due to Newton and was published in 1687. In geometrical optics, he explained amongst other things the decomposition of light and the theory of the rainbow; he invented the reflecting telescope known by his name, and the sextant. In physical optics, he suggested and elaborated the emission theory of light.

The above list does not exhaust the subjects he investigated, but it will serve to illustrate how marked was his influence on the history of mathematics. On his writings and on their effects, it will be enough to quote the remarks of two or three of those who were subsequently concerned with the subject-matter of the *Principia*. Lagrange described the *Principia* as the greatest production of the human mind, and said he felt dazed at such an illustration of what man's intellect might be capable. In describing the effect of his own writings and those of Laplace it was a favourite remark of his that Newton was not only the greatest genius that had ever existed, but he was also the most fortunate, for as there is but one universe, it can happen but to one man in the world's history to be the interpreter of its laws. Laplace, who is in general very sparing of his praise, makes of Newton the one exception, and the words in which he enumerates the causes which "will always assure to the *Principia* a pre-eminence above all the other productions of human genius" have been often quoted. Not less remarkable is the homage rendered by Gauss; for other great mathematicians or philosophers he used the epithets magnus, or clarus, or clarissimus: for Newton alone he kept the prefix summus. Finally Biot, who had made a special study of Newton's works, sums up his remarks by saying, "comme géomètre et comme expérimentateur Newton est sans égal; par la réunion de ces deux genres de génies à leur plus haut degré, il est sans exemple."

CHAPTER XVII.

LEIBNITZ AND THE MATHEMATICIANS OF THE FIRST HALF OF THE EIGHTEENTH CENTURY.[1]

I HAVE briefly traced in the last chapter the nature and extent of Newton's contributions to science. Modern analysis is, however, derived directly from the works of Leibnitz and the elder Bernoullis; and it is immaterial to us whether the fundamental ideas of it were obtained by them from Newton, or discovered independently. The English mathematicians of the years considered in this chapter continued to use the language and notation of Newton; they are thus somewhat distinct from their continental contemporaries, and I have therefore grouped them together in a section by themselves.

Leibnitz and the Bernoullis.

Leibnitz.[2]   *Gottfried Wilhelm Leibnitz* (or *Leibniz*) was born at Leipzig on June 21 (O.S.), 1646, and died at Hanover on November 14, 1716. His father died before he was six, and the teaching at the school to which he was then sent was inefficient, but his industry triumphed over all difficulties; by the time he was twelve he had taught himself to read Latin easily, and had begun Greek; and before he was twenty he had mastered the ordinary text-books on mathematics, philosophy, theology, and law. Refused the degree of doctor of laws at Leipzig by those who were jealous of his youth and learning, he moved to Nuremberg. An essay which he there wrote on the study of law was dedicated

---

[1]See Cantor, vol. iii; other authorities for the mathematicians of the period are mentioned in the footnotes.

[2]See the life of Leibnitz by G. E. Guhrauer, two volumes and a supplement, Breslau, 1842 and 1846. Leibnitz's mathematical papers have been collected and edited by C. J. Gerhardt in seven volumes, Berlin and Halle, 1849–63.

to the Elector of Mainz, and led to his appointment by the elector on a commission for the revision of some statutes, from which he was subsequently promoted to the diplomatic service. In the latter capacity he supported (unsuccessfully) the claims of the German candidate for the crown of Poland. The violent seizure of various small places in Alsace in 1670 excited universal alarm in Germany as to the designs of Louis XIV.; and Leibnitz drew up a scheme by which it was proposed to offer German co-operation, if France liked to take Egypt, and use the possession of that country as a basis for attack against Holland in Asia, provided France would agree to leave Germany undisturbed. This bears a curious resemblance to the similar plan by which Napoleon I. proposed to attack England. In 1672 Leibnitz went to Paris on the invitation of the French government to explain the details of the scheme, but nothing came of it.

At Paris he met Huygens who was then residing there, and their conversation led Leibnitz to study geometry, which he described as opening a new world to him; though as a matter of fact he had previously written some tracts on various minor points in mathematics, the most important being a paper on combinations written in 1668, and a description of a new calculating machine. In January, 1673, he was sent on a political mission to London, where he stopped some months and made the acquaintance of Oldenburg, Collins, and others; it was at this time that he communicated the memoir to the Royal Society in which he was found to have been forestalled by Mouton.

In 1673 the Elector of Mainz died, and in the following year Leibnitz entered the service of the Brunswick family; in 1676 he again visited London, and then moved to Hanover, where, till his death, he occupied the well-paid post of librarian in the ducal library. His pen was thenceforth employed in all the political matters which affected the Hanoverian family, and his services were recognized by honours and distinctions of various kinds; his memoranda on the various political, historical, and theological questions which concerned the dynasty during the forty years from 1673 to 1713 form a valuable contribution to the history of that time.

Leibnitz's appointment in the Hanoverian service gave him more time for his favourite pursuits. He used to assert that as the first-fruit of his increased leisure, he invented the differential and integral calculus in 1674, but the earliest traces of the use of it in his extant note-books do not occur till 1675, and it was not till 1677 that we find it developed into a consistent system; it was not published till 1684.

Most of his mathematical papers were produced within the ten years from 1682 to 1692, and many of them in a journal, called the *Acta Eruditorum*, founded by himself and Otto Mencke in 1682, which had a wide circulation on the continent.

Leibnitz occupies at least as large a place in the history of philosophy as he does in the history of mathematics. Most of his philosophical writings were composed in the last twenty or twenty-five years of his life; and the point as to whether his views were original or whether they were appropriated from Spinoza, whom he visited in 1676, is still in question among philosophers, though the evidence seems to point to the originality of Leibnitz. As to Leibnitz's system of philosophy it will be enough to say that he regarded the ultimate elements of the universe as individual percipient beings whom he called monads. According to him the monads are centres of force, and substance is force, while space, matter, and motion are merely phenomenal; finally, the existence of God is inferred from the existing harmony among the monads. His services to literature were almost as considerable as those to philosophy; in particular, I may single out his overthrow of the then prevalent belief that Hebrew was the primeval language of the human race.

In 1700 the academy of Berlin was created on his advice, and he drew up the first body of statutes for it. On the accession in 1714 of his master, George I., to the throne of England, Leibnitz was thrown aside as a useless tool; he was forbidden to come to England; and the last two years of his life were spent in neglect and dishonour. He died at Hanover in 1716. He was overfond of money and personal distinctions; was unscrupulous, as perhaps might be expected of a professional diplomatist of that time; but possessed singularly attractive manners, and all who once came under the charm of his personal presence remained sincerely attached to him. His mathematical reputation was largely augmented by the eminent position that he occupied in diplomacy, philosophy, and literature; and the power thence derived was considerably increased by his influence in the management of the *Acta Eruditorum*.

The last years of his life—from 1709 to 1716—were embittered by the long controversy with John Keill, Newton, and others, as to whether he had discovered the differential calculus independently of Newton's previous investigations, or whether he had derived the fundamental idea from Newton, and merely invented another notation for it. The controversy[1] occupies a place in the scientific history of the early years of the eighteenth century quite disproportionate to its true importance, but it so materially affected the history of mathematics in western Europe, that I feel obliged to give the leading facts, though I am reluctant to take up so much space with questions of a personal character.

The ideas of the infinitesimal calculus can be expressed either in the notation of fluxions or in that of differentials. The former was used by Newton in 1666, but no distinct account of it was printed till 1693. The earliest use of the latter in the note-books of Leibnitz may be probably referred to 1675, it was employed in the letter sent to Newton in 1677, and an account of it was printed in the memoir of 1684 described below. There is no question that the differential notation is due to Leibnitz, and the sole question is as to whether the general idea of the calculus was taken from Newton or discovered independently.

The case in favour of the independent invention by Leibnitz rests on the ground that he published a description of his method some years before Newton printed anything on fluxions, that he always alluded to the discovery as being his own invention, and that for some years this statement was unchallenged; while of course there must be a strong presumption that he acted in good faith. To rebut this case it is necessary to shew (i) that he saw some of Newton's papers on the subject in or before 1675, or at least 1677, and (ii) that he thence derived the fundamental ideas of the calculus. The fact that his claim was unchallenged for some years is, in the particular circumstances of the case, immaterial.

That Leibnitz saw some of Newton's manuscripts was always intrinsically probable; but when, in 1849, C. J. Gerhardt[2] examined Leibnitz's papers he found among them a manuscript copy, the existence of which had been previously unsuspected, in Leibnitz's handwriting, of extracts from Newton's *De Analysi per Equationes Numero Termino-*

---

[1] The case in favour of the independent invention by Leibnitz is stated in Gerhardt's *Leibnizens mathematische Schriften*; and in the third volume of M. Cantor's *Geschichte der Mathematik*. The arguments on the other side are given in H. Sloman's *Leibnitzens Anspruch auf die Erfindung der Differenzialrechnung*, Leipzig, 1857, of which an English translation, with additions by Dr. Sloman, was published at Cambridge in 1860. A summary of the evidence will be found in G. A. Gibson's memoir, *Proceedings of the Edinburgh Mathematical Society*, vol. xiv, 1896, pp. 148–174. The history of the invention of the calculus is given in an article on it in the ninth edition of the *Encyclopaedia Britannica*, and in P. Mansion's *Esquisse de l'histoire du calcul infinitésimal*, Gand, 1887.

[2] Gerhardt, *Leibnizens mathematische Schriften*, vol. i, p. 7.

rum Infinitas (which was printed in the *De Quadratura Curvarum* in 1704), together with notes on their expression in the differential notation. The question of the date at which these extracts were made is therefore all-important. Tschirnhausen seems to have possessed a copy of Newton's *De Analysi* in 1675, and as in that year he and Leibnitz were engaged together on a piece of work, it is not impossible that these extracts were made then. It is also possible that they may have been made in 1676, for Leibnitz discussed the question of analysis by infinite series with Collins and Oldenburg in that year, and it is *a priori* probable that they would have then shewn him the manuscript of Newton on that subject, a copy of which was possessed by one or both of them. On the other hand it may be supposed that Leibnitz made the extracts from the printed copy in or after 1704. Leibnitz shortly before his death admitted in a letter to Conti that in 1676 Collins had shewn him some Newtonian papers, but implied that they were of little or no value,—presumably he referred to Newton's letters of June 13 and Oct. 24, 1676, and to the letter of Dec. 10, 1672, on the method of tangents, extracts from which accompanied[1] the letter of June 13,— but it is remarkable that, on the receipt of these letters, Leibnitz should have made no further inquiries, unless he was already aware from other sources of the method followed by Newton.

Whether Leibnitz made no use of the manuscript from which he had copied extracts, or whether he had previously invented the calculus, are questions on which at this distance of time no direct evidence is available. It is, however, worth noting that the unpublished Portsmouth Papers shew that when, in 1711, Newton went carefully into the whole dispute, he picked out this manuscript as the one which had probably somehow fallen into the hands of Leibnitz.[2] At that time there was no direct evidence that Leibnitz had seen this manuscript before it was printed in 1704, and accordingly Newton's conjecture was not published; but Gerhardt's discovery of the copy made by Leibnitz tends to confirm the accuracy of Newton's judgment in the matter. It is said by those who question Leibnitz's good faith that to a man of his ability the manuscript, especially if supplemented by the letter of Dec. 10, 1672, would supply sufficient hints to give him a clue to the methods of the calculus, though as the fluxional notation is not employed in it anyone who used it would have to invent a notation; but this is denied

---

[1] Gerhardt, vol. i, p. 91.

[2] *Catalogue of Portsmouth Papers*, pp. xvi, xvii, 7, 8.

by others.

There was at first no reason to suspect the good faith of Leibnitz; and it was not until the appearance in 1704 of an anonymous review of Newton's tract on quadrature, in which it was implied that Newton had borrowed the idea of the fluxional calculus from Leibnitz, that any responsible mathematician[1] questioned the statement that Leibnitz had invented the calculus independently of Newton. It is universally admitted that there was no justification or authority for the statements made in this review, which was rightly attributed to Leibnitz. But the subsequent discussion led to a critical examination of the whole question, and doubt was expressed as to whether Leibnitz had not derived the fundamental idea from Newton. The case against Leibnitz as it appeared to Newton's friends was summed up in the *Commercium Epistolicum* issued in 1712, and detailed references are given for all the facts mentioned.

No such summary (with facts, dates, and references) of the case for Leibnitz was issued by his friends; but John Bernoulli attempted to indirectly weaken the evidence by attacking the personal character of Newton: this was in a letter dated June 7, 1713. The charges were false, and, when pressed for an explanation of them, Bernoulli most solemnly denied having written the letter. In accepting the denial Newton added in a private letter to him the following remarks, which are interesting as giving Newton's account of why he was at last induced to take any part in the controversy. "I have never," said he, "grasped at fame among foreign nations, but I am very desirous to preserve my character for honesty, which the author of that epistle, as if by the authority of a great judge, had endeavoured to wrest from me. Now that I am old, I have little pleasure in mathematical studies, and I have never tried to propagate my opinions over the world, but have rather taken care not to involve myself in disputes on account of them."

Leibnitz's defence or explanation of his silence is given in the following letter, dated April 9, 1716, from him to Conti. "Pour répondre de point en point à l'ouvrage publié contre moi, il falloit un autre ouvrage aussi grand pour le moins que celui-là: il falloit entrer dans un grand détail de quantité de minutiés passées il y a trente à quarante ans, dont je ne me souvenois guère: il me falloit chercher mes vieilles lettres, dont plusieurs se sont perdues, outre que le plus souvent je n'ai point gardé

---

[1] In 1699 Duillier had accused Leibnitz of plagiarism from Newton, but Duillier was not a person of much importance.

les minutes des miennes: et les autres sont ensevelies dans un grand tas de papiers, que je ne pouvois débrouiller qu'avec du temps et de la patience; mais je n'en avois guère le loisir, étant chargé présentement d'occupations d'une toute autre nature."

The death of Leibnitz in 1716 only put a temporary stop to the controversy which was bitterly debated for many years later. The question is one of difficulty; the evidence is conflicting and circumstantial; and every one must judge for himself which opinion seems most reasonable. Essentially it is a case of Leibnitz's word against a number of suspicious details pointing against him. His unacknowledged possession of a copy of part of one of Newton's manuscripts may be explicable; but the fact that on more than one occasion he deliberately altered or added to important documents (ex. gr. the letter of June 7, 1713, in the *Charta Volans*, and that of April 8, 1716, in the *Acta Eruditorum*), before publishing them, and, what is worse, that a material date in one of his manuscripts has been falsified[1] (1675 being altered to 1673), makes his own testimony on the subject of little value. It must be recollected that what he is alleged to have received was rather a number of suggestions than an account of the calculus; and it is possible that as he did not publish his results of 1677 until 1684, and that as the notation and subsequent development of it were all of his own invention, he may have been led, thirty years later, to minimize any assistance which he had obtained originally, and finally to consider that it was immaterial. During the eighteenth century the prevalent opinion was against Leibnitz, but to-day the majority of writers incline to think it more likely that the inventions were independent.

If we must confine ourselves to one system of notation then there can be no doubt that that which was invented by Leibnitz is better fitted for most of the purposes to which the infinitesimal calculus is applied than that of fluxions, and for some (such as the calculus of variations) it is indeed almost essential. It should be remembered, however, that at the beginning of the eighteenth century the methods of the infinitesimal calculus had not been systematized, and either notation was equally good. The development of that calculus was the main work of the mathematicians of the first half of the eighteenth century. The differential form was adopted by continental mathematicians. The application of it by Euler, Lagrange, and Laplace to the principles of

---

[1] Cantor, who advocates Leibnitz's claims, thinks that the falsification must be taken to be Leibnitz's act: see Cantor, vol. iii, p. 176.

mechanics laid down in the *Principia* was the great achievement of the last half of that century, and finally demonstrated the superiority of the differential to the fluxional calculus. The translation of the *Principia* into the language of modern analysis, and the filling in of the details of the Newtonian theory by the aid of that analysis, were effected by Laplace.

The controversy with Leibnitz was regarded in England as an attempt by foreigners to defraud Newton of the credit of his invention, and the question was complicated on both sides by national jealousies. It was therefore natural, though it was unfortunate, that in England the geometrical and fluxional methods as used by Newton were alone studied and employed. For more than a century the English school was thus out of touch with continental mathematicians. The consequence was that, in spite of the brilliant band of scholars formed by Newton, the improvements in the methods of analysis gradually effected on the continent were almost unknown in Britain. It was not until 1820 that the value of analytical methods was fully recognized in England, and that Newton's countrymen again took any large share in the development of mathematics.

Leaving now this long controversy I come to the discussion of the mathematical papers produced by Leibnitz, all the more important of which were published in the *Acta Eruditorum*. They are mainly concerned with applications of the infinitesimal calculus and with various questions on mechanics.

The only papers of first-rate importance which he produced are those on the differential calculus. The earliest of these was one published in the *Acta Eruditorum* for October, 1684, in which he enunciated a general method for finding maxima and minima, and for drawing tangents to curves. One inverse problem, namely, to find the curve whose subtangent is constant, was also discussed. The notation is the same as that with which we are familiar, and the differential coefficients of x^n and of products and quotients are determined. In 1686 he wrote a paper on the principles of the new calculus. In both of these papers the principle of continuity is explicitly assumed, while his treatment of the subject is based on the use of infinitesimals and not on that of the limiting value of ratios. In answer to some objections which were raised in 1694 by Bernard Nieuwentyt, who asserted that dy/dx stood for an unmeaning quantity like $0/0$, Leibnitz explained, in the same way as Barrow had previously done, that the value of dy/dx in geometry could be expressed as the ratio of two finite quantities. I think that Leibnitz's statement of the objects and methods of the infinitesimal calculus as contained in these papers, which are the three most important memoirs on it that he produced, is somewhat obscure, and his attempt to place the subject on a metaphysical basis did not tend to clearness; but the fact that all the results of modern mathematics are expressed in the language invented by Leibnitz has proved the best monument of his work. Like Newton, he treated integration not only as a summation, but as the inverse of differentiation.

In 1686 and 1692 he wrote papers on osculating curves. These, however, contain some bad blunders, as, for example, the assertion that an osculating circle will necessarily cut a curve in four consecutive points: this error was pointed out by John Bernoulli, but in his article of 1692 Leibnitz defended his original assertion, and insisted that a circle could never cross a curve where it touched it.

In 1692 Leibnitz wrote a memoir in which he laid the foundation of the theory of envelopes. This was further developed in another paper in 1694, in which he introduced for the first time the terms "co-ordinates" and "axes of co-ordinates."

Leibnitz also published a good many papers on mechanical subjects; but some of them contain mistakes which shew that he did not understand the principles of the subject. Thus, in 1685, he wrote a memoir to find the pressure exerted by a sphere of weight W placed between two inclined planes of complementary inclinations, placed so that the lines of greatest slope are perpendicular to the line of the intersection of the planes. He asserted that the pressure on each plane must consist of two components, "unum quo decliviter descendere tendit, alterum quo planum declive premit." He further said that for metaphysical reasons the sum of the two pressures must be equal to W. Hence, if R and R' be the required pressures, and α and $\frac{1}{2}\pi - \alpha$ the inclinations of the planes, he finds that

$$R = \frac{1}{2}W(1 - \sin\alpha + \cos\alpha) \quad \text{and} \quad R' = \frac{1}{2}W(1 - \cos\alpha + \sin\alpha).$$

The true values are $R = W\cos\alpha$ and $R' = W\sin\alpha$. Nevertheless some of his papers on mechanics are valuable. Of these the most important were two, in 1689 and 1694, in which he solved the problem of finding an isochronous curve; one, in 1697, on the curve of quickest descent (this was the problem sent as a challenge to Newton); and two, in 1691 and 1692, in which he stated the intrinsic equation of the curve assumed by a flexible rope suspended from two points, that is, the catenary, but gave no proof. This last problem had been originally proposed by Galileo.

In 1689, that is, two years after the *Principia* had been published, he wrote on the movements of the planets which he stated were produced by a motion of the ether. Not only were the equations of motion which he obtained wrong, but his deductions from them were not even in accordance with his own axioms. In another memoir in 1706, that is, nearly twenty years after the *Principia* had been written, he admitted that he had made some mistakes in his former paper, but adhered to his previous conclusions, and summed the matter up by saying "it is certain that gravitation generates a new force at each instant to the centre, but the centrifugal force also generates another away from the centre.... The centrifugal force may be considered in two aspects according as the movement is treated as along the tangent to the curve or as along the arc of the circle itself." It seems clear from this paper that he did not really understand the principles of dynamics, and it is hardly necessary to consider his work on the subject in further detail. Much of it is vitiated by a constant confusion between momentum and kinetic energy: when the force is "passive" he uses the first, which he calls the *vis mortua*, as the measure of a force; when the force is "active" he uses the latter, the double of which he calls the *vis viva*.

The series quoted by Leibnitz comprise those for e^x, $\log(1+x)$, $\sin x$, vers x, and $\tan^{-1} x$; all of these had been previously published, and he rarely, if ever, added any demonstrations. Leibnitz (like Newton) recognised the importance of James Gregory's remarks on the necessity of examining whether infinite series are convergent or divergent, and proposed a test to distinguish series whose terms are alternately positive and negative. In 1693 he explained the method of expansion by indeterminate coefficients, though his applications were not free from error.

To sum the matter up briefly, it seems to me that Leibnitz's work exhibits great skill in analysis, but much of it is unfinished, and when he leaves his symbols and attempts to interpret his results he frequently commits blunders. No doubt the demands of politics, philosophy, and literature on his time may have prevented him from elaborating any problem completely or writing a systematic exposition of his views, though they are no excuse for the mistakes of principle which occur in his papers. Some of his memoirs contain suggestions of methods which have now become valuable means of analysis, such as the use of determinants and of indeterminate coefficients; but when a writer of manifold interests like Leibnitz throws out innumerable suggestions, some of them are likely to turn out valuable; and to enumerate these (which he did not work out) without reckoning the others (which are wrong) gives a false impression of the value of his work. But in spite of this, his title to fame rests on a sure basis, for by his advocacy of the differential calculus his name is inseparably connected with one of the chief instruments of analysis, as that of Descartes—another philosopher—is similarly connected with analytical geometry.

Leibnitz was but one amongst several continental writers whose papers in the *Acta Eruditorum* familiarised mathematicians with the use of the differential calculus. Among the most important of these were James and John Bernoulli, both of whom were warm friends and admirers of Leibnitz, and to their devoted advocacy his reputation is largely due. Not only did they take a prominent part in nearly every mathematical question then discussed, but nearly all the leading mathematicians on the continent during the first half of the eighteenth century came directly or indirectly under the influence of one or both of them.

The Bernoullis[1] (or as they are sometimes, and perhaps more correctly, called, the Bernouillis) were a family of Dutch origin, who were driven from Holland by the Spanish persecutions, and finally settled at Bâle in Switzerland. The first member of the family who attained distinction in mathematics was James.

James Bernoulli.[2]   *Jacob* or *James Bernoulli* was born at Bâle on December 27, 1654; in 1687 he was appointed to a chair of mathematics in the university there; and occupied it until his death on August 16, 1705.

He was one of the earliest to realize how powerful as an instrument of analysis was the infinitesimal calculus, and he applied it to several problems, but he did not himself invent any new processes. His great influence was uniformly and successfully exerted in favour of the use of the differential calculus, and his lessons on it, which were written in the form of two essays in 1691 and are published in the second volume of his works, shew how completely he had even then grasped the principles of the new analysis. These lectures, which contain the earliest use of the term integral, were the first published attempt to construct an integral

---

[1]See the account in the *Allgemeine deutsche Biographie*, vol. ii, Leipzig, 1875, pp. 470–483.

[2]See the *éloge* by B. de Fontenelle, Paris, 1766; also Montucla's *Histoire*, vol. ii. A collected edition of the works of James Bernoulli was published in two volumes at Geneva in 1744, and an account of his life is prefixed to the first volume.

calculus; for Leibnitz had treated each problem by itself, and had not laid down any general rules on the subject.

The most important discoveries of James Bernoulli were his solution of the problem to find an isochronous curve; his proof that the construction for the catenary which had been given by Leibnitz was correct, and his extension of this to strings of variable density and under a central force; his determination of the form taken by an elastic rod fixed at one end and acted on by a given force at the other, the *elastica*; also of a flexible rectangular sheet with two sides fixed horizontally and filled with a heavy liquid, the *lintearia*; and, lastly, of a sail filled with wind, the *velaria*. In 1696 he offered a reward for the general solution of isoperimetrical figures, that is, of figures of a given species and given perimeter which shall include a maximum area: his own solution, published in 1701, is correct as far as it goes. In 1698 he published an essay on the differential calculus and its applications to geometry. He here investigated the chief properties of the equiangular spiral, and especially noticed the manner in which various curves deduced from it reproduced the original curve: struck by this fact he begged that, in imitation of Archimedes, an equiangular spiral should be engraved on his tombstone with the inscription *eadem numero mutata resurgo*. He also brought out in 1695 an edition of Descartes's *Géométrie*. In his *Ars Conjectandi*, published in 1713, he established the fundamental principles of the calculus of probabilities; in the course of the work he defined the numbers known by his name[1] and explained their use, he also gave some theorems on finite differences. His higher lectures were mostly on the theory of series; these were published by Nicholas Bernoulli in 1713.

John Bernoulli.[2]   *John Bernoulli*, the brother of James Bernoulli, was born at Bâle on August 7, 1667, and died there on January 1, 1748. He occupied the chair of mathematics at Groningen from 1695 to 1705; and at Bâle, where he succeeded his brother, from 1705 to 1748. To all who did not acknowledge his merits in a manner commensurate with his own view of them he behaved most unjustly: as an illustration

---

[1] A bibliography of Bernoulli's Numbers was given by G. S. Ely, in the American *Journal of Mathematics*, 1882, vol. v, pp. 228–235.

[2] D'Alembert wrote a eulogistic *éloge* on the work and influence of John Bernoulli, but he explicitly refused to deal with his private life or quarrels; see also Montucla's *Histoire*, vol. ii. A collected edition of the works of John Bernoulli was published at Geneva in four volumes in 1742, and his correspondence with Leibnitz was published in two volumes at the same place in 1745.

of his character it may be mentioned that he attempted to substitute for an incorrect solution of his own on the problem of isoperimetrical curves another stolen from his brother James, while he expelled his son Daniel from his house for obtaining a prize from the French Academy which he had expected to receive himself. He was, however, the most successful teacher of his age, and had the faculty of inspiring his pupils with almost as passionate a zeal for mathematics as he felt himself. The general adoption on the continent of the differential rather than the fluxional notation was largely due to his influence.

Leaving out of account his innumerable controversies, the chief discoveries of John Bernoulli were the exponential calculus, the treatment of trigonometry as a branch of analysis, the conditions for a geodesic, the determination of orthogonal trajectories, the solution of the brachistochrone, the statement that a ray of light pursues such a path that $\Sigma\mu ds$ is a minimum, and the enunciation of the principle of virtual work. I believe that he was the first to denote the accelerating effect of gravity by an algebraical sign g, and he thus arrived at the formula $v^2 = 2gh$: the same result would have been previously expressed by the proportion $v_1^2 : v_2^2 = h_1 : h_2$. The notation ϕx to indicate a function[1] of x was introduced by him in 1718, and displaced the notation X or ξ proposed by him in 1698; but the general adoption of symbols like f, F, ϕ, ψ, ... to represent functions, seems to be mainly due to Euler and Lagrange.

Several members of the same family, but of a younger generation, enriched mathematics by their teaching and writings. The most important of these were the three sons of John; namely, Nicholas, Daniel, and John the younger; and the two sons of John the younger, who bore the names of John and James. To make the account complete I add here their respective dates. *Nicholas Bernoulli*, the eldest of the three sons of John, was born on Jan. 27, 1695, and was drowned at Petrograd, where he was professor, on July 26, 1726. *Daniel Bernoulli*, the second son of John, was born on Feb. 9, 1700, and died on March 17, 1782; he was professor first at Petrograd and afterwards at Bâle, and shares with Euler the unique distinction of having gained the prize proposed annually by the French Academy no less than ten times: I refer to him again a few pages later. *John Bernoulli*, the younger, a brother of Nicholas and Daniel, was born on May 18, 1710, and died in 1790; he also was

---

[1] On the meaning assigned at first to the word *function* see a note by M. Cantor, *L'Intermédiaire des mathématiciens*, January 1896, vol. iii, pp. 22–23.

a professor at Bâle. He left two sons, *John* and *James*: of these, the former, who was born on Dec. 4, 1744, and died on July 10, 1807, was astronomer-royal and director of mathematical studies at Berlin; while the latter, who was born on Oct. 17, 1759, and died in July 1789, was successively professor at Bâle, Verona, and Petrograd.

The development of analysis on the continent.

Leaving for a moment the English mathematicians of the first half of the eighteenth century we come next to a number of continental writers who barely escape mediocrity, and to whom it will be necessary to devote but few words. Their writings mark the steps by which analytical geometry and the differential and integral calculus were perfected and made familiar to mathematicians. Nearly all of them were pupils of one or other of the two elder Bernoullis, and they were so nearly contemporaries that it is difficult to arrange them chronologically. The most eminent of them are *Cramer, de Gua, de Montmort, Fagnano, l'Hospital, Nicole, Parent, Riccati, Saurin,* and *Varignon.*

L'Hospital.   *Guillaume François Antoine l'Hospital, Marquis de St.-Mesme,* born at Paris in 1661, and died there on Feb. 2, 1704, was among the earliest pupils of John Bernoulli, who, in 1691, spent some months at l'Hospital's house in Paris for the purpose of teaching him the new calculus. It seems strange, but it is substantially true, that a knowledge of the infinitesimal calculus and the power of using it was then confined to Newton, Leibnitz, and the two elder Bernoullis—and it will be noticed that they were the only mathematicians who solved the more difficult problems then proposed as challenges. There was at that time no text-book on the subject, and the credit of putting together the first treatise which explained the principles and use of the method is due to l'Hospital; it was published in 1696 under the title *Analyse des infiniment petits.* This contains a partial investigation of the limiting value of the ratio of functions which for a certain value of the variable take the indeterminate form 0 : 0, a problem solved by John Bernoulli in 1704. This work had a wide circulation; it brought the differential notation into general use in France, and helped to make it known in Europe. A supplement, containing a similar treatment of the integral calculus, together with additions to the differential calculus which had been made in the following half century, was published at Paris, 1754–56, by L. A. de Bougainville.

L'Hospital took part in most of the challenges issued by Leibnitz, the Bernoullis, and other continental mathematicians of the time; in particular he gave a solution of the brachistochrone, and investigated the form of the solid of least resistance of which Newton in the *Principia* had stated the result. He also wrote a treatise on analytical conics, which was published in 1707, and for nearly a century was deemed a standard work on the subject.

Varignon.[1]   *Pierre Varignon*, born at Caen in 1654, and died in Paris on Dec. 22, 1722, was an intimate friend of Newton, Leibnitz, and the Bernoullis, and, after l'Hospital, was the earliest and most powerful advocate in France of the use of the differential calculus. He realized the necessity of obtaining a test for examining the convergency of series, but the analytical difficulties were beyond his powers. He simplified the proofs of many of the leading propositions in mechanics, and in 1687 recast the treatment of the subject, basing it on the composition of forces. His works were published at Paris in 1725.

De Montmort. Nicole.   *Pierre Raymond de Montmort*, born at Paris on Oct. 27, 1678, and died there on Oct. 7, 1719, was interested in the subject of finite differences. He determined in 1713 the sum of n terms of a finite series of the form

$$na + \frac{n(n-1)}{1\cdot 2}\Delta a + \frac{n(n-1)(n-2)}{1\cdot 2\cdot 3}\Delta^2 a + \ldots;$$

a theorem which seems to have been independently rediscovered by Chr. Goldbach in 1718. *François Nicole*, who was born at Paris on Dec. 23, 1683, and died there on Jan. 18, 1758, published his *Traité du calcul des differences finies* in 1717; it contains rules both for forming differences and for effecting the summation of given series. Besides this, in 1706 he wrote a work on roulettes, especially spherical epicycloids; and in 1729 and 1731 he published memoirs on Newton's essay on curves of the third degree.

Parent. Saurin. De Gua.   *Antoine Parent*, born at Paris on Sept. 16, 1666, and died there on Sept. 26, 1716, wrote in 1700 on analytical geometry of three dimensions. His works were collected and published in three volumes at Paris in 1713. *Joseph Saurin*, born at Courtaison in 1659, and died at Paris on Dec. 29, 1737, was the first to show how the tangents at the multiple points of curves could be determined by analysis. *Jean Paul de Gua de Malves* was born at Carcassonne in 1713, and died at Paris on June 2, 1785. He published in

---

[1]See the *éloge* by B. de Fontenelle, Paris, 1766.

1740 a work on analytical geometry in which he applied it, without the aid of the differential calculus, to find the tangents, asymptotes, and various singular points of an algebraical curve; and he further shewed how singular points and isolated loops were affected by conical projection. He gave the proof of Descartes's rule of signs which is to be found in most modern works. It is not clear whether Descartes ever proved it strictly, and Newton seems to have regarded it as obvious.

Cramer. *Gabriel Cramer*, born at Geneva in 1704, and died at Bagnols in 1752, was professor at Geneva. The work by which he is best known is his treatise on algebraic curves[1] published in 1750, which, as far as it goes, is fairly complete; it contains the earliest demonstration that a curve of the nth degree is in general determined if $\frac{1}{2}n(n+3)$ points on it be given. This work is still sometimes read. Besides this, he edited the works of the two elder Bernoullis; and wrote on the physical cause of the spheroidal shape of the planets and the motion of their apses, 1730, and on Newton's treatment of cubic curves, 1746.

Riccati. *Jacopo Francesco, Count Riccati*, born at Venice on May 28, 1676, and died at Trèves on April 15, 1754, did a great deal to disseminate a knowledge of the Newtonian philosophy in Italy. Besides the equation known by his name, certain cases of which he succeeded in integrating, he discussed the question of the possibility of lowering the order of a given differential equation. His works were published at Trèves in four volumes in 1758. He had two sons who wrote on several minor points connected with the integral calculus and differential equations, and applied the calculus to several mechanical questions: these were *Vincenzo*, who was born in 1707 and died in 1775, and *Giordano*, who was born in 1709 and died in 1790.

Fagnano. *Giulio Carlo, Count Fagnano*, and *Marquis de Toschi*, born at Sinigaglia on Dec. 6, 1682, and died on Sept. 26, 1766, may be said to have been the first writer who directed attention to the theory of elliptic functions. Failing to rectify the ellipse or hyperbola, Fagnano attempted to determine arcs whose difference should be rectifiable. He also pointed out the remarkable analogy existing between the integrals which represent the arc of a circle and the arc of a lemniscate. Finally he proved the formula

$$\pi = 2i \log\{(1 - i)/(1 + i)\},$$

[1] See Cantor, chapter cxvi.

where i stands for $\sqrt{-1}$. His works were collected and published in two volumes at Pesaro in 1750.

It was inevitable that some mathematicians should object to methods of analysis founded on the infinitesimal calculus. The most prominent of these were *Viviani, De la Hire*, and *Rolle*, whose names were mentioned at the close of chapter XV.

So far no one of the school of Leibnitz and the two elder Bernoullis had shewn any exceptional ability, but by the action of a number of second-rate writers the methods and language of analytical geometry and the differential calculus had become well known by about 1740. The close of this school is marked by the appearance of *Clairaut, D'Alembert*, and *Daniel Bernoulli*. Their lives overlap the period considered in the next chapter, but, though it is difficult to draw a sharp dividing line which shall separate by a definite date the mathematicians there considered from those whose writings are discussed in this chapter, I think that on the whole the works of these three writers are best treated here.

Clairaut.   *Alexis Claude Clairaut* was born at Paris on May 13, 1713, and died there on May 17, 1765. He belongs to the small group of children who, though of exceptional precocity, survive and maintain their powers when grown up. As early as the age of twelve he wrote a memoir on four geometrical curves; but his first important work was a treatise on tortuous curves, published when he was eighteen—a work which procured for him admission to the French Academy. In 1731 he gave a demonstration of the fact noted by Newton that all curves of the third order were projections of one of five parabolas.

In 1741 Clairaut went on a scientific expedition to measure the length of a meridian degree on the earth's surface, and on his return in 1743 he published his *Théorie de la figure de la terre*. This is founded on a paper by Maclaurin, wherein it had been shewn that a mass of homogeneous fluid set in rotation about a line through its centre of mass would, under the mutual attraction of its particles, take the form of a spheroid. This work of Clairaut treated of heterogeneous spheroids and contains the proof of his formula for the accelerating effect of gravity in a place of latitude l, namely,

$$g = G \left\{ 1 + \left(\tfrac{5}{2}m - \epsilon \right) \sin^2 l \right\},$$

where G is the value of equatorial gravity, m the ratio of the centrifugal force to gravity at the equator, and ϵ the ellipticity of a meridian section of the earth. In 1849 Stokes[1] shewed that the same result was true whatever was the interior constitution or density of the earth, provided the surface was a spheroid of equilibrium of small ellipticity.

Impressed by the power of geometry as shewn in the writings of Newton and Maclaurin, Clairaut abandoned analysis, and his next work, the *Théorie de la lune*, published in 1752, is strictly Newtonian in character. This contains the explanation of the motion of the apse which had previously puzzled astronomers, and which Clairaut had at first deemed so inexplicable that he was on the point of publishing a new hypothesis as to the law of attraction when it occurred to him to carry the approximation to the third order, and he thereupon found that the result was in accordance with the observations. This was followed in 1754 by some lunar tables. Clairaut subsequently wrote various papers on the orbit of the moon, and on the motion of comets as affected by the perturbation of the planets, particularly on the path of Halley's comet.

His growing popularity in society hindered his scientific work: "engagé," says Bossut, "à des soupers, à des veilles, entraîné par un goût vif pour les femmes, voulant allier le plaisir à ses travaux ordinaires, il perdit le repos, la santé, enfin la vie à l'âge de cinquante-deux ans."

D'Alembert.[2]   *Jean-le-Rond D'Alembert* was born at Paris on November 16, 1717, and died there on October 29, 1783. He was the illegitimate child of the chevalier Destouches. Being abandoned by his mother on the steps of the little church of St. Jean-le-Rond, which then nestled under the great porch of Notre-Dame, he was taken to the parish commissary, who, following the usual practice in such cases, gave him the Christian name of Jean-le-Rond; I do not know by what authority he subsequently assumed the right to prefix *de* to his name. He was boarded out by the parish with the wife of a glazier in a small way of business who lived near the cathedral, and here he found a real home, though a humble one. His father appears to have looked after him, and paid for his going to a school where he obtained a fair mathematical education.

An essay written by him in 1738 on the integral calculus, and another in 1740 on "ducks and drakes" or ricochets, attracted some at-

---

[1]See *Cambridge Philosophical Transactions*, vol. viii, pp. 672–695.

[2]Bertrand, Condorcet, and J. Bastien have left sketches of D'Alembert's life. His literary works have been published, but there is no complete edition of his scientific writings. Some papers and letters, discovered comparatively recently, were published by C. Henry at Paris in 1887.

tention, and in the same year he was elected a member of the French Academy; this was probably due to the influence of his father. It is to his credit that he absolutely refused to leave his adopted mother, with whom he continued to live until her death in 1757. It cannot be said that she sympathised with his success, for at the height of his fame she remonstrated with him for wasting his talents on such work: "Vous ne serez jamais qu'un philosophe," said she, "et qu'est-ce qu'un philosophe? c'est un fou qui se tourmente pendant sa vie, pour qu'on parle de lui lorsqu'il n'y sera plus."

Nearly all his mathematical works were produced during the years 1743 to 1754. The first of these was his *Traité de dynamique*, published in 1743, in which he enunciates the principle known by his name, namely, that the "internal forces of inertia" (that is, the forces which resist acceleration) must be equal and opposite to the forces which produce the acceleration. This may be inferred from Newton's second reading of his third law of motion, but the full consequences had not been realized previously. The application of this principle enables us to obtain the differential equations of motion of any rigid system.

In 1744 D'Alembert published his *Traité de l'équilibre et du mouvement des fluides*, in which he applies his principle to fluids; this led to partial differential equations which he was then unable to solve. In 1745 he developed that part of the subject which dealt with the motion of air in his *Théorie générale des vents*, and this again led him to partial differential equations. A second edition of this in 1746 was dedicated to Frederick the Great of Prussia, and procured an invitation to Berlin and the offer of a pension; he declined the former, but subsequently, after some pressing, pocketed his pride and the latter. In 1747 he applied the differential calculus to the problem of a vibrating string, and again arrived at a partial differential equation.

His analysis had three times brought him to an equation of the form

$$\frac{\partial^2 u}{\partial t^2} = \frac{\partial^2 u}{\partial x^2},$$

and he now succeeded in shewing that it was satisfied by

$$u = \phi(x + t) + \psi(x - t),$$

where ϕ and ψ are arbitrary functions. It may be interesting to give his solution which was published in the transactions of the Berlin Academy for 1747. He begins by saying that, if $\dfrac{\partial u}{\partial x}$ be denoted by p and $\dfrac{\partial u}{\partial t}$ by

q, then
$$du = pdx + qdt.$$

But, by the given equation, $\dfrac{\partial q}{\partial t} = \dfrac{\partial p}{\partial x}$, and therefore $pdt + qdx$ is also an exact differential: denote it by dv.

Therefore $$dv = pdt + qdx.$$

Hence $du + dv = (pdx + qdt) + (pdt + qdx) = (p + q)(dx + dt)$,

and $\quad du - dv = (pdx + qdt) - (pdt + qdx) = (p - q)(dx - dt)$.

Thus $u + v$ must be a function of $x + t$, and $u - v$ must be a function of $x - t$. We may therefore put

$$u + v = 2\phi(x + t),$$

and $$u - v = 2\psi(x - t).$$

Hence $$u = \phi(x + t) + \psi(x - t).$$

D'Alembert added that the conditions of the physical problem of a vibrating string demand that, when $x = 0$, u should vanish for all values of t. Hence identically

$$\phi(t) + \psi(-t) = 0.$$

Assuming that both functions can be expanded in integral powers of t, this requires that they should contain only odd powers. Hence

$$\psi(-t) = -\phi(t) = \phi(-t).$$

Therefore
$$u = \phi(x + t) + \phi(x - t).$$

Euler now took the matter up and shewed that the equation of the form of the string was $\dfrac{\partial^2 u}{\partial t^2} = a^2 \dfrac{\partial^2 u}{\partial x^2}$, and that the general integral was $u = \phi(x + at) + \psi(x - at)$, where ϕ and ψ are arbitrary functions.

The chief remaining contributions of D'Alembert to mathematics were on physical astronomy, especially on the precession of the equinoxes, and on variations in the obliquity of the ecliptic. These were collected in his *Système du monde*, published in three volumes in 1754.

During the latter part of his life he was mainly occupied with the great French encyclopaedia. For this he wrote the introduction, and numerous philosophical and mathematical articles; the best are those on geometry and on probabilities. His style is brilliant, but not polished, and faithfully reflects his character, which was bold, honest, and frank. He defended a severe criticism which he had offered on some mediocre work by the remark, "j'aime mieux être incivil qu'ennuyé"; and with his dislike of sycophants and bores it is not surprising that during his life he had more enemies than friends.

Daniel Bernoulli.[1]   *Daniel Bernoulli,* whose name I mentioned above, and who was by far the ablest of the younger Bernoullis, was a contemporary and intimate friend of Euler, whose works are mentioned in the next chapter. Daniel Bernoulli was born on Feb. 9, 1700, and died at Bâle, where he was professor of natural philosophy, on March 17, 1782. He went to Petrograd in 1724 as professor of mathematics, but the roughness of the social life was distasteful to him, and he was not sorry when a temporary illness in 1733 allowed him to plead his health as an excuse for leaving. He then returned to Bâle, and held successively chairs of medicine, metaphysics, and natural philosophy there.

His earliest mathematical work was the *Exercitationes,* published in 1724, which contains a solution of the differential equation proposed by Riccati. Two years later he pointed out for the first time the frequent desirability of resolving a compound motion into motions of translation and motions of rotation. His chief work is his *Hydrodynamica,* published in 1738; it resembles Lagrange's *Mécanique analytique* in being arranged so that all the results are consequences of a single principle, namely, in this case, the conservation of energy. This was followed by a memoir on the theory of the tides, to which, conjointly with memoirs by Euler and Maclaurin, a prize was awarded by the French Academy: these three memoirs contain all that was done on this subject between the publication of Newton's *Principia* and the investigations of Laplace. Bernoulli also wrote a large number of papers on various mechanical questions, especially on problems connected with vibrating strings, and

---

[1]The only account of Daniel Bernoulli's life with which I am acquainted is the *éloge* by his friend Condorcet. *Marie Jean Antoine Nicolas Caritat, Marquis de Condorcet,* was born in Picardy on Sept. 17, 1743, and fell a victim to the republican terrorists on March 28, 1794. He was secretary to the Academy, and is the author of numerous *éloges.* He is perhaps more celebrated for his studies in philosophy, literature, and politics than in mathematics, but his mathematical treatment of probabilities, and his discussion of differential equations and finite differences, shew an ability which might have put him in the first rank had he concentrated his attention on mathematics. He sacrificed himself in a vain effort to guide the revolutionary torrent into a constitutional channel.

the solutions given by Taylor and by D'Alembert. He is the earliest writer who attempted to formulate a kinetic theory of gases, and he applied the idea to explain the law associated with the names of Boyle and Mariotte.

The English mathematicians of the eighteenth century.

I have reserved a notice of the English mathematicians who succeeded Newton, in order that the members of the English school may be all treated together. It was almost a matter of course that the English should at first have adopted the notation of Newton in the infinitesimal calculus in preference to that of Leibnitz, and consequently the English school would in any case have developed on somewhat different lines to that on the continent, where a knowledge of the infinitesimal calculus was derived solely from Leibnitz and the Bernoullis. But this separation into two distinct schools became very marked owing to the action of Leibnitz and John Bernoulli, which was naturally resented by Newton's friends; and so for forty or fifty years, to the disadvantage of both sides, the quarrel raged. The leading members of the English school were *Cotes, Demoivre, Ditton, David Gregory, Halley, Maclaurin, Simpson,* and *Taylor.* I may, however, again remind my readers that as we approach modern times the number of capable mathematicians in Britain, France, Germany, and Italy becomes very considerable, but that in a popular sketch like this book it is only the leading men whom I propose to mention.

To David Gregory, Halley, and Ditton I need devote but few words.

David Gregory. *David Gregory,* the nephew of the James Gregory mentioned above, born at Aberdeen on June 24, 1661, and died at Maidenhead on Oct. 10, 1708, was appointed professor at Edinburgh in 1684, and in 1691 was on Newton's recommendation elected Savilian professor at Oxford. His chief works are one on geometry, issued in 1684; one on optics, published in 1695, which contains [p. 98] the earliest suggestion of the possibility of making an achromatic combination of lenses; and one on the Newtonian geometry, physics, and astronomy, issued in 1702.

Halley. *Edmund Halley,* born in London in 1656, and died at Greenwich in 1742, was educated at St. Paul's School, London, and Queen's College, Oxford, in 1703 succeeded Wallis as Savilian professor, and subsequently in 1720 was appointed astronomer-royal in succession to Flamsteed, whose *Historia Coelestis Britannica* he edited; the first

and imperfect edition was issued in 1712. Halley's name will be recollected for the generous manner in which he secured the immediate
publication of Newton's *Principia* in 1687. Most of his original work
was on astronomy and allied subjects, and lies outside the limits of this
book; it may be, however, said that the work is of excellent quality,
and both Lalande and Mairan speak of it in the highest terms. Halley
conjecturally restored the eighth and lost book of the conics of Apollonius, and in 1710 brought out a magnificent edition of the whole work;
he also edited the works of Serenus, those of Menelaus, and some of the
minor works of Apollonius. He was in his turn succeeded at Greenwich
as astronomer-royal by Bradley.[1]

Ditton.   *Humphry Ditton* was born at Salisbury on May 29,
1675, and died in London in 1715 at Christ's Hospital, where he was
mathematical master. He does not seem to have paid much attention to
mathematics until he came to London about 1705, and his early death
was a distinct loss to English science. He published in 1706 a text book
on fluxions; this and another similar work by William Jones, which
was issued in 1711, occupied in England much the same place that
l'Hospital's treatise did in France. In 1709 Ditton issued an algebra,
and in 1712 a treatise on perspective. He also wrote numerous papers
in the *Philosophical Transactions*. He was the earliest writer to attempt
to explain the phenomenon of capillarity on mathematical principles;
and he invented a method for finding the longitude, which has been
since used on various occasions.

Taylor.[2]   *Brook Taylor,* born at Edmonton on August 18, 1685,
and died in London on December 29, 1731, was educated at St. John's
College, Cambridge, and was among the most enthusiastic of Newton's
admirers. From the year 1712 onwards he wrote numerous papers in the
Philosophical Transactions, in which, among other things, he discussed
the motion of projectiles, the centre of oscillation, and the forms taken
by liquids when raised by capillarity. In 1719 he resigned the secretary-

---

[1] *James Bradley*, born in Gloucestershire in 1692, and died in 1762, was the most
distinguished astronomer of the first half of the eighteenth century. Among his more
important discoveries were the explanation of astronomical aberration (1729), the
cause of nutation (1748), and his empirical formula for corrections for refraction. It
is perhaps not too much to say that he was the first astronomer who made the art
of observing part of a methodical science.

[2] An account of his life by Sir William Young is prefixed to the *Contemplatio
Philosophica*. This was printed at London in 1793 for private circulation and is now
extremely rare.

ship of the Royal Society and abandoned the study of mathematics. His earliest work, and that by which he is generally known, is his *Methodus Incrementorum Directa et Inversa*, published in London in 1715. This contains [prop. 7] a proof of the well-known theorem

$$f(x + h) = f(x) + hf'(x) + \frac{h^2}{\underline{|2}} f''(x) + \dots,$$

by which a function of a single variable can be expanded in powers of it. He does not consider the convergency of the series, and the proof which involves numerous assumptions is not worth reproducing. The work also includes several theorems on interpolation. Taylor was the earliest writer to deal with theorems on the change of the independent variable; he was perhaps the first to realize the possibility of a calculus of operation, and just as he denotes the nth differential coefficient of y by y_n, so he uses y_{-1} to represent the integral of y; lastly, he is usually recognized as the creator of the theory of finite differences.

The applications of the calculus to various questions given in the *Methodus* have hardly received that attention they deserve. The most important of them is the theory of the transverse vibrations of strings, a problem which had baffled previous investigators. In this investigation Taylor shews that the number of half-vibrations executed in a second is

$$\pi \sqrt{DP/LN},$$

where L is the length of the string, N its weight, P the weight which stretches it, and D the length of a seconds pendulum. This is correct, but in arriving at it he assumes that every point of the string will pass through its position of equilibrium at the same instant, a restriction which D'Alembert subsequently shewed to be unnecessary. Taylor also found the form which the string assumes at any instant.

The *Methodus* also contains the earliest determination of the differential equation of the path of a ray of light when traversing a heterogeneous medium; and, assuming that the density of the air depends only on its distance from the earth's surface, Taylor obtained by means of quadratures the approximate form of the curve. The form of the catenary and the determination of the centres of oscillation and percussion are also discussed.

A treatise on perspective by Taylor, published in 1719, contains the earliest general enunciation of the principle of vanishing points; though the idea of vanishing points for horizontal and parallel lines in a picture hung in a vertical plane had been enunciated by Guido Ubaldi in his *Perspectivae Libri*, Pisa, 1600, and by Stevinus in his *Sciagraphia*, Leyden, 1608.

Cotes.   *Roger Cotes* was born near Leicester on July 10, 1682, and died at Cambridge on June 5, 1716. He was educated at Trinity College, Cambridge, of which society he was a fellow, and in 1706 was elected to the newly-created Plumian chair of astronomy in the university of Cambridge. From 1709 to 1713 his time was mainly occupied in editing the second edition of the *Principia*. The remark of Newton that if only Cotes had lived "we might have known something" indicates the opinion of his abilities held by most of his contemporaries.

Cotes's writings were collected and published in 1722 under the titles *Harmonia Mensurarum* and *Opera Miscellanea*. His lectures on hydrostatics were published in 1738. A large part of the *Harmonia Mensurarum* is given up to the decomposition and integration of rational algebraical expressions. That part which deals with the theory of partial fractions was left unfinished, but was completed by Demoivre. Cotes's theorem in trigonometry, which depends on forming the quadratic factors of $x^n - 1$, is well known. The proposition that "if from a fixed point O a line be drawn cutting a curve in Q_1, Q_2, ..., Q_n, and a point P be taken on the line so that the reciprocal of OP is the arithmetic mean of the reciprocals of OQ_1, OQ_2, ..., OQ_n, then the locus of P will be a straight line" is also due to Cotes. The title of the book was derived from the latter theorem. The *Opera Miscellanea* contains a paper on the method for determining the most probable result from a number of observations. This was the earliest attempt to frame a theory of errors. It also contains essays on Newton's *Methodus Differentialis*, on the construction of tables by the method of differences, on the descent of a body under gravity, on the cycloidal pendulum, and on projectiles.

Demoivre.   *Abraham Demoivre* (more correctly written as *de Moivre*) was born at Vitry on May 26, 1667, and died in London on November 27, 1754. His parents came to England when he was a boy, and his education and friends were alike English. His interest in the higher mathematics is said to have originated in his coming by chance across a copy of Newton's *Principia*. From the *éloge* on him delivered in 1754 before the French Academy it would seem that his work as a teacher of mathematics had led him to the house of the Earl of Devonshire at the instant when Newton, who had asked permission to present a copy of his work to the earl, was coming out. Taking up the book, and charmed by the far-reaching conclusions and the apparent simplicity of the reasoning, Demoivre thought nothing would be easier than to master the subject, but to his surprise found that to follow the argument overtaxed his powers. He, however, bought a copy, and as he had but little leisure he tore out the pages in order to carry one or two of them loose in his pocket so that he could study them in the intervals of his work as a teacher. Subsequently he joined the Royal Society, and became intimately connected with Newton, Halley, and other mathematicians of the English school. The manner of his death has a certain interest for psychologists. Shortly before it he declared that it was necessary for him to sleep some ten minutes or a quarter of an hour longer each day than the preceding one. The day after he had thus reached a total of something over twenty-three hours he slept up to the limit of twenty-four hours, and then died in his sleep.

He is best known for having, together with Lambert, created that part of trigonometry which deals with imaginary quantities. Two theorems on this part of the subject are still connected with his name, namely, that which asserts that $\cos nx + i \sin nx$ is one of the values of $(\cos x + i \sin x)^n$, and that which gives the various quadratic factors of $x^{2n} - 2px^n + 1$. His chief works, other than numerous papers in the *Philosophical Transactions*, were *The Doctrine of Chances*, published in 1718, and the *Miscellanea Analytica*, published in 1730. In the former the theory of recurring series was first given, and the theory of partial fractions which Cotes's premature death had left unfinished was completed, while the rule for finding the probability of a compound event was enunciated. The latter book, besides the trigonometrical propositions mentioned above, contains some theorems in astronomy, but they are treated as problems in analysis.

Maclaurin.[1] *Colin Maclaurin*, who was born at Kilmodan in Argyllshire in February 1698, and died at York on June 14, 1746, was educated at the university of Glasgow; in 1717 he was elected, at the early age of nineteen, professor of mathematics at Aberdeen; and in 1725 he was appointed the deputy of the mathematical professor at Edinburgh, and ultimately succeeded him. There was some difficulty in securing a stipend for a deputy, and Newton privately wrote offering to bear the cost so as to enable the university to secure the services of Maclaurin. Maclaurin took an active part in opposing the advance of the Young Pretender in 1745; on the approach of the Highlanders he

---

[1] A sketch of Maclaurin's life is prefixed to his posthumous account of Newton's discoveries, London, 1748.

fled to York, but the exposure in the trenches at Edinburgh and the privations he endured in his escape proved fatal to him.

His chief works are his *Geometria Organica*, London, 1720; his *De Linearum Geometricarum Proprietatibus*, London, 1720; his *Treatise on Fluxions*, Edinburgh, 1742; his *Algebra*, London, 1748; and his *Account of Newton's Discoveries*, London, 1748.

The first section of the first part of the *Geometria Organica* is on conics; the second on nodal cubics; the third on other cubics and on quartics; and the fourth section is on general properties of curves. Newton had shewn that, if two angles bounded by straight lines turn round their respective summits so that the point of intersection of two of these lines moves along a straight line, the other point of intersection will describe a conic; and, if the first point move along a conic, the second will describe a quartic. Maclaurin gave an analytical discussion of the general theorem, and shewed how by this method various curves could be practically traced. This work contains an elaborate discussion on curves and their pedals, a branch of geometry which he had created in two papers published in the *Philosophical Transactions* for 1718 and 1719.

The second part of the work is divided into three sections and an appendix. The first section contains a proof of Cotes's theorem above alluded to; and also the analogous theorem (discovered by himself) that, if a straight line $OP_1P_2\ldots$ drawn through a fixed point O cut a curve of the nth degree in n points P_1, P_2, $\ldots$, and if the tangents at P_1, P_2, $\ldots$ cut a fixed line Ox in points A_1, A_2, $\ldots$, then the sum of the reciprocals of the distances OA_1, OA_2, $\ldots$ is constant for all positions of the line $OP_1P_2\ldots$. These two theorems are generalizations of those given by Newton on diameters and asymptotes. Either is deducible from the other. In the second and third sections these theorems are applied to conics and cubics; most of the harmonic properties connected with a quadrilateral inscribed in a conic are determined; and in particular the theorem on an inscribed hexagon which is known by the name of Pascal is deduced. Pascal's essay was not published till 1779, and the earliest printed enunciation of his theorem was that given by Maclaurin. Amongst other propositions he shews that, if a quadrilateral be inscribed in a cubic, and if the points of intersection of the opposite sides also lie on the curve, then the tangents to the cubic at any two opposite angles of the quadrilateral will meet on the curve. In the fourth section he considers some theorems on central force. The fifth section contains some theorems on the description of curves through given points.

One of these (which includes Pascal's as a particular case) is that if a polygon be deformed so that while each of its sides passes through a fixed point its angles (save one) describe respectively curves of the mth, nth, pth,... degrees, then shall a remaining angle describe a curve of the degree $2mnp...$; but if the given points be collinear, the resulting curve will be only of the degree $mnp...$. This essay was reprinted with additions in the *Philosophical Transactions* for 1735.

The *Treatise of Fluxions*, published in 1742, was the first logical and systematic exposition of the method of fluxions. The cause of its publication was an attack by Berkeley on the principles of the infinitesimal calculus. In it [art. 751, p. 610] Maclaurin gave a proof of the theorem that

$$f(x) = f(0) + xf'(0) + \frac{x^2}{|2} f''(0) + \dots.$$

This was obtained in the manner given in many modern text-books by assuming that $f(x)$ can be expanded in a form like

$$f(x) = A_0 + A_1 x + A_2 x^2 + \dots,$$

then, on differentiating and putting $x = 0$ in the successive results, the values of A_0, A_1,... are obtained; but he did not investigate the convergency of the series. The result had been previously given in 1730 by James Stirling in his *Methodus Differentialis* [p. 102], and of course is at once deducible from Taylor's theorem. Maclaurin also here enunciated [art. 350, p. 289] the important theorem that, if $\phi(x)$ be positive and decrease as x increases from $x = a$ to $x = \infty$, then the series

$$\phi(a) + \phi(a + 1) + \phi(a + 2) + \dots$$

is convergent or divergent as the integral from $x = a$ to $x = \infty$ of $\phi(x)$ is finite or infinite. The theorem had been given by Euler[1] in 1732, but in so awkward a form that its value escaped general attention. Maclaurin here also gave the correct theory of maxima and minima, and rules for finding and discriminating multiple points.

This treatise is, however, especially valuable for the solutions it contains of numerous problems in geometry, statics, the theory of attractions, and astronomy. To solve these Maclaurin reverted to classical methods, and so powerful did these processes seem, when used by him, that Clairaut, after reading the work, abandoned analysis, and attacked

---

[1]See Cantor, vol. iii, p. 663.

the problem of the figure of the earth again by pure geometry. At a later time this part of the book was described by Lagrange as the "chef-d'œuvre de géométrie qu'on peut comparer à tout ce qu'Archimède nous a laissé de plus beau et de plus ingénieux." Maclaurin also determined the attraction of a homogeneous ellipsoid at an internal point, and gave some theorems on its attraction at an external point; in attacking these questions he introduced the conception of level surfaces, that is, surfaces at every point of which the resultant attraction is perpendicular to the surface. No further advance in the theory of attractions was made until Lagrange in 1773 introduced the idea of the potential. Maclaurin also shewed that a spheroid was a possible form of equilibrium of a mass of homogeneous liquid rotating about an axis passing through its centre of mass. Finally he discussed the tides; this part had been previously published (in 1740) and had received a prize from the French Academy.

Among Maclaurin's minor works is his *Algebra*, published in 1748, and founded on Newton's *Universal Arithmetic*. It contains the results of some early papers of Maclaurin; notably of two, written in 1726 and 1729, on the number of imaginary roots of an equation, suggested by Newton's theorem; and of one, written in 1729, containing the well-known rule for finding equal roots by means of the derived equation. In this book negative quantities are treated as being not less real than positive quantities. To this work a treatise, entitled *De Linearum Geometricarum Proprietatibus Generalibus*, was added as an appendix; besides the paper of 1720 above alluded to, it contains some additional and elegant theorems. Maclaurin also produced in 1728 an exposition of the Newtonian philosophy, which is incorporated in the posthumous work printed in 1748. Almost the last paper he wrote was one printed in the *Philosophical Transactions* for 1743 in which he discussed from a mathematical point of view the form of a bee's cell.

Maclaurin was one of the most able mathematicians of the eighteenth century, but his influence on the progress of British mathematics was on the whole unfortunate. By himself abandoning the use both of analysis and of the infinitesimal calculus, he induced Newton's countrymen to confine themselves to Newton's methods, and it was not until about 1820, when the differential calculus was introduced into the Cambridge curriculum, that English mathematicians made any general use of the more powerful methods of modern analysis.

Stewart.   Maclaurin was succeeded in his chair at Edinburgh by his pupil *Matthew Stewart*, born at Rothesay in 1717 and died at Edinburgh on January 23, 1785, a mathematician of considerable power, to whom I allude in passing, for his theorems on the problem of three bodies, and for his discussion, treated by transversals and involution, of the properties of the circle and straight line.

Simpson.[1]   The last member of the English school whom I need mention here is *Thomas Simpson*, who was born in Leicestershire on August 20, 1710, and died on May 14, 1761. His father was a weaver, and he owed his education to his own efforts. His mathematical interests were first aroused by the solar eclipse which took place in 1724, and with the aid of a fortune-telling pedlar he mastered Cocker's *Arithmetic* and the elements of algebra. He then gave up his weaving and became an usher at a school, and by constant and laborious efforts improved his mathematical education, so that by 1735 he was able to solve several questions which had been recently proposed and which involved the infinitesimal calculus. He next moved to London, and in 1743 was appointed professor of mathematics at Woolwich, a post which he continued to occupy till his death.

The works published by Simpson prove him to have been a man of extraordinary natural genius and extreme industry. The most important of them are his *Fluxions*, 1737 and 1750, with numerous applications to physics and astronomy; his *Laws of Chance* and his *Essays*, 1740; his theory of *Annuities and Reversions* (a branch of mathematics that is due to James Dodson, died in 1757, who was a master at Christ's Hospital, London), with tables of the value of lives, 1742; his *Dissertations*, 1743, in which the figure of the earth, the force of attraction at the surface of a nearly spherical body, the theory of the tides, and the law of astronomical refraction are discussed; his *Algebra*, 1745; his *Geometry*, 1747; his *Trigonometry*, 1748, in which he introduced the current abbreviations for the trigonometrical functions; his *Select Exercises*, 1752, containing the solutions of numerous problems and a theory of gunnery; and lastly, his *Miscellaneous Tracts*, 1754.

The work last mentioned consists of eight memoirs, and these contain his best known investigations. The first three papers are on various problems in astronomy; the fourth is on the theory of mean observations; the fifth and sixth on problems in fluxions and algebra; the seventh contains a general solution of the isoperimetrical problem; the eighth contains a discussion of the third and ninth sections of the *Prin-*

---

[1] A sketch of Simpson's life, with a bibliography of his writings, by J. Bevis and C. Hutton, was published in London in 1764. A short memoir is also prefixed to the later editions of his work on fluxions.

cipia, and their application to the lunar orbit. In this last memoir Simpson obtained a differential equation for the motion of the apse of the lunar orbit similar to that arrived at by Clairaut, but instead of solving it by successive approximations, he deduced a general solution by indeterminate coefficients. The result agrees with that given by Clairaut. Simpson solved this problem in 1747, two years later than the publication of Clairaut's memoir, but the solution was discovered independently of Clairaut's researches, of which Simpson first heard in 1748.

CHAPTER XVIII.

LAGRANGE, LAPLACE, AND THEIR CONTEMPORARIES.[1]
CIRC. 1740–1830.

THE last chapter contains the history of two separate schools—the continental and the British. In the early years of the eighteenth century the English school appeared vigorous and fruitful, but decadence rapidly set in, and after the deaths of Maclaurin and Simpson no British mathematician appeared who is at all comparable to the continental mathematicians of the latter half of the eighteenth century. This fact is partly explicable by the isolation of the school, partly by its tendency to rely too exclusively on geometrical and fluxional methods. Some attention was, however, given to practical science, but, except for a few remarks at the end of this chapter, I do not think it necessary to discuss English mathematics in detail, until about 1820, when analytical methods again came into vogue.

On the continent, under the influence of John Bernoulli, the calculus had become an instrument of great analytical power expressed in an admirable notation—and for practical applications it is impossible to over-estimate the value of a good notation. The subject of mechanics remained, however, in much the condition in which Newton had left it, until D'Alembert, by making use of the differential calculus, did something to extend it. Universal gravitation as enunciated in the *Principia* was accepted as an established fact, but the geometrical methods adopted in proving it were difficult to follow or to use in anal-

---

[1] A fourth volume of M. Cantor's *History*, covering the period from 1759 to 1799, was brought out in 1907. It contains memoirs by S. Günther on the mathematics of the period; by F. Cajori on arithmetic, algebra, and numbers; by E. Netto on series, imaginaries, &c.; by V. von Braunmühl on trigonometry; by V. Bobynin and G. Loria on pure geometry; by V. Kommerell on analytical geometry; by G. Vivanti on the infinitesimal calculus; and by C. R. Wallner on differential equations.

ogous problems; Maclaurin, Simpson, and Clairaut may be regarded as the last mathematicians of distinction who employed them. Lastly, the Newtonian theory of light was generally received as correct.

The leading mathematicians of the era on which we are now entering are Euler, Lagrange, Laplace, and Legendre. Briefly we may say that Euler extended, summed up, and completed the work of his predecessors; while Lagrange with almost unrivalled skill developed the infinitesimal calculus and theoretical mechanics, and presented them in forms similar to those in which we now know them. At the same time Laplace made some additions to the infinitesimal calculus, and applied that calculus to the theory of universal gravitation; he also created a calculus of probabilities. Legendre invented spherical harmonic analysis and elliptic integrals, and added to the theory of numbers. The works of these writers are still standard authorities. I shall content myself with a mere sketch of the chief discoveries embodied in them, referring any one who wishes to know more to the works themselves. Lagrange, Laplace, and Legendre created a French school of mathematics of which the younger members are divided into two groups; one (including Poisson and Fourier) began to apply mathematical analysis to physics, and the other (including Monge, Carnot, and Poncelet) created modern geometry. Strictly speaking, some of the great mathematicians of recent times, such as Gauss and Abel, were contemporaries of the mathematicians last named; but, except for this remark, I think it convenient to defer any consideration of them to the next chapter.

The development of analysis and mechanics.

Euler.[1]   *Leonhard Euler* was born at Bâle on April 15, 1707, and died at Petrograd on September 7, 1783. He was the son of a Lutheran minister who had settled at Bâle, and was educated in his native town under the direction of John Bernoulli, with whose sons Daniel and Nicholas he formed a lifelong friendship. When, in 1725, the younger Bernoullis went to Russia, on the invitation of the empress, they procured a place there for Euler, which in 1733 he exchanged for the chair of mathematics, then vacated by Daniel Bernoulli. The

---

[1] The chief facts in Euler's life are given by N. Fuss, and a list of Euler's writings is prefixed to his *Correspondence*, 2 vols., Petrograd, 1843; see also *Index Operum Euleri* by J. G. Hagen, Berlin, 1896. Euler's earlier works are discussed by Cantor, chapters cxi, cxiii, cxv, and cxvii. No complete edition of Euler's writings has been published, though the work has been begun twice.

severity of the climate affected his eyesight, and in 1735 he lost the use of one eye completely. In 1741 he moved to Berlin at the request, or rather command, of Frederick the Great; here he stayed till 1766, when he returned to Russia, and was succeeded at Berlin by Lagrange. Within two or three years of his going back to Petrograd he became blind; but in spite of this, and although his house, together with many of his papers, were burnt in 1771, he recast and improved most of his earlier works. He died of apoplexy in 1783. He was married twice.

I think we may sum up Euler's work by saying that he created a good deal of analysis, and revised almost all the branches of pure mathematics which were then known, filling up the details, adding proofs, and arranging the whole in a consistent form. Such work is very important, and it is fortunate for science when it falls into hands as competent as those of Euler.

Euler wrote an immense number of memoirs on all kinds of mathematical subjects. His chief works, in which many of the results of earlier memoirs are embodied, are as follows.

In the first place, he wrote in 1748 his *Introductio in Analysin Infinitorum*, which was intended to serve as an introduction to pure analytical mathematics. This is divided into two parts.

The first part of the *Analysis Infinitorum* contains the bulk of the matter which is to be found in modern text-books on algebra, theory of equations, and trigonometry. In the algebra he paid particular attention to the expansion of various functions in series, and to the summation of given series; and pointed out explicitly that an infinite series cannot be safely employed unless it is convergent. In the trigonometry, much of which is founded on F. C. Mayer's *Arithmetic of Sines*, which had been published in 1727, Euler developed the idea of John Bernoulli, that the subject was a branch of analysis and not a mere appendage of astronomy or geometry. He also introduced (contemporaneously with Simpson) the current abbreviations for the trigonometrical functions, and shewed that the trigonometrical and exponential functions were connected by the relation $\cos\theta + i\sin\theta = e^{i\theta}$.

Here, too [pp. 85, 90, 93], we meet the symbol e used to denote the base of the Napierian logarithms, namely, the incommensurable number $2.71828\ldots$, and the symbol π used to denote the incommensurable number $3.14159\ldots$. The use of a single symbol to denote the number $2.71828\ldots$ seems to be due to Cotes, who denoted it by M; Euler in 1731 denoted it by e. To the best of my knowledge, Newton had been the first to employ the literal exponential notation, and Euler, using the form a^z, had taken a as the base of any system of logarithms. It is probable that the choice of e for a particular base was determined by its being the vowel consecutive to a. The use of a single symbol to denote the number 3.14159... appears to have been introduced about the beginning of the eighteenth century. W. Jones in 1706 represented it by π, a symbol which had been used by Oughtred in 1647, and by Barrow a few years later, to denote the periphery of a circle. John Bernoulli represented the number by c; Euler in 1734 denoted it by p, and in a letter of 1736 (in which he enunciated the theorem that the sum of the squares of the reciprocals of the natural numbers is $\pi^2/6$) he used the letter c; Chr. Goldbach in 1742 used π; and after the publication of Euler's *Analysis* the symbol π was generally employed.

The numbers e and π would enter into mathematical analysis from whatever side the subject was approached. The latter represents among other things the ratio of the circumference of a circle to its diameter, but it is a mere accident that that is taken for its definition. De Morgan in the *Budget of Paradoxes* tells an anecdote which illustrates how little the usual definition suggests its real origin. He was explaining to an actuary what was the chance that at the end of a given time a certain proportion of some group of people would be alive; and quoted the actuarial formula involving π, which, in answer to a question, he explained stood for the ratio of the circumference of a circle to its diameter. His acquaintance, who had so far listened to the explanation with interest, interrupted him and explained, "My dear friend, that must be a delusion; what can a circle have to do with the number of people alive at the end of a given time?"

The second part of the *Analysis Infinitorum* is on analytical geometry. Euler commenced this part by dividing curves into algebraical and transcendental, and established a variety of propositions which are true for all algebraical curves. He then applied these to the general equation of the second degree in two dimensions, shewed that it represents the various conic sections, and deduced most of their properties from the general equation. He also considered the classification of cubic, quartic, and other algebraical curves. He next discussed the question as to what surfaces are represented by the general equation of the second degree in three dimensions, and how they may be discriminated one from the other: some of these surfaces had not been previously investigated. In the course of this analysis he laid down the rules for the transformation of co-ordinates in space. Here also we find the earliest attempt to bring the curvature of surfaces within the domain of mathematics, and the first complete discussion of tortuous curves.

The *Analysis Infinitorum* was followed in 1755 by the *Institutiones Calculi Differentialis*, to which it was intended as an introduction. This is the first text-book on the differential calculus which has any claim to be regarded as complete, and it may be said that until recently many modern treatises on the subject are based on it; at the same time it should be added that the exposition of the principles of the subject is often prolix and obscure, and sometimes not altogether accurate.

This series of works was completed by the publication in three volumes in 1768 to 1770 of the *Institutiones Calculi Integralis*, in which the results of several of Euler's earlier memoirs on the same subject and on differential equations are included. This, like the similar treatise on the differential calculus, summed up what was then known on the subject, but many of the theorems were recast and the proofs improved. The Beta and Gamma[1] functions were invented by Euler and are discussed here, but only as illustrations of methods of reduction and integration. His treatment of elliptic integrals is superficial; it was suggested by a theorem, given by John Landen in the *Philosophical Transactions* for 1775, connecting the arcs of a hyperbola and an ellipse. Euler's works that form this trilogy have gone through numerous subsequent editions.

The classic problems on isoperimetrical curves, the brachistochrone in a resisting medium, and the theory of geodesics (all of which had been suggested by his master, John Bernoulli) had engaged Euler's attention at an early date; and in solving them he was led to the calculus of variations. The idea of this was given in his *Curvarum Maximi Minimive Proprietate Gaudentium Inventio*, published in 1741 and extended in 1744, but the complete development of the new calculus was first effected by Lagrange in 1759. The method used by Lagrange is described in Euler's integral calculus, and is the same as that given in most modern text-books on the subject.

In 1770 Euler published his *Vollständige Anleitung zur Algebra*. A French translation, with numerous and valuable additions by Lagrange, was brought out in 1774; and a treatise on arithmetic by Euler was appended to it. The first volume treats of determinate algebra. This contains one of the earliest attempts to place the fundamental processes on a scientific basis: the same subject had attracted D'Alembert's attention. This work also includes the proof of the binomial theorem for an

---

[1] The history of the Gamma function is given in a monograph by Brunel in the *Mémoires de la société des sciences*, Bordeaux, 1886.

unrestricted real index which is still known by Euler's name; the proof is founded on the principle of the permanence of equivalent forms, but Euler made no attempt to investigate the convergency of the series: that he should have omitted this essential step is the more curious as he had himself recognized the necessity of considering the convergency of infinite series: Vandermonde's proof given in 1764 suffers from the same defect.

The second volume of the algebra treats of indeterminate or Diophantine algebra. This contains the solutions of some of the problems proposed by Fermat, and which had hitherto remained unsolved.

As illustrating the simplicity and directness of Euler's methods I give the substance of his demonstration,[1] alluded to above, that all even perfect numbers are included in Euclid's formula, $2^{n-1}p$, where p stands for $2^n - 1$ and is a prime.[2] Let N be an even perfect number. N is even, hence it can be written in the form $2^{n-1}a$, where a is not divisible by 2. N is perfect, that is, is equal to the sum of all its integral subdivisors; therefore (if the number itself be reckoned as one of its divisors) it is equal to half the sum of all its integral divisors, which we may denote by ΣN. Since $2N = \Sigma N$, we have

$$2 \times 2^{n-1}\alpha = \Sigma 2^{n-1}\alpha = \Sigma 2^{n-1} \times \Sigma\alpha.$$
$$\therefore 2^n\alpha = (1 + 2 + \ldots + 2^{n-1})\Sigma\alpha = (2^n - 1)\Sigma\alpha,$$

therefore $\alpha : \Sigma\alpha = 2^n - 1 : 2^n = p : p + 1$. Hence $\alpha = \lambda p$, and $\Sigma\alpha = \lambda(p + 1)$; and since the ratio $p : p + 1$ is in its lowest terms, λ must be a positive integer. Now, unless $\lambda = 1$, we have $1, \lambda, p$, and λp as factors of λp; moreover, if p be not prime, there will be other factors also. Hence, unless $\lambda = 1$ and p be a prime, we have

$$\Sigma\lambda p = 1 + \lambda + p + \lambda p + \ldots = (\lambda + 1)(p + 1) + \ldots$$

But this is inconsistent with the result $\Sigma\lambda p = \Sigma\alpha = \lambda(p + 1)$. Hence λ must be equal to 1 and p must be a prime. Therefore $\alpha = p$, therefore $N = 2^{n-1}\alpha = 2^{n-1}(2^n - 1)$. I may add the corollary that since p is a prime, it follows that n is a prime; and the determination of what values of n (less than 257) make p prime falls under Mersenne's rule.

---

[1] *Commentationes Arithmeticae Collectae*, Petrograd, 1849, vol. ii, p. 514, art. 107. Sylvester published an analysis of the argument in *Nature*, December 15, 1887, vol. xxxvii, p. 152.

[2] Euc. ix, 36; see above, page 252.

The four works mentioned above comprise most of what Euler produced in pure mathematics. He also wrote numerous memoirs on nearly all the subjects of applied mathematics and mathematical physics then studied: the chief novelties in them are as follows.

In the mechanics of a rigid system he determined the general equations of motion of a body about a fixed point, which are ordinarily written in the form

$$A\frac{d\omega_1}{dt} - (B - C)\omega_2\omega_3 = L :$$

and he gave the general equations of motion of a free body, which are usually presented in the form

$$\frac{d}{dt}(mu) - mv\theta_3 + mw\theta_2 = X, \text{ and } \frac{dh'_1}{dt} - h'_2\theta_3 + h'_3\theta_2 = L.$$

He also defended and elaborated the theory of "least action" which had been propounded by Maupertuis in 1751 in his *Essai de cosmologie*[p. 70].

In hydrodynamics Euler established the general equations of motion, which are commonly expressed in the form

$$\frac{1}{\rho}\frac{dp}{dx} = X - \frac{du}{dt} - u\frac{du}{dx} - v\frac{du}{dy} - w\frac{du}{dz}.$$

At the time of his death he was engaged in writing a treatise on hydromechanics in which the treatment of the subject would have been completely recast.

His most important works on astronomy are his *Theoria Motuum Planetarum et Cometarum*, published in 1744; his *Theoria Motus Lunaris*, published in 1753; and his *Theoria Motuum Lunae*, published in 1772. In these he attacked the problem of three bodies: he supposed the body considered (*ex. gr.* the moon) to carry three rectangular axes with it in its motion, the axes moving parallel to themselves, and to these axes all the motions were referred. This method is not convenient, but it was from Euler's results that Mayer[1] constructed the lunar tables for which his widow in 1770 received £5000 from the English parliament, and in recognition of Euler's services a sum of £300 was also voted as an honorarium to him.

---

[1] *Johann Tobias Mayer*, born in Würtemberg in 1723, and died in 1762, was director of the English observatory at Göttingen. Most of his memoirs, other than his lunar tables, were published in 1775 under the title *Opera Inedita*.

Euler was much interested in optics. In 1746 he discussed the relative merits of the emission and undulatory theories of light; he on the whole preferred the latter. In 1770–71 he published his optical researches in three volumes under the title *Dioptrica*.

He also wrote an elementary work on physics and the fundamental principles of mathematical philosophy. This originated from an invitation he received when he first went to Berlin to give lessons on physics to the princess of Anhalt-Dessau. These lectures were published in 1768–1772 in three volumes under the title *Lettres ... sur quelques sujets de physique ...*, and for half a century remained a standard treatise on the subject.

Of course Euler's magnificent works were not the only text-books containing original matter produced at this time. Amongst numerous writers I would specially single out *Daniel Bernoulli, Simpson, Lambert, Bézout, Trembley,* and *Arbogast,* as having influenced the development of mathematics. To the two first-mentioned I have already alluded in the last chapter.

Lambert.[1]   *Johann Heinrich Lambert* was born at Mülhausen on August 28, 1728, and died at Berlin on September 25, 1777. He was the son of a small tailor, and had to rely on his own efforts for his education; from a clerk in some ironworks he got a place in a newspaper office, and subsequently, on the recommendation of the editor, he was appointed tutor in a private family, which secured him the use of a good library and sufficient leisure to use it. In 1759 he settled at Augsburg, and in 1763 removed to Berlin where he was given a small pension, and finally made editor of the Prussian astronomical almanack.

Lambert's most important works were one on optics, issued in 1759, which suggested to Arago the lines of investigation he subsequently pursued; a treatise on perspective, published in 1759 (to which in 1768 an appendix giving practical applications were added); and a treatise on comets, printed in 1761, containing the well-known expression for the area of a focal sector of a conic in terms of the chord and the bounding radii. Besides these he communicated numerous papers to the Berlin Academy. Of these the most important are his memoir in 1768 on transcendental magnitudes, in which he proved that π is incommensurable (the proof is given in Legendre's *Géométrie*, and is there extended to

---

[1]See *Lambert nach seinem Leben und Wirken,* by D. Huber, Bâle, 1829. Most of Lambert's memoirs are collected in his *Beiträge zum Gebrauche der Mathematik,* published in four volumes, Berlin, 1765–1772.

π^2): his paper on trigonometry, read in 1768, in which he developed Demoivre's theorems on the trigonometry of complex variables, and introduced the hyperbolic sine and cosine[1] denoted by the symbols $\sinh x, \cosh x$: his essay entitled analytical observations, published in 1771, which is the earliest attempt to form functional equations by expressing the given properties in the language of the differential calculus, and then integrating his researches on non-Euclidean geometry: lastly, his paper on vis viva, published in 1783, in which for the first time he expressed Newton's second law of motion in the notation of the differential calculus.

Bézout. Trembley. Arbogast.   Of the other mathematicians above mentioned I here add a few words. *Étienne Bézout*, born at Nemours on March 31, 1730, and died on September 27, 1783, besides numerous minor works, wrote a *Théorie générale des équations algébriques*, published at Paris in 1779, which in particular contained much new and valuable matter on the theory of elimination and symmetrical functions of the roots of an equation: he used determinants in a paper in the *Histoire de L'académie royale*, 1764, but did not treat of the general theory. *Jean Trembley*, born at Geneva in 1749, and died on September 18, 1811, contributed to the development of differential equations, finite differences, and the calculus of probabilities. *Louis François Antoine Arbogast*, born in Alsace on October 4, 1759, and died at Strassburg, where he was professor, on April 8, 1803, wrote on series and the derivatives known by his name: he was the first writer to separate the symbols of operation from those of quantity.

I do not wish to crowd my pages with an account of those who have not distinctly advanced the subject, but I have mentioned the above writers because their names are still well known. We may, however, say that the discoveries of Euler and Lagrange in the subjects which they treated were so complete and far-reaching that what their less gifted contemporaries added is not of sufficient importance to require mention in a book of this nature.

Lagrange.[2]   *Joseph Louis Lagrange*, the greatest mathematician

---

[1] These functions are said to have been previously suggested by F. C. Mayer, see *Die Lehre von den Hyperbelfunktionen* by S. Günther, Halle, 1881, and *Beiträge zur Geschichte der neueren Mathematik*, Ansbach, 1881.

[2] Summaries of the life and works of Lagrange are given in the *English Cyclopaedia* and the *Encyclopaedia Britannica* (ninth edition), of which I have made considerable use: the former contains a bibliography of his writings. Lagrange's works, edited by MM. J. A. Serret and G. Darboux, were published in 14 volumes, Paris, 1867–1892.

of the eighteenth century, was born at Turin on January 25, 1736, and died at Paris on April 10, 1813. His father, who had charge of the Sardinian military chest, was of good social position and wealthy, but before his son grew up he had lost most of his property in speculations, and young Lagrange had to rely for his position on his own abilities. He was educated at the college of Turin, but it was not until he was seventeen that he shewed any taste for mathematics—his interest in the subject being first excited by a memoir by Halley,[1] across which he came by accident. Alone and unaided he threw himself into mathematical studies; at the end of a year's incessant toil he was already an accomplished mathematician, and was made a lecturer in the artillery school.

The first fruit of Lagrange's labours here was his letter, written when he was still only nineteen, to Euler, in which he solved the isoperimetrical problem which for more than half a century had been a subject of discussion. To effect the solution (in which he sought to determine the form of a function so that a formula in which it entered should satisfy a certain condition) he enunciated the principles of the calculus of variations. Euler recognized the generality of the method adopted, and its superiority to that used by himself; and with rare courtesy he withheld a paper he had previously written, which covered some of the same ground, in order that the young Italian might have time to complete his work, and claim the undisputed invention of the new calculus. The name of this branch of analysis was suggested by Euler. This memoir at once placed Lagrange in the front rank of mathematicians then living.

In 1758 Lagrange established with the aid of his pupils a society, which was subsequently incorporated as the Turin Academy, and in the five volumes of its transactions, usually known as the *Miscellanea Taurinensia*, most of his early writings are to be found. Many of these are elaborate memoirs. The first volume contains a memoir on the theory of the propagation of sound; in this he indicates a mistake made by Newton, obtains the general differential equation for the motion, and integrates it for motion in a straight line. This volume also contains the complete solution of the problem of a string vibrating transversely; in this paper he points out a lack of generality in the solutions previously given by Taylor, D'Alembert, and Euler, and arrives at the conclusion

Delambre's account of his life is printed in the first volume.

[1] On the excellence of the modern algebra in certain optical problems, *Philosophical Transactions*, 1693, vol. xviii, p. 960.

that the form of the curve at any time t is given by the equation $y = a \sin mx \sin nt$. The article concludes with a masterly discussion of echoes, beats, and compound sounds. Other articles in this volume are on recurring series, probabilities, and the calculus of variations.

The second volume contains a long paper embodying the results of several memoirs in the first volume on the theory and notation of the calculus of variations; and he illustrates its use by deducing the principle of least action, and by solutions of various problems in dynamics.

The third volume includes the solution of several dynamical problems by means of the calculus of variations; some papers on the integral calculus; a solution of Fermat's problem mentioned above, to find a number x which will make $(x^2n + 1)$ a square where n is a given integer which is not a square; and the general differential equations of motion for three bodies moving under their mutual attractions.

In 1761 Lagrange stood without a rival as the foremost mathematician living; but the unceasing labour of the preceding nine years had seriously affected his health, and the doctors refused to be responsible for his reason or life unless he would take rest and exercise. Although his health was temporarily restored his nervous system never quite recovered its tone, and henceforth he constantly suffered from attacks of profound melancholy.

The next work he produced was in 1764 on the libration of the moon, and an explanation as to why the same face was always turned to the earth, a problem which he treated by the aid of virtual work. His solution is especially interesting as containing the germ of the idea of generalized equations of motion, equations which he first formally proved in 1780.

He now started to go on a visit to London, but on the way fell ill at Paris. There he was received with marked honour, and it was with regret he left the brilliant society of that city to return to his provincial life at Turin. His further stay in Piedmont was, however, short. In 1766 Euler left Berlin, and Frederick the Great immediately wrote expressing the wish of "the greatest king in Europe" to have "the greatest mathematician in Europe" resident at his court. Lagrange accepted the offer and spent the next twenty years in Prussia, where he produced not only the long series of memoirs published in the Berlin and Turin transactions, but his monumental work, the *Mécanique analytique*. His residence at Berlin commenced with an unfortunate mistake. Finding most of his colleagues married, and assured by their wives that it was the only way to be happy, he married; his wife soon died, but the union was not a happy one.

Lagrange was a favourite of the king, who used frequently to discourse to him on the advantages of perfect regularity of life. The lesson went home, and thenceforth Lagrange studied his mind and body as though they were machines, and found by experiment the exact amount of work which he was able to do without breaking down. Every night he set himself a definite task for the next day, and on completing any branch of a subject he wrote a short analysis to see what points in the demonstrations or in the subject-matter were capable of improvement. He always thought out the subject of his papers before he began to compose them, and usually wrote them straight off without a single erasure or correction.

His mental activity during these twenty years was amazing. Not only did he produce his splendid *Mécanique analytique*, but he contributed between one and two hundred papers to the Academies of Berlin, Turin, and Paris. Some of these are really treatises, and all without exception are of a high order of excellence. Except for a short time when he was ill he produced on an average about one memoir a month. Of these I note the following as among the most important.

First, his contributions to the fourth and fifth volumes, 1766–1773, of the *Miscellanea Taurinensia*; of which the most important was the one in 1771, in which he discussed how numerous astronomical observations should be combined so as to give the most probable result. And later, his contributions to the first two volumes, 1784–1785, of the transactions of the Turin Academy; to the first of which he contributed a paper on the pressure exerted by fluids in motion, and to the second an article on integration by infinite series, and the kind of problems for which it is suitable.

Most of the memoirs sent to Paris were on astronomical questions, and among these I ought particularly to mention his memoir on the Jovian system in 1766, his essay on the problem of three bodies in 1772, his work on the secular equation of the moon in 1773, and his treatise on cometary perturbations in 1778. These were all written on subjects proposed by the French Academy, and in each case the prize was awarded to him.

The greater number of his papers during this time were, however, contributed to the Berlin Academy. Several of them deal with questions on *algebra*. In particular I may mention the following. (i) His discussion of the solution in integers of indeterminate quadratics, 1769, and generally of indeterminate equations, 1770. (ii) His tract on the theory of elimination, 1770. (iii) His memoirs on a general process for solving an algebraical equation of any degree, 1770 and 1771; this method fails for equations of an order above the fourth, because it then involves the solution of an equation of higher dimensions than the one proposed, but it gives all the solutions of his predecessors as modifications of a single principle. (iv) The complete solution of a binomial equation of any degree; this is contained in the memoirs last mentioned. (v) Lastly, in 1773, his treatment of determinants of the second and third order, and of invariants.

Several of his early papers also deal with questions connected with the neglected but singularly fascinating subject of the *theory of numbers*. Among these are the following. (i) His proof of the theorem that every integer which is not a square can be expressed as the sum of two, three, or four integral squares, 1770. (ii) His proof of Wilson's theorem that if n be a prime, then $\lfloor n - 1 + 1$ is always a multiple of n, 1771. (iii) His memoirs of 1773, 1775, and 1777, which give the demonstrations of several results enunciated by Fermat, and not previously proved. (iv) And, lastly, his method for determining the factors of numbers of the form $x^2 + ay^2$.

There are also numerous articles on various points of *analytical geometry*. In two of them, written rather later, in 1792 and 1793, he reduced the equations of the quadrics (or conicoids) to their canonical forms.

During the years from 1772 to 1785 he contributed a long series of memoirs which created the science of *differential equations*, at any rate as far as partial differential equations are concerned. I do not think that any previous writer had done anything beyond considering equations of some particular form. A large part of these results were collected in the second edition of Euler's integral calculus which was published in 1794.

Lagrange's papers on *mechanics* require no separate mention here as the results arrived at are embodied in the *Mécanique analytique* which is described below.

Lastly, there are numerous memoirs on problems in *astronomy*. Of these the most important are the following. (i) On the attraction of ellipsoids, 1773: this is founded on Maclaurin's work. (ii) On the secular equation of the moon, 1773; also noticeable for the earliest introduction of the idea of the potential. The potential of a body at any point is the sum of the mass of every element of the body when divided by its distance from the point. Lagrange shewed that if the potential of a body at an external point were known, the attraction in any direction could be at once found. The theory of the potential was elaborated in a paper sent to Berlin in 1777. (iii) On the motion of the nodes of a planet's orbit, 1774. (iv) On the stability of the planetary orbits, 1776. (v) Two memoirs in which the method of determining the orbit of a comet from three observations is completely worked out, 1778 and 1783: this has not indeed proved practically available, but his system of calculating the perturbations by means of mechanical quadratures has formed the basis of most subsequent researches on the subject. (vi) His determination of the secular and periodic variations of the elements of the planets, 1781–1784: the upper limits assigned for these agree closely with those obtained later by Leverrier, and Lagrange proceeded as far as the knowledge then possessed of the masses of the planets permitted. (vii) Three memoirs on the method of interpolation, 1783, 1792, and 1793: the part of finite differences dealing therewith is now in the same stage as that in which Lagrange left it.

Over and above these various papers he composed his great treatise, the *Mécanique analytique*. In this he lays down the law of virtual work, and from that one fundamental principle, by the aid of the calculus of variations, deduces the whole of mechanics, both of solids and fluids. The object of the book is to shew that the subject is implicitly included in a single principle, and to give general formulae from which any particular result can be obtained. The method of generalized co-ordinates by which he obtained this result is perhaps the most brilliant result of his analysis. Instead of following the motion of each individual part of a material system, as D'Alembert and Euler had done, he shewed that, if we determine its configuration by a sufficient number of variables whose number is the same as that of the degrees of freedom possessed by the system, then the kinetic and potential energies of the system can be expressed in terms of these variables, and the differential equations of motion thence deduced by simple differentiation. For example, in dynamics of a rigid system he replaces the consideration of the particular problem by the general equation which is now usually written in the form

$$\frac{d}{dt}\frac{\partial T}{\partial \theta} - \frac{\partial T}{\partial \theta} + \frac{\partial V}{\partial \theta} = 0.$$

Amongst other theorems here given are the proposition that the kinetic energy imparted by given impulses to a material system under given constraints is a maximum, and a more general statement of the principle of least action than had been given by Maupertuis or Euler. All the analysis is so elegant that Sir William Rowan Hamilton said the work could be only described as a scientific poem. Lagrange held that mechanics was really a branch of pure mathematics analogous to a geometry of four dimensions, namely, the time and the three co-ordinates of the point in space;[1] and it is said that he prided himself that from the beginning to the end of the work there was not a single diagram. At first no printer could be found who would publish the book; but Legendre at last persuaded a Paris firm to undertake it, and it was issued in 1788.

In 1787 Frederick died, and Lagrange, who had found the climate of Berlin trying, gladly accepted the offer of Louis XVI. to migrate to Paris. He received similar invitations from Spain and Naples. In France he was received with every mark of distinction, and special apartments in the Louvre were prepared for his reception. At the beginning of his residence here he was seized with an attack of melancholy, and even the printed copy of his *Mécanique* on which he had worked for a quarter of a century lay for more than two years unopened on his desk. Curiosity as to the results of the French revolution first stirred him out of his lethargy, a curiosity which soon turned to alarm as the revolution developed. It was about the same time, 1792, that the unaccountable sadness of his life and his timidity moved the compassion of a young girl who insisted on marrying him, and proved a devoted wife to whom he became warmly attached. Although the decree of October 1793, which ordered all foreigners to leave France, specially exempted him by name, he was preparing to escape when he was offered the presidency of the commission for the reform of weights and measures. The choice of the units finally selected was largely due to him, and it was mainly owing to his influence that the decimal subdivision was accepted by the commission of 1799.

Though Lagrange had determined to escape from France while there was yet time, he was never in any danger; and the different revolutionary governments (and, at a later time, Napoleon) loaded him with honours and distinctions. A striking testimony to the respect in which he was held was shown in 1796 when the French commissary in Italy was ordered to attend in full state on Lagrange's father, and tender the congratulations of the republic on the achievements of his son, who "had done honour to all mankind by his genius, and whom it was the

[1]On the development of this idea, see H. Minkowski, *Raum und Zeit*, Leipzig, 1909.

special glory of Piedmont to have produced." It may be added that Napoleon, when he attained power, warmly encouraged scientific studies in France, and was a liberal benefactor of them.

In 1795 Lagrange was appointed to a mathematical chair at the newly-established École normale, which enjoyed only a brief existence of four months. His lectures here were quite elementary, and contain nothing of any special importance, but they were published because the professors had to "pledge themselves to the representatives of the people and to each other neither to read nor to repeat from memory," and the discourses were ordered to be taken down in shorthand in order to enable the deputies to see how the professors acquitted themselves.

On the establishment of the École polytechnique in 1797 Lagrange was made a professor; and his lectures there are described by mathematicians who had the good fortune to be able to attend them, as almost perfect both in form and matter. Beginning with the merest elements, he led his hearers on until, almost unknown to themselves, they were themselves extending the bounds of the subject: above all he impressed on his pupils the advantage of always using general methods expressed in a symmetrical notation.

His lectures on the differential calculus form the basis of his *Théorie des fonctions analytiques* which was published in 1797. This work is the extension of an idea contained in a paper he had sent to the Berlin Memoirs in 1772, and its object is to substitute for the differential calculus a group of theorems based on the development of algebraic functions in series. A somewhat similar method had been previously used by John Landen in his *Residual Analysis*, published in London in 1758. Lagrange believed that he could thus get rid of those difficulties, connected with the use of infinitely large and infinitely small quantities, to which some philosophers objected in the usual treatment of the differential calculus. The book is divided into three parts: of these, the first treats of the general theory of functions, and gives an algebraic proof of Taylor's theorem, the validity of which is, however, open to question; the second deals with applications to geometry; and the third with applications to mechanics. Another treatise on the same lines was his *Leçons sur le calcul des fonctions*, issued in 1804. These works may be considered as the starting-point for the researches of Cauchy, Jacobi, and Weierstrass, and are interesting from the historical point of view.

Lagrange, however, did not himself object to the use of infinitesimals in the differential calculus; and in the preface to the second edition of the *Mécanique*, which was issued in 1811, he justifies their employment, and concludes by saying that "when we have grasped the spirit of the infinitesimal method, and have verified the exactness of its results either by the geometrical method of prime and ultimate ratios, or by the analytical method of derived functions, we may employ infinitely small quantities as a sure and valuable means of shortening and simplifying our proofs."

His *Résolution des équations numériques*, published in 1798, was also the fruit of his lectures at the Polytechnic. In this he gives the method of approximating to the real roots of an equation by means of continued fractions, and enunciates several other theorems. In a note at the end he shows how Fermat's theorem that $a^{p-1} - 1 \equiv 0 (\bmod p)$, where p is a prime and a is prime to p, may be applied to give the complete algebraical solution of any binomial equation. He also here explains how the equation whose roots are the squares of the differences of the roots of the original equation may be used so as to give considerable information as to the position and nature of those roots.

The theory of the planetary motions had formed the subject of some of the most remarkable of Lagrange's Berlin papers. In 1806 the subject was reopened by Poisson, who, in a paper read before the French Academy, showed that Lagrange's formulae led to certain limits for the stability of the orbits. Lagrange, who was present, now discussed the whole subject afresh, and in a memoir communicated to the Academy in 1808 explained how, by the variation of arbitrary constants, the periodical and secular inequalities of any system of mutually interacting bodies could be determined.

In 1810 Lagrange commenced a thorough revision of the *Mécanique analytique,* but he was able to complete only about two-thirds of it before his death.

In appearance he was of medium height, and slightly formed, with pale blue eyes and a colourless complexion. In character he was nervous and timid, he detested controversy, and to avoid it willingly allowed others to take the credit for what he had himself done.

Lagrange's interests were essentially those of a student of pure mathematics: he sought and obtained far-reaching abstract results, and was content to leave the applications to others. Indeed, no inconsiderable part of the discoveries of his great contemporary, Laplace, consists of the application of the Lagrangian formulae to the facts of nature; for example, Laplace's conclusions on the velocity of sound and the secular acceleration of the moon are implicitly involved in Lagrange's results. The only difficulty in understanding Lagrange is that of the subject-

matter and the extreme generality of his processes; but his analysis is
"as lucid and luminous as it is symmetrical and ingenious."

A recent writer speaking of Lagrange says truly that he took a
prominent part in the advancement of almost every branch of pure
mathematics. Like Diophantus and Fermat, he possessed a special genius for the theory of numbers, and in this subject he gave solutions of
many of the problems which had been proposed by Fermat, and added
some theorems of his own. He developed the calculus of variations. To
him, too, the theory of differential equations is indebted for its position as a science rather than a collection of ingenious artifices for the
solution of particular problems. To the calculus of finite differences he
contributed the formula of interpolation which bears his name. But
above all he impressed on mechanics (which it will be remembered he
considered a branch of pure mathematics) that generality and completeness towards which his labours invariably tended.

Laplace.[1]   *Pierre Simon Laplace* was born at Beaumont-en-Auge
in Normandy on March 23, 1749, and died at Paris on March 5, 1827.
He was the son of a small cottager or perhaps a farm-labourer, and owed
his education to the interest excited in some wealthy neighbours by his
abilities and engaging presence. Very little is known of his early years,
for when he became distinguished he had the pettiness to hold himself
aloof both from his relatives and from those who had assisted him.
It would seem that from a pupil he became an usher in the school at
Beaumont; but, having procured a letter of introduction to D'Alembert,
he went to Paris to push his fortune. A paper on the principles of
mechanics excited D'Alembert's interest, and on his recommendation
a place in the military school was offered to Laplace.

Secure of a competency, Laplace now threw himself into original
research, and in the next seventeen years, 1771–1787, he produced much
of his original work in astronomy. This commenced with a memoir,
read before the French Academy in 1773, in which he shewed that
the planetary motions were stable, and carried the proof as far as the
cubes of the eccentricities and inclinations. This was followed by several
papers on points in the integral calculus, finite differences, differential
equations, and astronomy.

[1] The following account of Laplace's life and writings is mainly founded on the
articles in the *English Cyclopaedia* and the *Encyclopaedia Britannica*. Laplace's
works were published in seven volumes by the French government in 1843–7; and a
new edition with considerable additional matter was issued at Paris in six volumes,
1878–84.

During the years 1784–1787 he produced some memoirs of exceptional power, Prominent among these is one read in 1784, and reprinted in the third volume of the *Mécanique céleste,* in which he completely determined the attraction of a spheroid on a particle outside it. This is memorable for the introduction into analysis of spherical harmonics or Laplace's coefficients, as also for the development of the use of the potential—a name first given by Green in 1828.

If the co-ordinates of two points be (r, μ, ω) and (r', μ', ω'), and if $r' \nleq r$, then the reciprocal of the distance between them can be expanded in powers of r/r', and the respective coefficients are Laplace's coefficients. Their utility arises from the fact that every function of the co-ordinates of a point on a sphere can be expanded in a series of them. It should be stated that the similar coefficients for space of two dimensions, together with some of their properties, had been previously given by Legendre in a paper sent to the French Academy in 1783. Legendre had good reason to complain of the way in which he was treated in this matter.

This paper is also remarkable for the development of the idea of the potential, which was appropriated from Lagrange,[1] who had used it in his memoirs of 1773, 1777, and 1780. Laplace shewed that the potential always satisfies the differential equation

$$\nabla^2 V = \frac{\partial^2 V}{\partial x^2} + \frac{\partial^2 V}{\partial y^2} + \frac{\partial^2 V}{\partial z^2} = 0,$$

and on this result his subsequent work on attractions was based. The quantity $\nabla^2 V$ has been termed the concentration of V, and its value at any point indicates the excess of the value of V there over its mean value in the neighbourhood of the point. Laplace's equation, or the more general form $\nabla^2 V = -4\pi\rho$, appears in all branches of mathematical physics. According to some writers this follows at once from the fact that $\nabla^2 V$ is a scalar operator; or the equation may represent analytically some general law of nature which has not been yet reduced to words; or possibly it might be regarded by a Kantian as the outward sign of one of the necessary forms through which all phenomena are perceived.

This memoir was followed by another on planetary inequalities, which was presented in three sections in 1784, 1785, and 1786. This deals mainly with the explanation of the "great inequality" of Jupiter

---

[1] See the *Bulletin* of the New York Mathematical Society, 1892, vol. i, pp. 66–74.

and Saturn. Laplace shewed by general considerations that the mutual action of two planets could never largely affect the eccentricities and inclinations of their orbits; and that the peculiarities of the Jovian system were due to the near approach to commensurability of the mean motions of Jupiter and Saturn: further developments of these theorems on planetary motion were given in his two memoirs of 1788 and 1789. It was on these data that Delambre computed his astronomical tables.

The year 1787 was rendered memorable by Laplace's explanation and analysis of the relation between the lunar acceleration and the secular changes in the eccentricity of the earth's orbit: this investigation completed the proof of the stability of the whole solar system on the assumption that it consists of a collection of rigid bodies moving in a vacuum. All the memoirs above alluded to were presented to the French Academy, and they are printed in the *Mémoires présentés par divers savans.*

Laplace now set himself the task to write a work which should "offer a complete solution of the great mechanical problem presented by the solar system, and bring theory to coincide so closely with observation that empirical equations should no longer find a place in astronomical tables." The result is embodied in the *Exposition du système du monde* and the *Mécanique céleste.*

The former was published in 1796, and gives a general explanation of the phenomena, but omits all details. It contains a summary of the history of astronomy: this summary procured for its author the honour of admission to the forty of the French Academy; it is commonly esteemed one of the masterpieces of French literature, though it is not altogether reliable for the later periods of which it treats.

The nebular hypothesis was here enunciated.[1] According to this hypothesis the solar system has been evolved from a quantity of incandescent gas rotating round an axis through its centre of mass. As it cooled the gas contracted and successive rings broke off from its outer edge. These rings in their turn cooled, and finally condensed into the planets, while the sun represents the central core which is still left. On this view we should expect that the more distant planets would be older than those nearer the sun. The subject is one of great difficulty, and though it seems certain that the solar system has a common origin, there are various features which appear almost inexplicable on the

---

[1] On hypotheses as to the origin of the solar system, see H. Poincaré, *Hypothèses cosmogoniques,* Paris, 1911.

nebular hypothesis as enunciated by Laplace.

Another theory which avoids many of the difficulties raised by Laplace's hypothesis has recently found favour. According to this, the origin of the solar system is to be found in the gradual aggregation of meteorites which swarm through our system, and perhaps through space. These meteorites which are normally cold may, by repeated collisions, be heated, melted, or even vaporized, and the resulting mass would, by the effect of gravity, be condensed into planet-like bodies—the larger aggregations so formed becoming the chief bodies of the solar system. To account for these collisions and condensations it is supposed that a vast number of meteorites were at some distant epoch situated in a spiral nebula, and that condensations and collisions took place at certain knots or intersections of orbits. As the resulting planetary masses cooled, moons or rings would be formed either by collisions of outlying parts or in the manner suggested in Laplace's hypothesis. This theory seems to be primarily due to Sir Norman Lockyer. It does not conflict with any of the known facts of cosmical science, but as yet our knowledge of the facts is so limited that it would be madness to dogmatize on the subject. Recent investigations have shown that our moon broke off from the earth while the latter was in a plastic condition owing to tidal friction. Hence its origin is neither nebular nor meteoric.

Probably the best modern opinion inclines to the view that nebular condensation, meteoric condensation, tidal friction, and possibly other causes as yet unsuggested, have all played their part in the evolution of the system.

The idea of the nebular hypothesis had been outlined by Kant[1] in 1755, and he had also suggested meteoric aggregations and tidal friction as causes affecting the formation of the solar system: it is probable that Laplace was not aware of this.

According to the rule published by Titius of Wittemberg in 1766—but generally known as Bode's law, from the fact that attention was called to it by Johann Elert Bode in 1778—the distances of the planets from the sun are nearly in the ratio of the numbers $0 + 4$, $3 + 4$, $6 + 4$, $12 + 4$, &c., the $(n + 2)$th term being $(2^n \times 3) + 4$. It would be an interesting fact if this could be deduced from the nebular, meteoric, or any other hypotheses, but so far as I am aware only one writer has made any serious attempt to do so, and his conclusion seems to be that the law is not sufficiently exact to be more than a convenient means of

[1] See *Kant's Cosmogony*, edited by W. Hastie, Glasgow, 1900.

remembering the general result.

Laplace's analytical discussion of the solar system is given in his *Mécanique céleste* published in five volumes. An analysis of the contents is given in the *English Cyclopaedia*. The first two volumes, published in 1799, contain methods for calculating the motions of the planets, determining their figures, and resolving tidal problems. The third and fourth volumes, published in 1802 and 1805, contain applications of these methods, and several astronomical tables. The fifth volume, published in 1825, is mainly historical, but it gives as appendices the results of Laplace's latest researches. Laplace's own investigations embodied in it are so numerous and valuable that it is regrettable to have to add that many results are appropriated from writers with scanty or no acknowledgment, and the conclusions—which have been described as the organized result of a century of patient toil—are frequently mentioned as if they were due to Laplace.

The matter of the *Mécanique céleste* is excellent, but it is by no means easy reading. Biot, who assisted Laplace in revising it for the press, says that Laplace himself was frequently unable to recover the details in the chain of reasoning, and, if satisfied that the conclusions were correct, he was content to insert the constantly recurring formula, "Il est aisé à voir." The *Mécanique céleste* is not only the translation of the *Principia* into the language of the differential calculus, but it completes parts of which Newton had been unable to fill in the details. F. F. Tisserand's recent work may be taken as the modern presentation of dynamical astronomy on classical lines, but Laplace's treatise will always remain a standard authority.

Laplace went in state to beg Napoleon to accept a copy of his work, and the following account of the interview is well authenticated, and so characteristic of all the parties concerned that I quote it in full. Someone had told Napoleon that the book contained no mention of the name of God; Napoleon, who was fond of putting embarrassing questions, received it with the remark, "M. Laplace, they tell me you have written this large book on the system of the universe, and have never even mentioned its Creator." Laplace, who, though the most supple of politicians, was as stiff as a martyr on every point of his philosophy, drew himself up and answered bluntly, "Je n'avais pas besoin de cette hypothèse-là." Napoleon, greatly amused, told this reply to Lagrange, who exclaimed, "Ah! c'est une belle hypothèse; ça explique beaucoup de choses."

In 1812 Laplace issued his *Théorie analytique des probabilités*.[1] The theory is stated to be only common sense expressed in mathematical language. The method of estimating the ratio of the number of favourable cases to the whole number of possible cases had been indicated by Laplace in a paper written in 1779. It consists in treating the successive values of any function as the coefficients in the expansion of another function with reference to a different variable. The latter is therefore called the generating function of the former. Laplace then shews how, by means of interpolation, these coefficients may be determined from the generating function. Next he attacks the converse problem, and from the coefficients he finds the generating function; this is effected by the solution of an equation in finite differences. The method is cumbersome, and in consequence of the increased power of analysis is now rarely used.

This treatise includes an exposition of the method of least squares, a remarkable testimony to Laplace's command over the processes of analysis. The method of least squares for the combination of numerous observations had been given empirically by Gauss and Legendre, but the fourth chapter of this work contains a formal proof of it, on which the whole of the theory of errors has been since based. This was effected only by a most intricate analysis specially invented for the purpose, but the form in which it is presented is so meagre and unsatisfactory that in spite of the uniform accuracy of the results it was at one time questioned whether Laplace had actually gone through the difficult work he so briefly and often incorrectly indicates.

In 1819 Laplace published a popular account of his work on probability. This book bears the same relation to the *Théorie des probabilités* that the *Système du monde* does to the *Mécanique céleste*.

Amongst the minor discoveries of Laplace in pure mathematics I may mention his discussion (simultaneously with Vandermonde) of the general theory of determinants in 1772; his proof that every equation of an even degree must have at least one real quadratic factor; his reduction of the solution of linear differential equations to definite integrals; and his solution of the linear partial differential equation of the second order. He was also the first to consider the difficult problems involved in equations of mixed differences, and to prove that the solution of an equation in finite differences of the first degree and the second order

---

[1] A summary of Laplace's reasoning is given in the article on Probability in the *Encyclopaedia Metropolitana*.

might be always obtained in the form of a continued fraction. Besides these original discoveries he determined, in his theory of probabilities, the values of a number of the more common definite integrals; and in the same book gave the general proof of the theorem enunciated by Lagrange for the development of any implicit function in a series by means of differential coefficients.

In theoretical physics the theory of capillary attraction is due to Laplace, who accepted the idea propounded by Hauksbee in the *Philosophical Transactions* for 1709, that the phenomenon was due to a force of attraction which was insensible at sensible distances. The part which deals with the action of a solid on a liquid and the mutual action of two liquids was not worked out thoroughly, but ultimately was completed by Gauss: Neumann later filled in a few details. In 1862 Lord Kelvin (Sir William Thomson) shewed that, if we assume the molecular constitution of matter, the laws of capillary attraction can be deduced from the Newtonian law of gravitation.

Laplace in 1816 was the first to point out explicitly why Newton's theory of vibratory motion gave an incorrect value for the velocity of sound. The actual velocity is greater than that calculated by Newton in consequence of the heat developed by the sudden compression of the air which increases the elasticity and therefore the velocity of the sound transmitted. Laplace's investigations in practical physics were confined to those carried on by him jointly with Lavoisier in the years 1782 to 1784 on the specific heat of various bodies.

Laplace seems to have regarded analysis merely as a means of attacking physical problems, though the ability with which he invented the necessary analysis is almost phenomenal. As long as his results were true he took but little trouble to explain the steps by which he arrived at them; he never studied elegance or symmetry in his processes, and it was sufficient for him if he could by any means solve the particular question he was discussing.

It would have been well for Laplace's reputation if he had been content with his scientific work, but above all things he coveted social fame. The skill and rapidity with which he managed to change his politics as occasion required would be amusing had they not been so servile. As Napoleon's power increased Laplace abandoned his republican principles (which, since they had faithfully reflected the opinions of the party in power, had themselves gone through numerous changes) and begged the first consul to give him the post of minister of the interior. Napoleon, who desired the support of men of science, agreed to the proposal; but a little less than six weeks saw the close of Laplace's political career. Napoleon's memorandum on his dismissal is as follows: "Géomètre de premier rang, Laplace ne tarda pas à se montrer administrateur plus que médiocre; dès son premier travail nous reconnûmes que nous nous étions trompé. Laplace ne saisissait aucune question sous son véritable point de vue: il cherchait des subtilités partout, n'avait que des idées problématiques, et portait enfin l'esprit des 'infiniment petits' jusque dans l'administration."

Although Laplace was removed from office it was desirable to retain his allegiance. He was accordingly raised to the senate, and to the third volume of the *Mécanique céleste* he prefixed a note that of all the truths therein contained the most precious to the author was the declaration he thus made of his devotion towards the peacemaker of Europe. In copies sold after the restoration this was struck out. In 1814 it was evident that the empire was falling; Laplace hastened to tender his services to the Bourbons, and on the restoration was rewarded with the title of marquis: the contempt that his more honest colleagues felt for his conduct in the matter may be read in the pages of Paul Louis Courier. His knowledge was useful on the numerous scientific commissions on which he served, and probably accounts for the manner in which his political insincerity was overlooked; but the pettiness of his character must not make us forget how great were his services to science.

That Laplace was vain and selfish is not denied by his warmest admirers; his conduct to the benefactors of his youth and his political friends was ungrateful and contemptible; while his appropriation of the results of those who were comparatively unknown seems to be well established and is absolutely indefensible—of those whom he thus treated three subsequently rose to distinction (Legendre and Fourier in France and Young in England) and never forgot the injustice of which they had been the victims. On the other side it may be said that on some questions he shewed independence of character, and he never concealed his views on religion, philosophy, or science, however distasteful they might be to the authorities in power: it should be also added that towards the close of his life, and especially to the work of his pupils, Laplace was both generous and appreciative, and in one case suppressed a paper of his own in order that a pupil might have the sole credit of the investigation.

Legendre.   *Adrian Marie Legendre* was born at Toulouse on September 18, 1752, and died at Paris on January 10, 1833. The leading events of his life are very simple and may be summed up briefly. He was educated at the Mazarin College in Paris, appointed professor at the military school in Paris in 1777, was a member of the Anglo-French commission of 1787 to connect Greenwich and Paris geodetically; served on several of the public commissions from 1792 to 1810; was made a professor at the Normal school in 1795; and subsequently held a few minor government appointments. The influence of Laplace was steadily exerted against his obtaining office or public recognition, and Legendre, who was a timid student, accepted the obscurity to which the hostility of his colleague condemned him.

Legendre's analysis is of a high order of excellence, and is second only to that produced by Lagrange and Laplace, though it is not so original. His chief works are his *Géométrie*, his *Théorie des nombres*, his *Exercices de calcul intégral,* and his *Fonctions elliptiques.* These include the results of his various papers on these subjects. Besides these he wrote a treatise which gave the rule for the method of least squares, and two groups of memoirs, one on the theory of attractions, and the other on geodetical operations.

The memoirs on attractions are analyzed and discussed in Todhunter's *History of the Theories of Attraction.* The earliest of these memoirs, presented in 1783, was on the attraction of spheroids. This contains the introduction of Legendre's coefficients, which are sometimes called circular (or zonal) harmonics, and which are particular cases of Laplace's coefficients; it also includes the solution of a problem in which the potential is used. The second memoir was communicated in 1784, and is on the form of equilibrium of a mass of rotating liquid which is approximately spherical. The third, written in 1786, is on the attraction of confocal ellipsoids. The fourth is on the figure which a fluid planet would assume, and its law of density.

His papers on geodesy are three in number, and were presented to the Academy in 1787 and 1788. The most important result is that by which a spherical triangle may be treated as plane, provided certain corrections are applied to the angles. In connection with this subject he paid considerable attention to geodesics.

The method of least squares was enunciated in his *Nouvelles méthodes* published in 1806, to which supplements were added in 1810 and 1820. Gauss independently had arrived at the same result, had used it in 1795, and published it and the law of facility in 1809. Laplace was the earliest writer to give a proof of it; this was in 1812.

Of the other books produced by Legendre, the one most widely known is his *Éléments de géométrie* which was published in 1794, and was at one time widely adopted on the continent as a substitute for Euclid. The later editions contain the elements of trigonometry, and proofs of the irrationality of π and π^2. An appendix on the difficult question of the theory of parallel lines was issued in 1803, and is bound up with most of the subsequent editions.

His *Théorie des nombres* was published in 1798, and appendices were added in 1816 and 1825; the third edition, issued in two volumes in 1830, includes the results of his various later papers, and still remains a standard work on the subject. It may be said that he here carried the subject as far as was possible by the application of ordinary algebra; but he did not realize that it might be regarded as a higher arithmetic, and so form a distinct subject in mathematics.

The law of quadratic reciprocity, which connects any two odd primes, was first proved in this book, but the result had been enunciated in a memoir of 1785. Gauss called the proposition "the gem of arithmetic," and no less than six separate proofs are to be found in his works. The theorem is as follows. If p be a prime and n be prime to p, then we know that the remainder when $n^{(p-1)/2}$ is divided by p is either $+1$ or -1. Legendre denoted this remainder by (n/p). When the remainder is $+1$ it is possible to find a square number which when divided by p leaves a remainder n, that is, n is a quadratic residue of p; when the remainder is -1 there exists no such square number, and n is a non-residue of p. The law of quadratic reciprocity is expressed by the theorem that, if a and b be any odd primes, then

$$(a/b)(b/a) = (-1)^{(a-1)(b-1)/4};$$

thus, if b be a residue of a, then a is also a residue of b, unless both of the primes a and b are of the form $4m + 3$. In other words, if a and b be odd primes, we know that

$$a^{(b-1)/2} \equiv \pm 1 (\bmod\ b), \text{ and } b^{(a-1)/2} \equiv \pm 1 (\bmod\ a);$$

and, by Legendre's law, the two ambiguities will be either both positive or both negative, unless a and b are both of the form $4m + 3$. Thus, if one odd prime be a non-residue of another, then the latter will be a non-residue of the former. Gauss and Kummer have subsequently proved similar laws of cubic and biquadratic reciprocity; and an important branch of the theory of numbers has been based on these researches.

This work also contains the useful theorem by which, when it is possible, an indeterminate equation of the second degree can be reduced to the form $ax^2 + by^2 + cz^2 = 0$. Legendre here discussed the forms of numbers which can be expressed as the sum of three squares; and he proved [art. 404] that the number of primes less than n is approximately $n/(\log_e n - 1.08366)$.

The *Exercices de calcul intégral* was published in three volumes, 1811, 1817, 1826. Of these the third and most of the first are devoted to elliptic functions; the bulk of this being ultimately included in the *Fonctions elliptiques*. The contents of the remainder of the treatise are of a miscellaneous character; they include integration by series, definite integrals, and in particular an elaborate discussion of the Beta and the Gamma functions.

The *Traité des fonctions elliptiques* was issued in two volumes in 1825 and 1826, and is the most important of Legendre's works. A third volume was added a few weeks before his death, and contains three memoirs on the researches of Abel and Jacobi. Legendre's investigations had commenced with a paper written in 1786 on elliptic arcs, but here and in his other papers he treated the subject merely as a problem in the integral calculus, and did not see that it might be considered as a higher trigonometry, and so constitute a distinct branch of analysis. Tables of the elliptic integrals were constructed by him. The modern treatment of the subject is founded on that of Abel and Jacobi. The superiority of their methods was at once recognized by Legendre, and almost the last act of his life was to recommend those discoveries which he knew would consign his own labours to comparative oblivion.

This may serve to remind us of a fact which I wish to specially emphasize, namely, that Gauss, Abel, Jacobi, and some others of the mathematicians alluded to in the next chapter, were contemporaries of the members of the French school.

Pfaff. I may here mention another writer who also made a special study of the integral calculus. This was *Johann Friederich Pfaff*, born at Stuttgart on Dec. 22, 1765, and died at Halle on April 21, 1825, who was described by Laplace as the most eminent mathematician in Germany at the beginning of this century, a description which, had it not been for Gauss's existence, would have been true enough.

Pfaff was the precursor of the German school, which under Gauss and his followers largely determined the lines on which mathematics developed during the nineteenth century. He was an intimate friend of Gauss, and in fact the two mathematicians lived together at Helmstadt during the year 1798, after Gauss had finished his university course. Pfaff's chief work was his (unfinished) *Disquisitiones Analyticae* on the integral calculus, published in 1797; and his most important memoirs were either on the calculus or on differential equations: on the latter subject his paper read before the Berlin Academy in 1814 is noticeable.

The creation of modern geometry.

While Euler, Lagrange, Laplace, and Legendre were perfecting analysis, the members of another group of French mathematicians were extending the range of geometry by methods similar to those previously used by Desargues and Pascal. The revival of the study of synthetic geometry is largely due to Poncelet, but the subject is also associated with the names of Monge and L. Carnot; its great development in more recent times is mainly due to Steiner, von Staudt, and Cremona.

Monge.[1]   *Gaspard Monge* was born at Beaune on May 10, 1746, and died at Paris on July 28, 1818. He was the son of a small pedlar, and was educated in the schools of the Oratorians, in one of which he subsequently became an usher. A plan of Beaune which he had made fell into the hands of an officer who recommended the military authorities to admit him to their training-school at Mézières. His birth, however, precluded his receiving a commission in the army, but his attendance at an annexe of the school where surveying and drawing were taught was tolerated, though he was told that he was not sufficiently well born to be allowed to attempt problems which required calculation. At last his opportunity came. A plan of a fortress having to be drawn from the data supplied by certain observations, he did it by a geometrical construction. At first the officer in charge refused to receive it, because etiquette required that not less than a certain time should be used in making such drawings, but the superiority of the method over that then taught was so obvious that it was accepted; and in 1768 Monge was made professor, on the understanding that the results of his descriptive geometry were to be a military secret confined to officers above a certain rank.

In 1780 he was appointed to a chair of mathematics in Paris, and this with some provincial appointments which he held gave him a comfortable income. The earliest paper of any special importance which he communicated to the French Academy was one in 1781, in which he discussed the lines of curvature drawn on a surface. These had been

---

[1]On the authorities for Monge's life and works, see the note by H. Brocard in *L'Intermédiaire des mathématiciens*, 1906, vol. xiii, pp. 118, 119.

first considered by Euler in 1760, and defined as those normal sections whose curvature was a maximum or a minimum. Monge treated them as the locus of those points on the surface at which successive normals intersect, and thus obtained the general differential equation. He applied his results to the central quadrics in 1795. In 1786 he published his well-known work on statics.

Monge eagerly embraced the doctrines of the revolution. In 1792 he became minister of the marine, and assisted the committee of public safety in utilizing science for the defence of the republic. When the Terrorists obtained power he was denounced, and escaped the guillotine only by a hasty flight. On his return in 1794 he was made a professor at the short-lived Normal school, where he gave lectures on descriptive geometry; the notes of these were published under the regulation above alluded to. In 1796 he went to Italy on the roving commission which was sent with orders to compel the various Italian towns to offer pictures, sculpture, or other works of art that they might possess, as a present or in lieu of contributions to the French republic for removal to Paris. In 1798 he accepted a mission to Rome, and after executing it joined Napoleon in Egypt. Thence after the naval and military victories of England he escaped to France.

Monge then settled down at Paris, and was made professor at the Polytechnic school, where he gave lectures on descriptive geometry; these were published in 1800 in the form of a text-book entitled *Géométrie descriptive*. This work contains propositions on the form and relative position of geometrical figures deduced by the use of transversals. The theory of perspective is considered; this includes the art of representing in two dimensions geometrical objects which are of three dimensions, a problem which Monge usually solved by the aid of two diagrams, one being the plan and the other the elevation. Monge also discussed the question as to whether, if in solving a problem certain subsidiary quantities introduced to facilitate the solution become imaginary, the validity of the solution is thereby impaired, and he shewed that the result would not be affected. On the restoration he was deprived of his offices and honours, a degradation which preyed on his mind and which he did not long survive.

Most of his miscellaneous papers are embodied in his works, *Application de l'algèbre à la géométrie*, published in 1805, and *Application de l'analyse à la géométrie*, the fourth edition of which, published in 1819, was revised by him just before his death. It contains among other results his solution of a partial differential equation of the second order.

Carnot.[1]   *Lazare Nicholas Marguerite Carnot*, born at Nolay on May 13, 1753, and died at Magdeburg on Aug. 22, 1823, was educated at Burgundy, and obtained a commission in the engineer corps of Condé. Although in the army, he continued his mathematical studies in which he felt great interest. His first work, published in 1784, was on machines; it contains a statement which foreshadows the principle of energy as applied to a falling weight, and the earliest proof of the fact that kinetic energy is lost in the collision of imperfectly elastic bodies. On the outbreak of the revolution in 1789 he threw himself into politics. In 1793 he was elected on the committee of public safety, and the victories of the French army were largely due to his powers of organization and enforcing discipline. He continued to occupy a prominent place in every successive form of government till 1796 when, having opposed Napoleon's *coup d'état*, he had to fly from France. He took refuge in Geneva, and there in 1797 issued his *Réflexions sur la métaphysique du calcul infinitésimal*: in this he amplifies views previously expounded by Berkeley and Lagrange. In 1802 he assisted Napoleon, but his sincere republican convictions were inconsistent with the retention of office. In 1803 he produced his *Géométrie de position*. This work deals with projective rather than descriptive geometry, it also contains an elaborate discussion of the geometrical meaning of negative roots of an algebraical equation. In 1814 he offered his services to fight for France, though not for the empire; and on the restoration he was exiled.

Poncelet.[2]   *Jean Victor Poncelet*, born at Metz on July 1, 1788, and died at Paris on Dec. 22, 1867, held a commission in the French engineers. Having been made a prisoner in the French retreat from Moscow in 1812 he occupied his enforced leisure by writing the *Traité des propriétés projectives des figures*, published in 1822, which was long one of the best known text-books on modern geometry. By means of projection, reciprocation, and homologous figures, he established all the chief properties of conics and quadrics. He also treated the theory of polygons. His treatise on practical mechanics in 1826, his memoir on water-mills in 1826, and his report on the English machinery and tools exhibited at the International Exhibition held in London in 1851 deserve mention. He contributed numerous articles to Crelle's journal; the most valuable of these deal with the explanation, by the aid of the

---

[1] See the *éloge* by Arago, which, like most obituary notices, is a panegyric rather than an impartial biography.

[2] See *La Vie et les ouvrages de Poncelet*, by I. Didion and C. Dupin, Paris, 1869.

doctrine of continuity, of imaginary solutions in geometrical problems.

The development of mathematical physics.

It will be noticed that Lagrange, Laplace, and Legendre mostly occupied themselves with analysis, geometry, and astronomy. I am inclined to regard Cauchy and the French mathematicians of the present day as belonging to a different school of thought to that considered in this chapter, and I place them amongst modern mathematicians, but I think that Fourier, Poisson, and the majority of their contemporaries, are the lineal successors of Lagrange and Laplace. If this view be correct, we may say that the successors of Lagrange and Laplace devoted much of their attention to the application of mathematical analysis to physics. Before considering these mathematicians I may mention the distinguished English experimental physicists who were their contemporaries, and whose merits have only recently received an adequate recognition. Chief among these are Cavendish and Young.

Cavendish.[1] The Honourable *Henry Cavendish* was born at Nice on October 10, 1731, and died in London on February 4, 1810. His tastes for scientific research and mathematics were formed at Cambridge, where he resided from 1749 to 1753. He created experimental electricity, and was one of the earliest writers to treat chemistry as an exact science. I mention him here on account of his experiment in 1798 to determine the density of the earth, by estimating its attraction as compared with that of two given lead balls: the result is that the mean density of the earth is about five and a half times that of water. This experiment was carried out in accordance with a suggestion which had been first made by John Mitchell (1724–1793), a fellow of Queens' College, Cambridge, who had died before he was able to carry it into effect.

Rumford.[2] *Sir Benjamin Thomson, Count Rumford,* born at Concord on March 26, 1753, and died at Auteuil on August 21, 1815, was of English descent, and fought on the side of the loyalists in the American War of Secession: on the conclusion of peace he settled in England, but subsequently entered the service of Bavaria, where his

---

[1] An account of his life by G. Wilson will be found in the first volume of the publications of the Cavendish Society, London, 1851. His *Electrical Researches* were edited by J. C. Maxwell, and published at Cambridge in 1879.

[2] An edition of Rumford's works, edited by George Ellis, accompanied by a biography, was published by the American Academy of Sciences at Boston in 1872.

powers of organization proved of great value in civil as well as military affairs. At a later period he again resided in England, and when there founded the Royal Institution. The majority of his papers were communicated to the Royal Society of London; of these the most important is his memoir in which he showed that heat and work are mutually convertible.

Young.[1]   Among the most eminent physicists of his time was *Thomas Young,* who was born at Milverton on June 13, 1773, and died in London on May 10, 1829. He seems as a boy to have been somewhat of a prodigy, being well read in modern languages and literature, as well as in science; he always kept up his literary tastes, and it was he who in 1819 first suggested the key to decipher the Egyptian hieroglyphics, which J. F. Champollion used so successfully. Young was destined to be a doctor, and after attending lectures at Edinburgh and Göttingen entered at Emmanuel College, Cambridge, from which he took his degree in 1799; and to his stay at the University he attributed much of his future distinction. His medical career was not particularly successful, and his favourite maxim that a medical diagnosis is only a balance of probabilities was not appreciated by his patients, who looked for certainty in return for their fee. Fortunately his private means were ample. Several papers contributed to various learned societies from 1798 onwards prove him to have been a mathematician of considerable power; but the researches which have immortalised his name are those by which he laid down the laws of interference of waves and of light, and was thus able to suggest the means by which the chief difficulties then felt in the way of the acceptance of the undulatory theory of light could be overcome.

Dalton.[2]   Another distinguished writer of the same period was *John Dalton,* who was born in Cumberland on September 5, 1766, and died at Manchester on July 27, 1844. Dalton investigated the tension of vapours, and the law of the expansion of a gas under changes of temperature. He also founded the atomic theory in chemistry.

It will be gathered from these notes that the English school of physicists at the beginning of this century were mostly concerned with the experimental side of the subject. But in fact no satisfactory theory could

---

[1]Young's collected works and a memoir on his life were published by G. Peacock, four volumes, London, 1855.

[2]See the *Memoir of Dalton,* by R. A. Smith, London, 1856; and W. C. Henry's memoir in the *Cavendish Society Transactions,* London, 1854.

be formed without some similar careful determination of the facts. The most eminent French physicists of the same time were Fourier, Poisson, Ampère, and Fresnel. Their method of treating the subject is more mathematical than that of their English contemporaries, and the two first named were distinguished for general mathematical ability.

Fourier.[1]   The first of these French physicists was *Jean Baptiste Joseph Fourier*, who was born at Auxerre on March 21, 1768, and died at Paris on May 16, 1830. He was the son of a tailor, and was educated by the Benedictines. The commissions in the scientific corps of the army were, as is still the case in Russia, reserved for those of good birth, and being thus ineligible he accepted a military lectureship on mathematics. He took a prominent part in his own district in promoting the revolution, and was rewarded by an appointment in 1795 in the Normal school, and subsequently by a chair in the Polytechnic school.

Fourier went with Napoleon on his Eastern expedition in 1798, and was made governor of Lower Egypt. Cut off from France by the English fleet, he organised the workshops on which the French army had to rely for their munitions of war. He also contributed several mathematical papers to the Egyptian Institute which Napoleon founded at Cairo, with a view of weakening English influence in the East. After the British victories and the capitulation of the French under General Menou in 1801, Fourier returned to France, and was made prefect of Grenoble, and it was while there that he made his experiments on the propagation of heat. He moved to Paris in 1816. In 1822 he published his *Théorie analytique de la chaleur*, in which he bases his reasoning on Newton's law of cooling, namely, that the flow of heat between two adjacent molecules is proportional to the infinitely small difference of their temperatures. In this work he shows that any function of a variable, whether continuous or discontinuous, can be expanded in a series of sines of multiples of the variable—a result which is constantly used in modern analysis. Lagrange had given particular cases of the theorem, and had implied that the method was general, but he had not pursued the subject. Dirichlet was the first to give a satisfactory demonstration of it.

Fourier left an unfinished work on determinate equations which was edited by Navier, and published in 1831; this contains much original matter, in particular there is a demonstration of Fourier's theorem on

[1]An edition of his works, edited by G. Darboux, was published in two volumes, Paris, 1888, 1890.

the position of the roots of an algebraical equation. Lagrange had shewn
how the roots of an algebraical equation might be separated by means
of another equation whose roots were the squares of the differences
of the roots of the original equation. Budan, in 1807 and 1811, had
enunciated the theorem generally known by the name of Fourier, but
the demonstration was not altogether satisfactory. Fourier's proof is the
same as that usually given in text-books on the theory of equations.
The final solution of the problem was given in 1829 by Jacques Charles
François Sturm (1803–1855).

Sadi Carnot.[1] Among Fourier's contemporaries who were interested in the theory of heat the most eminent was *Sadi Carnot*, a son
of the eminent geometrician mentioned above. Sadi Carnot was born
at Paris in 1796, and died there of cholera in August 1832; he was an
officer in the French army. In 1824 he issued a short work entitled
Réflexions sur la puissance motrice du feu, in which he attempted to
determine in what way heat produced its mechanical effect. He made
the mistake of assuming that heat was material, but his essay may be
taken as initiating the modern theory of thermodynamics.

Poisson.[2] *Siméon Denis Poisson*, born at Pithiviers on June 21,
1781, and died at Paris on April 25, 1840, is almost equally distinguished for his applications of mathematics to mechanics and to physics. His father had been a private soldier, and on his retirement was
given some small administrative post in his native village; when the
revolution broke out he appears to have assumed the government of
the place, and, being left undisturbed, became a person of some local
importance. The boy was put out to nurse, and he used to tell how one
day his father, coming to see him, found that the nurse had gone out,
on pleasure bent, having left him suspended by a small cord attached
to a nail fixed in the wall. This, she explained, was a necessary pre-
caution to prevent him from perishing under the teeth of the various
animals and animalculae that roamed on the floor. Poisson used to
add that his gymnastic efforts carried him incessantly from one side to
the other, and it was thus in his tenderest infancy that he commenced
those studies on the pendulum that were to occupy so large a part of
his mature age.

---

[1]A sketch of S. Carnot's life and an English translation of his *Réflexions* was
published by R. H. Thurston, London and New York, 1890.

[2]Memoirs of Poisson will be found in the *Encyclopaedia Britannica*, the *Transactions of the Royal Astronomical Society*, vol. v, and Arago's *Éloges*, vol. ii; the
latter contains a bibliography of Poisson's papers and works.

He was educated by his father, and destined much against his will to be a doctor. His uncle offered to teach him the art, and began by making him prick the veins of cabbage-leaves with a lancet. When perfect in this, he was allowed to put on blisters; but in almost the first case he did this by himself, the patient died in a few hours, and though all the medical practitioners of the place assured him that "the event was a very common one," he vowed he would have nothing more to do with the profession.

Poisson, on his return home after this adventure, discovered amongst the official papers sent to his father a copy of the questions set at the Polytechnic school, and at once found his career. At the age of seventeen he entered the Polytechnic, and his abilities excited the interest of Lagrange and Laplace, whose friendship he retained to the end of their lives. A memoir on finite differences which he wrote when only eighteen was reported on so favourably by Legendre that it was ordered to be published in the *Recueil des savants étrangers*. As soon as he had finished his course he was made a lecturer at the school, and he continued through his life to hold various government scientific posts and professorships. He was somewhat of a socialist, and remained a rigid republican till 1815, when, with a view to making another empire impossible, he joined the legitimists. He took, however, no active part in politics, and made the study of mathematics his amusement as well as his business.

His works and memoirs are between three and four hundred in number. The chief treatises which he wrote were his *Traité de mécanique*,[1] published in two volumes, 1811 and 1833, which was long a standard work; his *Théorie nouvelle de l'action capillaire*, 1831; his *Théorie mathématique de la chaleur*, 1835, to which a supplement was added in 1837; and his *Recherches sur la probabilité des jugements*, 1837. He had intended, if he had lived, to write a work which should cover all mathematical physics and in which the results of the three books last named would have been incorporated.

Of his memoirs in pure mathematics the most important are those on definite integrals, and Fourier's series, their application to physical

---

[1] Among Poisson's contemporaries who studied mechanics and of whose works he made use I may mention *Louis Poinsot*, who was born in Paris on Jan. 3, 1777, and died there on Dec. 5, 1859. In his *Statique*, published in 1803, he treated the subject without any explicit reference to dynamics. The theory of couples is largely due to him (1806), as also the motion of a body in space under the action of no forces.

problems constituting one of his chief claims to distinction; his essays on the calculus of variations; and his papers on the probability of the mean results of observations.[1]

Perhaps the most remarkable of his memoirs in applied mathematics are those on the theory of electrostatics and magnetism, which originated a new branch of mathematical physics; he supposed that the results were due to the attractions and repulsions of imponderable particles. The most important of those on physical astronomy are the two read in 1806 (printed in 1809) on the secular inequalities of the mean motions of the planets, and on the variation of arbitrary constants introduced into the solutions of questions on mechanics; in these Poisson discusses the question of the stability of the planetary orbits (which Lagrange had already proved to the first degree of approximation for the disturbing forces), and shews that the result can be extended to the third order of small quantities: these were the memoirs which led to Lagrange's famous memoir of 1808. Poisson also published a paper in 1821 on the libration of the moon; and another in 1827 on the motion of the earth about its centre of gravity. His most important memoirs on the theory of attraction are one in 1829 on the attraction of spheroids, and another in 1835 on the attraction of a homogeneous ellipsoid: the substitution of the correct equation involving the potential, namely, $\nabla^2 V = -4\pi\rho$, for Laplace's form of it, $\nabla^2 V = 0$, was first published[2] in 1813. Lastly, I may mention his memoir in 1825 on the theory of waves.

Ampère.[3]   *André Marie Ampère* was born at Lyons on January 22, 1775, and died at Marseilles on June 10, 1836. He was widely read in all branches of learning, and lectured and wrote on many of them, but after the year 1809, when he was made professor of analysis at the Polytechnic school in Paris, he confined himself almost entirely to mathematics and science. His papers on the connection between electricity and magnetism were written in 1820. According to his theory, propounded in 1826, a molecule of matter which can be magnetized is traversed by a closed electric current, and magnetization is produced by any cause which makes the direction of these currents in the different

---

[1] See the *Journal de l'école polytechnique* from 1813 to 1823, and the *Mémoires de l'académie* for 1823; the *Mémoires de l'académie*, 1833; and the *Connaissance des temps*, 1827 and following years. Most of his memoirs were published in the three periodicals here mentioned.

[2] In the *Bulletin des sciences* of the *Société philomatique*.

[3] See C. A. Valson's *Étude sur la vie et les ouvrages d'Ampère*, Lyons, 1885.

molecules of the body approach parallelism.

Fresnel.   Biot.   *Augustin Jean Fresnel,* born at Broglie on May 10, 1788, and died at Ville-d'Avray on July 14, 1827, was a civil engineer by profession, but he devoted his leisure to the study of physical optics. The undulatory theory of light, which Hooke, Huygens, and Euler had supported on *a priori* grounds, had been based on experiment by the researches of Young. Fresnel deduced the mathematical consequences of these experiments, and explained the phenomena of interference both of ordinary and polarized light. Fresnel's friend and contemporary, *Jean Baptiste Biot,* who was born at Paris on April 21, 1774, and died there in 1862, requires a word or two in passing. Most of his mathematical work was in connection with the subject of optics, and especially the polarization of light. His systematic works were produced within the years 1805 and 1817; a selection of his more valuable memoirs was published in Paris in 1858.

Arago.[1]   *François Jean Dominique Arago* was born at Estagel in the Pyrenees on February 26, 1786, and died in Paris on October 2, 1853. He was educated at the Polytechnic school, Paris, and we gather from his autobiography that however distinguished were the professors of that institution they were remarkably incapable of imparting their knowledge or maintaining discipline.

In 1804 Arago was made secretary to the observatory at Paris, and from 1806 to 1809 he was engaged in measuring a meridian arc in order to determine the exact length of a metre. He was then appointed to a leading post in the observatory, given a residence there, and made a professor at the Polytechnic school, where he enjoyed a marked success as a lecturer. He subsequently gave popular lectures on astronomy, which were both lucid and accurate—a combination of qualities which was rarer then than now. He reorganized the national observatory, the management of which had long been inefficient, but in doing this his want of tact and courtesy raised many unnecessary difficulties. He remained to the end a consistent republican, and after the *coup d'état* of 1852, though half blind and dying, he resigned his post as astronomer rather than take the oath of allegiance. It is to the credit of Napoleon III. that he gave directions that the old man should be in no way disturbed, and should be left free to say and do what he liked.

---

[1] Arago's works, which include *éloges* on many of the leading mathematicians of the last five or six centuries, have been edited by M. J. A. Barral, and published in fourteen volumes, Paris, 1856–57. An autobiography is prefixed to the first volume.

Arago's earliest physical researches were on the pressure of steam at different temperatures, and the velocity of sound, 1818 to 1822. His magnetic observations mostly took place from 1823 to 1826. He discovered what has been called rotatory magnetism, and the fact that most bodies could be magnetized; these discoveries were completed and explained by Faraday. He warmly supported Fresnel's optical theories, and the two philosophers conducted together those experiments on the polarization of light which led to the inference that the vibrations of the luminiferous ether were transverse to the direction of motion, and that polarization consisted in a resolution of rectilinear motion into components at right angles to each other. The subsequent invention of the polariscope and discovery of rotatory polarization are due to Arago. The general idea of the experimental determination of the velocity of light in the manner subsequently effected by Fizeau and Foucault was suggested by him in 1838, but his failing eyesight prevented his arranging the details or making the experiments.

It will be noticed that some of the last members of the French school were alive at a comparatively recent date, but nearly all their mathematical work was done before the year 1830. They are the direct successors of the French writers who flourished at the commencement of the nineteenth century, and seem to have been out of touch with the great German mathematicians of the early part of it, on whose researches much of the best work of that century is based; they are thus placed here, though their writings are in some cases of a later date than those of Gauss, Abel, and Jacobi.

The introduction of analysis into England.

The complete isolation of the English school and its devotion to geometrical methods are the most marked features in its history during the latter half of the eighteenth century; and the absence of any considerable contribution to the advancement of mathematical science was a natural consequence. One result of this was that the energy of English men of science was largely devoted to practical physics and practical astronomy, which were in consequence studied in Britain perhaps more than elsewhere.

Ivory.　Almost the only English mathematician at the beginning of this century who used analytical methods, and whose work requires mention here, is Ivory, to whom the celebrated theorem in attractions is due. *Sir James Ivory* was born in Dundee in 1765, and died on

September 21, 1842. After graduating at St. Andrews he became the managing partner in a flax-spinning company in Forfarshire, but continued to devote most of his leisure to mathematics. In 1804 he was made professor at the Royal Military College at Marlow, which was subsequently moved to Sandhurst; he was knighted in 1831. He contributed numerous papers to the *Philosophical Transactions*, the most remarkable being those on attractions. In one of these, in 1809, he shewed how the attraction of a homogeneous ellipsoid on an external point is a multiple of that of another ellipsoid on an internal point: the latter can be easily obtained. He criticized Laplace's solution of the method of least squares with unnecessary bitterness, and in terms which shewed that he had failed to understand it.

The Cambridge Analytical School. Towards the beginning of the last century the more thoughtful members of the Cambridge school of mathematics began to recognize that their isolation from their continental contemporaries was a serious evil. The earliest attempt in England to explain the notation and methods of the calculus as used on the continent was due to Woodhouse, who stands out as the apostle of the new movement. It is doubtful if he could have brought the analytical methods into vogue by himself; but his views were enthusiastically adopted by three students, Peacock, Babbage, and Herschel, who succeeded in carrying out the reforms he had suggested. In a book which will fall into the hands of few but English readers I may be pardoned for making space for a few remarks on these four mathematicians, though otherwise a notice of them would not be required in a work of this kind.[1] The original stimulus came from French sources, and I therefore place these remarks at the close of my account of the French school; but I should add that the English mathematicians of this century at once struck out a line independent of their French contemporaries.

Woodhouse. *Robert Woodhouse* was born at Norwich on April 28, 1773; was educated at Caius College, Cambridge, of which society he was subsequently a fellow; was Plumian professor in the university; and continued to live at Cambridge till his death on December 23, 1827.

Woodhouse's earliest work, entitled the *Principles of Analytical Calculation*, was published at Cambridge in 1803. In this he explained the differential notation and strongly pressed the employment of it; but he severely criticized the methods used by continental writers, and their

---

[1] The following account is condensed from my *History of the Study of Mathematics at Cambridge*, Cambridge, 1889.

constant assumption of non-evident principles. This was followed in 1809 by a trigonometry (plane and spherical), and in 1810 by a historical treatise on the calculus of variations and isoperimetrical problems. He next produced an astronomy; of which the first book (usually bound in two volumes), on practical and descriptive astronomy, was issued in 1812, and the second book, containing an account of the treatment of physical astronomy by Laplace and other continental writers, was issued in 1818. All these works deal critically with the scientific foundation of the subjects considered—a point which is not unfrequently neglected in modern text-books.

A man like Woodhouse, of scrupulous honour, universally respected, a trained logician, and with a caustic wit, was well fitted to introduce a new system; and the fact that when he first called attention to the continental analysis he exposed the unsoundness of some of the usual methods of establishing it, more like an opponent than a partisan, was as politic as it was honest. Woodhouse did not exercise much influence on the majority of his contemporaries, and the movement might have died away for the time being if it had not been for the advocacy of Peacock, Babbage, and Herschel, who formed an Analytical Society, with the object of advocating the general use in the university of analytical methods and of the differential notation.

Peacock.　*George Peacock*, who was the most influential of the early members of the new school, was born at Denton on April 9, 1791. He was educated at Trinity College, Cambridge, of which society he was subsequently a fellow and tutor. The establishment of the university observatory was mainly due to his efforts, and in 1836 he was appointed to the Lowndean professorship of astronomy and geometry. In 1839 he was made dean of Ely, and resided there till his death on Nov. 8, 1858. Although Peacock's influence on English mathematicians was considerable, he has left but few memorials of his work; but I may note that his report on progress in analysis, 1833, commenced those valuable summaries of current scientific progress which enrich many of the annual volumes of the *Transactions* of the British Association.

Babbage.　Another important member of the Analytical Society was *Charles Babbage*, who was born at Totnes on Dec. 26, 1792; he entered at Trinity College, Cambridge, in 1810; subsequently became Lucasian professor in the university; and died in London on Oct. 18, 1871. It was he who gave the name to the Analytical Society, which, he stated, was formed to advocate "the principles of pure *d*-ism as opposed to the *dot*-age of the university". In 1820 the Astronomical Society was founded mainly through his efforts, and at a later time, 1830 to 1832, he took a prominent part in the foundation of the British Association. He will be remembered for his mathematical memoirs on the calculus of functions, and his invention of an analytical machine which could not only perform the ordinary processes of arithmetic, but could tabulate the values of any function and print the results.

Herschel.   The third of those who helped to bring analytical methods into general use in England was the son of Sir William Herschel (1738–1822), the most illustrious astronomer of the latter half of the eighteenth century and the creator of modern stellar astronomy. *Sir John Frederick William Herschel* was born on March 7, 1792, educated at St. John's College, Cambridge, and died on May 11, 1871. His earliest original work was a paper on Cotes's theorem, and it was followed by others on mathematical analysis, but his desire to complete his father's work led ultimately to his taking up astronomy. His papers on light and astronomy contain a clear exposition of the principles which underlie the mathematical treatment of those subjects.

In 1813 the Analytical Society published a volume of memoirs, of which the preface and the first paper (on continued products) are due to Babbage; and three years later they issued a translation of Lacroix's *Traité élémentaire du calcul différentiel et du calcul intégral.* In 1817, and again in 1819, the differential notation was used in the university examinations, and after 1820 its use was well established. The Analytical Society followed up this rapid victory by the issue in 1820 of two volumes of examples illustrative of the new method; one by Peacock on the differential and integral calculus, and the other by Herschel on the calculus of finite differences. Since then English works on the infinitesimal calculus have abandoned the exclusive use of the fluxional notation. It should be noticed in passing that Lagrange and Laplace, like the majority of other modern writers, employ both the fluxional and the differential notation; it was the exclusive adoption of the former that was so hampering.

Amongst those who materially assisted in extending the use of the new analysis were William Whewell (1794–1866) and George Biddell Airy (1801–1892), both Fellows of Trinity College, Cambridge. The former issued in 1819 a work on mechanics, and the latter, who was a pupil of Peacock, published in 1826 his *Tracts*, in which the new method was applied with great success to various physical problems. The efforts of the society were supplemented by the rapid publication of good text-books in which analysis was freely used. The employment of analytical methods spread from Cambridge over the rest of Britain, and by 1830 these methods had come into general use there.

CHAPTER XIX.

MATHEMATICS OF THE NINETEENTH CENTURY.

THE nineteenth century saw the creation of numerous new departments of pure mathematics—notably of a theory of numbers, or higher arithmetic; of theories of forms and groups, or a higher algebra; of theories of functions of multiple periodicity, or a higher trigonometry; and of a general theory of functions, embracing extensive regions of higher analysis. Further, the developments of synthetic and analytical geometry created what practically were new subjects. The foundations of the subject and underlying assumptions (notably in arithmetic, geometry, and the calculus) were also subjected to a rigorous scrutiny. Lastly, the application of mathematics to physical problems revolutionized the foundations and treatment of that subject. Numerous Schools, Journals, and Teaching Posts were established, and the facilities for the study of mathematics were greatly extended.

Developments, such as these, may be taken as opening a new period in the history of the subject, and I recognize that in the future a writer who divides the history of mathematics as I have done would probably treat the mathematics of the seventeenth and eighteenth centuries as forming one period, and would treat the mathematics of the nineteenth century as commencing a new period. This, however, would imply a tolerably complete and systematic account of the development of the subject in the nineteenth century. But evidently it is impossible for me to discuss adequately the mathematics of a time so near to us, and the works of mathematicians some of whom are living and some of whom I have met and known. Hence I make no attempt to give a complete account of the mathematics of the nineteenth century, but as a sort of appendix to the preceding chapters I mention the more striking features in the history of recent pure mathematics, in which I include theoretical dynamics and astronomy; I do not, however, propose to discuss in general the recent application of mathematics to physics.

In only a few cases do I give an account of the life and works of the mathematicians mentioned; but I have added brief notes about some of those to whom the development of any branch of the subject is chiefly due, and an indication of that part of it to which they have directed most attention. Even with these limitations it has been very difficult to put together a connected account of the mathematics of recent times; and I wish to repeat explicitly that I do not suggest, nor do I wish my readers to suppose, that my notes on a subject give the names of all the chief writers who have studied it. In fact the quantity of matter produced has been so enormous that no one can expect to do more than make himself acquainted with the works produced in some special branch or branches. As an illustration of this remark I may add that the committee appointed by the Royal Society to report on a catalogue of periodical literature estimated, in 1900, that more than 1500 memoirs on pure mathematics were then issued annually, and more than 40,000 a year on scientific subjects.

Most histories of mathematics do not treat of the work produced during this century. The chief exceptions with which I am acquainted are R. d'Adhémar's *L'Œuvre mathématique du xixe siècle*; K. Fink's *Geschichte der Mathematik*, Tübingen, 1890; E. J. Gerhardt's *Geschichte der Mathematik in Deutschland*, Munich, 1877; S. Günther's *Verm. Unt. zur Geschichte der mathematischen Wissenschaften*, Leipzig, 1876, and *Ziele und Resultate der neueren mathematisch-historischen Forschung*, Erlangen, 1876; J. G. Hagen, *Synopsis der höheren Mathematik*, 3 volumes, Berlin, 1891, 1893, 1906; a short dissertation by H. Hankel, entitled *Die Entwickelung der Mathematik in den letzten Jahrhunderten*, Tübingen, 1885; a *Discours* on the professors at the Sorbonne by C. Hermite in the *Bulletin des sciences mathématiques*, 1890; F. C. Klein's *Lectures on Mathematics*, Evanston Colloquium, New York and London, 1894; E. Lampe's *Die reine Mathematik in den Jahren 1884–1899*, Berlin, 1899; the eleventh and twelfth volumes of Marie's *Histoire des sciences*, in which are some notes on mathematicians who were born in the last century; P. Painlevé's *Les Sciences mathématiques au xixe siècle*; a chapter by D. E. Smith in *Higher Mathematics*, by M. Merriman and R. S. Woodward, New York, 1900; and V. Volterra's lecture at the Rome Congress, 1908, "On the history of mathematics in Italy during the latter half of the nineteenth century."

A few histories of the development of particular subjects have been written—such as those by Isaac Todhunter on the theories of attraction and on the calculus of probabilities; those by T. Muir on determinants, that by A. von Braunmühl on trigonometry, that by R. Reiff on infinite series, that by G. Loria, *Il passato ed il presente delle principali teorie geometriche*, and that by F. Engel and P. Stäckel on the theory of parallels. The transactions of some of the scientific societies and academies also contain reports on the progress in different branches of the subject, while information on the memoirs by particular mathematicians is given in the invaluable volumes of J. C. Poggendorff's *Biographisch-literarisches Handwörterbuch zur Geschichte der exacten Wissenschaften*, Leipzig. The *Encyklopädie der mathematischen Wissenschaften*, which is now in course of issue, aims at representing the present state of knowledge in pure and applied mathematics, and doubtless in some branches of mathematics it will supersede these reports. The French translation of this encyclopaedia contains numerous and valuable additions. I have found these authorities and these reports useful, and I have derived further assistance in writing this chapter from the obituary notices in the proceedings of various learned Societies. I am also indebted to information kindly furnished me by various friends, and if I do not further dwell on this, it is only that I would not seem to make them responsible for my errors and omissions.

A period of exceptional intellectual activity in any subject is usually followed by one of comparative stagnation; and after the deaths of Lagrange, Laplace, Legendre, and Poisson, the French school, which had occupied so prominent a position at the beginning of this century, ceased for some years to produce much new work. Some of the mathematicians whom I intend to mention first, Gauss, Abel, and Jacobi, were contemporaries of the later years of the French mathematicians just named, but their writings appear to me to belong to a different school, and thus are properly placed at the beginning of a fresh chapter.

There is no mathematician of this century whose writings have had a greater effect than those of Gauss; nor is it on only one branch of the science that his influence has left a permanent mark. I cannot, therefore, commence my account of the mathematics of recent times better than by describing very briefly his more important researches.

Gauss.[1]   *Karl Friedrich Gauss* was born at Brunswick on April 23,

[1] Biographies of Gauss have been published by L. Hänselmann, Leipzig, 1878, and by S. von Walterhausen, Leipzig, 1856. The Royal Society of Göttingen undertook

1777, and died at Göttingen on February 23, 1855. His father was a bricklayer, and Gauss was indebted for a liberal education (much against the will of his parents, who wished to profit by his wages as a labourer) to the notice which his talents procured from the reigning duke. In 1792 he was sent to the Caroline College, and by 1795 professors and pupils alike admitted that he knew all that the former could teach him: it was while there that he investigated the method of least squares, and proved by induction the law of quadratic reciprocity. Thence he went to Göttingen, where he studied under Kästner: many of his discoveries in the theory of numbers were made while a student here. In 1798 he returned to Brunswick, where he earned a somewhat precarious livelihood by private tuition.

In 1799 Gauss published a demonstration that every integral algebraical function of one variable can be expressed as a product of real linear or quadratic factors. Hence every algebraical equation has a root of the form $a + bi$, a theorem of which he gave later two other distinct proofs. His *Disquisitiones Arithmeticae* appeared in 1801. A large part of this had been submitted as a memoir to the French Academy in the preceding year, and had been rejected in a most regrettable manner; Gauss was deeply hurt, and his reluctance to publish his investigations may be partly attributable to this unfortunate incident.

The next discovery of Gauss was in a totally different department of mathematics. The absence of any planet in the space between Mars and Jupiter, where Bode's law would have led observers to expect one, had been long remarked, but it was not till 1801 that any one of the numerous group of minor planets which occupy that space was observed. The discovery was made by G. Piazzi of Palermo; and was the more interesting as its announcement occurred simultaneously with a publication by Hegel in which he severely criticised astronomers for not paying more attention to philosophy,—a science, said he, which would at once have shewn them that there could not possibly be more than seven planets, and a study of which would therefore have prevented an absurd waste of time in looking for what in the nature of things could never be found. The new planet was named Ceres, but it was seen under conditions which appeared to render it impracticable to forecast its orbit. The observations were fortunately communicated to Gauss;

the issue of a collection of Gauss's works, and nine volumes are already published. Further additions are expected, and some hints of what may be expected have been given by F C. Klein.

he calculated its elements, and his analysis put him in the first rank of theoretical astronomers.

The attention excited by these investigations procured for him in 1807 the offer of a chair at Petrograd, which he declined. In the same year he was appointed director of the Göttingen Observatory and professor of Astronomy there. These offices he retained to his death; and after his appointment he never slept away from his Observatory except on one occasion when he attended a scientific congress at Berlin. His lectures were singularly lucid and perfect in form, and it is said that he used here to give the analysis by which he had arrived at his various results, and which is so conspicuously absent from his published demonstrations; but for fear his auditors should lose the thread of his discourse, he never willingly permitted them to take notes.

I have already mentioned Gauss's publications in 1799, 1801, and 1802. For some years after 1807 his time was mainly occupied by work connected with his Observatory. In 1809 he published at Hamburg his *Theoria Motus Corporum Coelestium*, a treatise which contributed largely to the improvement of practical astronomy, and introduced the principle of curvilinear triangulation; and on the same subject, but connected with observations in general, we have his memoir *Theoria Combinationis Observationum Erroribus Minimis Obnoxia*, with a second part and a supplement.

Somewhat later he took up the subject of geodesy, acting from 1821 to 1848 as scientific adviser to the Danish and Hanoverian Governments for the survey then in progress; his papers of 1843 and 1866, *Ueber Gegenstände der höhern Geodäsie*, contain his researches on the subject.

Gauss's researches on electricity and magnetism date from about the year 1830. His first paper on the theory of magnetism, entitled *Intensitas Vis Magneticae Terrestris ad Mensuram Absolutam Revocata*, was published in 1833. A few months afterwards he, together with W. E. Weber, invented the declination instrument and the bifilar magnetometer; and in the same year they erected at Göttingen a magnetic observatory free from iron (as Humboldt and Arago had previously done on a smaller scale) where they made magnetic observations, and in particular showed that it was practicable to send telegraphic signals. In connection with this Observatory Gauss founded an association with the object of securing continuous observations at fixed times. The volumes of their publications, *Resultate aus der Beobachtungen des magnetischen Vereins* for 1838 and 1839, contain two important memoirs by Gauss: one on the general theory of earth-magnetism, and the other on the theory of forces attracting according to the inverse square of the distance.

Gauss, like Poisson, treated the phenomena in electrostatics as due to attractions and repulsions between imponderable particles. Lord Kelvin, then William Thomson (1824–1907), of Glasgow, shewed in 1846 that the effects might also be supposed analogous to a flow of heat from various sources of electricity properly distributed.

In electrodynamics Gauss arrived (in 1835) at a result equivalent to that given by W. E. Weber of Göttingen in 1846, namely, that the attraction between two electrified particles e and e', whose distance apart is r, depends on their relative motion and position according to the formula

$$ee'r^{-2}\{1 + (r\ddot{r} - \tfrac{1}{2}\dot{r}^2)^2 c^{-2}\}.$$

Gauss, however, held that no hypothesis was satisfactory which rested on a formula and was not a consequence of a physical conjecture, and as he could not frame a plausible physical conjecture he abandoned the subject.

Such conjectures were proposed by Riemann in 1858, and by C. Neumann, now of Leipzig, and E. Betti (1823–1892) of Pisa in 1868, but Helmholtz in 1870, 1873, and 1874 showed that they were untenable. A simpler view which regards all electric and magnetic phenomena as stresses and motions of a material elastic medium had been outlined by Michael Faraday (1791–1867), and was elaborated by James Clerk Maxwell (1831–1879) of Cambridge in 1873; the latter, by the use of generalised co-ordinates, was able to deduce the consequences, and the agreement with experiment is close. Maxwell concluded by showing that if the medium were the same as the so-called luminiferous ether, the velocity of light would be equal to the ratio of the electromagnetic and electrostatic units, and subsequent experiments have tended to confirm this conclusion. The theories previously current had assumed the existence of a simple elastic solid or an action between matter and ether.

The above and other electric theories were classified by J. J. Thomson of Cambridge, in a report to the British Association in 1885, into those not founded on the principle of the conservation of energy (such as those of Ampère, Grassmann, Stefan, and Korteweg); those which rest on assumptions concerning the velocities and positions of electrified particles (such as those of Gauss, W. E. Weber, Riemann, and R. J. E. Clausius); those which require the existence of a kind of energy of which we have no other knowledge (such as the theory of C. Neumann); those which rest on dynamical considerations, but in which no account is taken of the action of the dielectric (such as the theory of F. E. Neumann); and, finally, those which rest on dynamical considerations and in which the action of the dielectric is considered (such as Maxwell's theory). In the report these theories are described, criticised, and compared with the results of experiments.

Gauss's researches on optics, and especially on systems of lenses, were published in 1840 in his *Dioptrische Untersuchungen*.

From this sketch it will be seen that the ground covered by Gauss's researches was extraordinarily wide, and it may be added that in many cases his investigations served to initiate new lines of work. He was, however, the last of the great mathematicians whose interests were nearly universal: since his time the literature of most branches of mathematics has grown so fast that mathematicians have been forced to specialise in some particular department or departments. I will now mention very briefly some of the most important of his discoveries in pure mathematics.

His most celebrated work in pure mathematics is the *Disquisitiones Arithmeticae*, which has proved a starting-point for several valuable investigations on the theory of numbers. This treatise and Legendre's *Théorie des nombres* remain standard works on the theory of numbers; but, just as in his discussion of elliptic functions Legendre failed to rise to the conception of a new subject, and confined himself to regarding their theory as a chapter in the integral calculus, so he treated the theory of numbers as a chapter in algebra. Gauss, however, realised that the theory of discrete magnitudes or higher arithmetic was of a different kind from that of continuous magnitudes or algebra, and he introduced a new notation and new methods of analysis, of which subsequent writers have generally availed themselves. The theory of numbers may be divided into two main divisions, namely, the theory of congruences and the theory of forms. Both divisions were discussed by Gauss. In particular the *Disquisitiones Arithmeticae* introduced the modern theory of congruences of the first and second orders, and to this Gauss reduced indeterminate analysis. In it also he discussed the solution of binomial equations of the form $x^n = 1$: this involves the celebrated theorem that it is possible to construct, by elementary geometry, regular polygons of $2^m(2^n + 1)$ sides, where m and n are integers and $2^n + 1$ is a prime—a discovery he had made in 1796. He developed the theory of ternary quadratic forms involving two indeterminates. He also investigated the theory of determinants, and it was on Gauss's results that Jacobi based his researches on that subject.

The theory of functions of double periodicity had its origin in the discoveries of Abel and Jacobi, which I describe later. Both these mathematicians arrived at the theta functions, which play so large a part in the theory of the subject. Gauss, however, had independently, and indeed at a far earlier date, discovered these functions and some of their properties, having been led to them by certain integrals which occurred in the *Determinatio Attractionis*, to evaluate which he invented the transformation now associated with the name of Jacobi. Though Gauss at a later time communicated the fact to Jacobi, he did not publish his researches; they occur in a series of note-books of a date not later than 1808, and are included in his collected works.

Of the remaining memoirs in pure mathematics the most remarkable are those on the theory of biquadratic residues (wherein the notion of complex numbers of the form $a + bi$ was first introduced into the theory of numbers), in which are included several tables, and notably one of the number of the classes of binary quadratic forms; that relating to the proof of the theorem that every algebraical equation has a real or imaginary root; that on the summation of series; and, lastly, one on interpolation. His introduction of rigorous tests for the convergency of infinite series is worthy of attention. Specially noticeable also are his investigations on hypergeometric series; these contain a discussion of the gamma function. This subject has since become one of considerable importance, and has been written on by (among others) Kummer and Riemann; later the original conceptions were greatly extended, and numerous memoirs on it and its extensions have appeared. I should also mention Gauss's theorems on the curvature of surfaces, wherein he devised a new and general method of treatment which has led to many new results. Finally, we have his important memoir on the conformal representation of one surface upon another, in which the results given by Lagrange for surfaces of revolution are generalised for all surfaces. It would seem also that Gauss had discovered some of the properties of quaternions, though these investigations were not published until a few years ago.

In the theory of attractions we have a paper on the attraction of homogeneous ellipsoids: the already-mentioned memoir of 1839, on the theory of forces attracting according to the inverse square of the distance; and the memoir, *Determinatio Attractionis*, in which it is shown that the secular variations, which the elements of the orbit of a planet experience from the attraction of another planet which disturbs it, are the same as if the mass of the disturbing planet were distributed over its orbit into an elliptic ring in such a manner that equal masses of the ring would correspond to arcs of the orbit described in equal times.

The great masters of modern analysis are Lagrange, Laplace, and Gauss, who were contemporaries. It is interesting to note the marked contrast in their styles. Lagrange is perfect both in form and matter, he is careful to explain his procedure, and though his arguments are general they are easy to follow. Laplace, on the other hand, explains nothing, is indifferent to style, and, if satisfied that his results are correct, is content to leave them either with no proof or with a faulty one. Gauss is as exact and elegant as Lagrange, but even more difficult to follow than Laplace, for he removes every trace of the analysis by which he reached his results, and studies to give a proof which, while rigorous, shall be as concise and synthetical as possible.

Dirichlet.[1]   One of Gauss's pupils to whom I may here allude is Lejeune Dirichlet, whose masterly exposition of the discoveries of Jacobi (who was his father-in-law) and of Gauss has unduly overshadowed his own original investigations on similar subjects. *Peter Gustav Lejeune Dirichlet* was born at Düren on February 13, 1805, and died at Göttingen on May 5, 1859. He held successively professorships at Breslau and Berlin, and on Gauss's death in 1855 was appointed to succeed him as professor of the higher mathematics at Göttingen. He intended to finish Gauss's incomplete works, for which he was admirably fitted, but his early death prevented this. He produced, however, several memoirs which have considerably facilitated the comprehension of some of Gauss's more abstruse methods. Of Dirichlet's original researches the most celebrated are those dealing with the establishment of Fourier's theorem, those in the theory of numbers on asymptotic laws (that is, laws which approximate more closely to accuracy as the numbers concerned become larger), and those on primes.

It is convenient to take Gauss's researches as the starting-point for the discussion of various subjects. Hence the length with which I have alluded to them.

---

[1]Dirichlet's works, edited by L. Kronecker, were issued in two volumes, Berlin, 1889, 1897. His lectures on the theory of numbers were edited by J. W. R. Dedekind, third edition, Brunswick, 1879–81. His investigations on the theory of the potential were edited by F. Grube, second edition, Leipzig, 1887. His researches on definite integrate have been edited by G. Arendt, Brunswick, 1904. There is a note on some of his researches by C. W. Borchardt in *Crelle's Journal*, vol. lvii, 1859, pp. 91–92.

The Theory of Numbers, or *Higher Arithmetic.*   The researches of Gauss on the theory of numbers were continued or supplemented by *Jacobi,* who first proved the law of cubic reciprocity; discussed the theory of residues; and, in his *Canon Arithmeticus,* gave a table of residues of prime roots. Dirichlet also paid some attention to this subject.

Eisenstein.[1]   The subject was next taken up by *Ferdinand Gotthold Eisenstein,* a professor at the University of Berlin, who was born at Berlin on April 16, 1823, and died there on October 11, 1852. The solution of the problem of the representation of numbers by binary quadratic forms is one of the great achievements of Gauss, and the fundamental principles upon which the treatment of such questions rest were given by him in the *Disquisitiones Arithmeticae.* Gauss there added some results relating to ternary quadratic forms, but the general extension from two to three indeterminates was the work of Eisenstein, who, in his memoir *Neue Theoreme der höheren Arithmetik,* defined the ordinal and generic characters of ternary quadratic forms of an uneven determinant; and, in the case of definite forms, assigned the weight of any order or genus; but he did not consider forms of an even determinant, nor give any demonstrations of his work.

Eisenstein also considered the theorems relating to the possibility of representing a number as a sum of squares, and showed that the general theorem was limited to eight squares. The solutions in the cases of two, four, and six squares may be obtained by means of elliptic functions, but the cases in which the number of squares is uneven involve special processes peculiar to the theory of numbers. Eisenstein gave the solution in the case of three squares. He also left a statement of the solution he had obtained in the case of five squares;[2] but his results were published without proofs, and apply only to numbers which are not divisible by a square.

Henry Smith.[3]   One of the most original mathematicians of the school founded by Gauss was Henry Smith. *Henry John Stephen Smith* was born in London on November 2, 1826, and died at Oxford on February 9, 1883. He was educated at Rugby, and at Balliol College,

---

[1] For a sketch of Eisenstein's life and researches see *Abhandlungen zur Geschichte der Mathematik,* 1895, p. 143 *et seq.*

[2] *Crelle's Journal,* vol. xxxv, 1847, p. 368.

[3] Smith's collected mathematical works, edited by J. W. L. Glaisher, and prefaced by a biographical sketch and other papers, were published in two volumes, Oxford, 1894. The following account is extracted from the obituary notice in the monthly notices of the Astronomical Society, 1884, pp. 138–149.

Oxford, of which latter society he was a fellow; and in 1861 he was elected Savilian professor of Geometry at Oxford, where he resided till his death.

The subject in connection with which Smith's name is specially associated is the theory of numbers, and to this he devoted the years from 1854 to 1864. The results of his historical researches were given in his report published in parts in the *Transactions* of the British Association from 1859 to 1865. This report contains an account of what had been done on the subject to that time together with some additional matter. The chief outcome of his own original work on the subject is included in two memoirs printed in the *Philosophical Transactions* for 1861 and 1867; the first being on linear indeterminate equations and congruences, and the second on the orders and genera of ternary quadratic forms. In the latter memoir demonstrations of Eisenstein's results and their extension to ternary quadratic forms of an even determinant were supplied, and a complete classification of ternary quadratic forms was given.

Smith, however, did not confine himself to the case of three indeterminates, but succeeded in establishing the principles on which the extension to the general case of n indeterminates depends, and obtained the general formulae—thus effecting the greatest advance made in the subject since the publication of Gauss's work. In the account of his methods and results which appeared in the *Proceedings* of the Royal Society,[1] Smith remarked that the theorems relating to the representation of numbers by four squares and other simple quadratic forms, are deducible by a uniform method from the principles there indicated, as also are the theorems relating to the representation of numbers by six and eight squares. He then proceeded to say that as the series of theorems relating to the representation of numbers by sums of squares ceases, for the reason assigned by Eisenstein, when the number of squares surpasses eight, it was desirable to complete it. The results for even squares were known. The principal theorems relating to the case of five squares had been given by Eisenstein, but he had considered only those numbers which are not divisible by a square, and he had not considered the case of seven squares. Smith here completed the enunciation of the theorems for the case of five squares, and added the corresponding theorems for the case of seven squares.

This paper was the occasion of a dramatic incident in the history of

[1] See vol. xiii, 1864, pp. 199–203, and vol. xvi, 1868, pp. 197–208.

mathematics. Fourteen years later, in ignorance of Smith's work, the demonstration and completion of Eisenstein's theorems for five squares were set by the French Academy as the subject of their "Grand prix des sciences mathématiques." Smith wrote out the demonstration of his general theorems so far as was required to prove the results in the special case of five squares, and only a month after his death, in March 1883, the prize was awarded to him, another prize being also awarded to H. Minkowski of Bonn. No episode could bring out in a more striking light the extent of Smith's researches than that a question, of which he had given the solution in 1867, as a corollary from general formulae which governed the whole class of investigations to which it belonged, should have been regarded by the French Academy as one whose solution was of such difficulty and importance as to be worthy of their great prize. It has been also a matter of comment that they should have known so little of contemporary English and German researches on the subject as to be unaware that the result of the problem they were proposing was then lying in their own library.

J. W. L. Glaisher of Cambridge has recently extended[1] these results, and investigated, by the aid of elliptic functions, the number of representations of a number as the sum of $2n$ squares where n is not greater than 9.

Among Smith's other investigations I may specially mention his geometrical memoir, *Sur quelques problèmes cubiques et biquadratiques*, for which in 1868 he was awarded the Steiner prize of the Berlin Academy. In a paper which he contributed to the *Atti* of the Accademia dei Lincei for 1877 he established a very remarkable analytical relation connecting the modular equation of order n, and the theory of binary quadratic forms belonging to the positive determinant n. In this paper the modular curve is represented analytically by a curve in such a manner as to present an actual geometrical image of the complete systems of the reduced quadratic forms belonging to the determinant, and a geometrical interpretation is given to the ideas of "class," "equivalence," and "reduced form." He was also the author of important papers in which he succeeded in extending to complex quadratic forms many of Gauss's investigations relating to real quadratic forms. He was led by his researches on the theory of numbers to the theory of elliptic functions, and the results he arrived at, especially on the theories of the theta and

---

[1] For a summary of his results see his paper in the *Proceedings of the London Mathematical Society*, 1907, vol. v, second series, pp. 479–490.

omega functions, are of importance.

Kummer. The theory of primes received a somewhat unexpected development by E. E. Kummer of Berlin, who was born in 1810 and died in 1893. In particular he treated higher complex members of the form $a + \Sigma bj$, where j is a complex root of $j^p - 1 = 0$, p being a prime. His theory brought out the unexpected result that the proposition that a number can be resolved into the product of powers of primes in one and only one way is not necessarily true of every complex number. This led to the theory of ideal primes, a theory which was developed later by J. W. R. Dedekind. Kummer also extended Gauss's theorems on quadratic residues to residues of a higher order, and wrote on the transformations of hypergeometric functions.

The theory of numbers, as treated to-day, may be said to originate with Gauss. I have already mentioned very briefly the investigations of *Jacobi, Dirichlet, Eisenstein, Henry Smith*, and *Kummer*. I content myself with adding some notes on the subsequent development of certain branches of the theory.[1]

The distribution of primes has been discussed in particular by *P. L. Tchebycheff*[2] (1821–1894) of Petrograd, *G. F. B. Riemann*, and *J. J. Sylvester*. Riemann's short tract on the number of primes which lie between two given numbers affords a striking instance of his analytical powers. Legendre had previously shown that the number of primes less than n is approximately $n/(\log_e n - 1.08366)$; but Riemann went farther, and this tract and a memoir by Tchebycheff contain nearly all that has been done yet in connection with a problem of so obvious a character, that it has suggested itself to all who have considered the theory of numbers, and yet which overtaxed the powers even of Lagrange and Gauss. In this paper also Riemann stated that all the roots of $\Gamma(\frac{1}{2}s+1)(s-1)\pi^{-\delta/2}\zeta(s)$ are of the form $\frac{1}{2}+it$ where t is real. It is believed that the theorem is true, but as yet it has defied all attempts to prove it. Riemann's work in this connection has proved the starting-point for researches by J. S. Hadamard, H. C. F. von Mangoldt, and other recent writers.

The partition of numbers, a problem to which Euler had paid considerable attention, has been treated by *A. Cayley, J. J. Sylvester*, and

---

[1] See H. J. S. Smith, *Report on the Theory of Numbers* in vol. i of his works, and O. Stolz, *Groessen und Zahlen*, Leipzig, 1891.

[2] Tchebycheff's collected works, edited by H. Markoff and N. Sonin, have been published in two volumes. A French translation was issued 1900, 1907.

P. A. MacMahon. The representation of numbers in special forms, the possible divisors of numbers of specified forms, and general theorems concerned with the divisors of numbers, have been discussed by *J. Liouville* (1809–1882), the editor from 1836 to 1874 of the well-known mathematical journal, and by *J. W. L. Glaisher* of Cambridge. The subject of quadratic binomials has been studied by *A. L. Cauchy*; of ternary and quadratic forms by *L. Kronecker*[1] (1823–1891) of Berlin; and of ternary forms by *C. Hermite* of Paris.

The most common text-books are, perhaps, that by O. Stolz of Innspruck, Leipzig, 1885–6; that by G. B. Mathews, Cambridge, 1892; that by E. Lucas, Paris, 1891; and those by P. Bachmann, Leipzig, 1892–1905. Possibly it may be found hereafter that the subject is approached better on other lines than those now usual.

The conception of *Number* has also been discussed at considerable length during the last quarter of the nineteenth century. Transcendent numbers had formed the subject of two memoirs by Liouville, but were subsequently treated as a distinct branch of mathematics, notably by L. Kronecker and G. Cantor. Irrational numbers and the nature of numbers have also been treated from first principles, in particular by K. Weierstrass, J. W. R. Dedekind,[2] H. C. R. Méray, G. Cantor, G. Peano, and B. A. W. Russell. This subject has attracted much attention of late years, and is now one of the most flourishing branches of modern mathematics. Transfinite, cardinal, and ordinal arithmetic, and the theory of sets of points, may be mentioned as prominent divisions. The theory of aggregates is related to this subject, and has been treated by G. Cantor, P. du Bois-Raymond, A. Schönflies, E. Zermelo, and B. A. W. Russell.

Elliptic and Abelian Functions, or *Higher Trigonometry.*[3]   The theory of functions of double and multiple periodicity is another subject to which much attention has been paid during this century. I have already mentioned that as early as 1808 Gauss had discovered the theta

---

[1]See the *Bulletin* of the New York (American) Mathematical Society, vol. i, 1891–2, pp. 173–184.

[2]Dedekind's Essays may serve as an introduction to the subject. They have been translated into English, Chicago, 1901.

[3]See the introduction to *Elliptische Functionen*, by A. Enneper, second edition (ed. by F. Müller), Halle, 1890; and *Geschichte der Theorie der elliptischen Transcendenten*, by L. Königsberger, Leipzig, 1879. On the history of Abelian functions see the *Transactions of the British Association*, vol. lxvii, London, 1897, pp. 246–286.

functions and some of their properties, but his investigations remained for many years concealed in his note-books; and it was to the researches made between 1820 and 1830 by Abel and Jacobi that the modern development of the subject is due. Their treatment of it has completely superseded that used by Legendre, and they are justly reckoned as the creators of this branch of mathematics.

Abel.[1]   *Niels Henrick Abel* was born at Findoe, in Norway, on August 5, 1802, and died at Arendal on April 6, 1829, at the age of twenty-six. His memoirs on elliptic functions, originally published in *Crelle's Journal* (of which he was one of the founders), treat the subject from the point of view of the theory of equations and algebraic forms, a treatment to which his researches naturally led him.

The important and very general result known as Abel's theorem, which was subsequently applied by Riemann to the theory of transcendental functions, was sent to the French Academy in 1826, but was not printed until 1841: its publication then was due to inquiries made by Jacobi, in consequence of a statement on the subject by B. Holmboe in his edition of Abel's works issued in 1839. It is far from easy to state Abel's theorem intelligently and yet concisely, but, broadly speaking, it may be described as a theorem for evaluating the sum of a number of integrals which have the same integrand, but different limits—these limits being the roots of an algebraic equation. The theorem gives the sum of the integrals in terms of the constants occurring in this equation and in the integrand. We may regard the inverse of the integral of this integrand as a new transcendental function, and if so the theorem furnishes a property of this function. For instance, if Abel's theorem be applied to the integrand $(1 - x^2)^{-1/2}$ it gives the addition theorem for the circular (or trigonometrical) functions.

The name of Abelian function has been given to the higher transcendents of multiple periodicity which were first discussed by Abel. The Abelian functions connected with a curve $f(x, y)$ are of the form $\int u\,dx$ where u is a rational function of x and y. The theory of Abelian functions has been studied by a very large number of modern writers.

Abel criticised the use of infinite series, and discovered the well-known theorem which furnishes a test for the validity of the result

---

[1]The life of Abel by C. A. Bjerknes was published at Stockholm in 1880, and another by L. de Pesloüan at Paris in 1906. Two editions of Abel's works have been published, of which the last, edited by Sylow and Lie, and issued at Christiania in two volumes in 1881, is the more complete. See also the Abel centenary volume, Christiania, 1902; and a memoir by G. Mittag-Leffler.

obtained by multiplying one infinite series by another. He also proved[1] the binomial theorem for the expansion of $(1 + x)^n$ when x and n are complex. As illustrating his fertility of ideas I may, in passing, notice his celebrated demonstration that it is impossible to express a root of the general quintic equation in terms of its coefficients by means of a finite number of radicals and rational functions; this theorem was the more important since it definitely limited a field of mathematics which had previously attracted numerous writers. I should add that this theorem had been enunciated as early as 1798 by Paolo Ruffini, an Italian physician practising at Modena; but I believe that the proof he gave was deficient in generality.

Jacobi.[2]   *Carl Gustav Jacob Jacobi*, born of Jewish parents at Potsdam on Dec. 10, 1804, and died at Berlin on Feb. 18, 1851, was educated at the University of Berlin, where he obtained the degree of Doctor of Philosophy in 1825. In 1827 he became extraordinary professor of Mathematics at Königsberg, and in 1829 was promoted to be an ordinary professor. This chair he occupied till 1842, when the Prussian Government gave him a pension, and he moved to Berlin, where he continued to live till his death in 1851. He was the greatest mathematical teacher of his generation, and his lectures, though somewhat unsystematic in arrangement, stimulated and influenced the more able of his pupils to an extent almost unprecedented at the time.

Jacobi's most celebrated investigations are those on elliptic functions, the modern notation in which is substantially due to him, and the theory of which he established simultaneously with Abel, but independently of him. Jacobi's results are given in his treatise on elliptic functions, published in 1829, and in some later papers in *Crelle's Journal*; they are earlier than Weierstrass's researches which are mentioned below. The correspondence between Legendre and Jacobi on elliptic functions has been reprinted in the first volume of Jacobi's collected works. Jacobi, like Abel, recognised that elliptic functions were not merely a group of theorems on integration, but that they were types of a new kind of function, namely, one of double periodicity; hence he paid

---

[1]See Abel, *Œuvres*, 1881, vol. i, pp. 219–250; and E. W. Barnes, *Quarterly Journal of Mathematics*, vol. xxxviii, 1907, pp. 108–116.

[2]See C. J. Gerhardt's *Geschichte der Mathematik in Deutschland*, Munich, 1877. Jacobi's collected works were edited by Dirichlet, three volumes, Berlin, 1846–71, and accompanied by a biography, 1852; a new edition, under the supervision of C. W. Borchardt and K. Weierstrass, was issued at Berlin in seven volumes, 1881–91. See also L. Königsberger's *C. G. J. Jacobi*, Leipzig, 1904.

particular attention to the theory of the theta function. The following passage,[1] in which he explains this view, is sufficiently interesting to deserve textual reproduction:—

E quo, cum universam, quae fingi potest, amplectatur periodicitatem analyticam elucet, functiones ellipticas non aliis adnumerari debere transcendentibus, quae quibusdam gaudent elegantiis, fortasse pluribus illas aut maioribus, sed speciem quandam iis inesse perfecti et absoluti.

Among Jacobi's other investigations I may specially single out his papers on Determinants, which did a great deal to bring them into general use; and particularly his introduction of the Jacobian, that is, of the functional determinant formed by the n^2 partial differential coefficients of the first order of n given functions of n independent variables. I ought also to mention his papers on Abelian transcendents; his investigations on the theory of numbers, to which I have already alluded; his important memoirs on the theory of differential equations, both ordinary and partial; his development of the calculus of variations; and his contributions to the problem of three bodies, and other particular dynamical problems. Most of the results of the researches last named are included in his *Vorlesungen über Dynamik.*

Riemann.[2]   *Georg Friedrich Bernhard Riemann* was born at Breselenz on Sept. 17, 1826, and died at Selasca on July 20, 1866. He studied at Göttingen under Gauss, and subsequently at Berlin under Jacobi, Dirichlet, Steiner, and Eisenstein, all of whom were professors there at the same time. In spite of poverty and sickness he struggled to pursue his researches. In 1857 he was made professor at Göttingen, general recognition of his powers soon followed, but in 1862 his health began to give way, and four years later he died, working, to the end, cheerfully and courageously.

Riemann must be esteemed one of the most profound and brilliant mathematicians of his time; he was a creative genius. The amount of matter he produced is small, but its originality and power are man-

[1] See Jacobi's collected works, vol. i, 1881, p. 87.

[2] Riemann's collected works, edited by H. Weber and prefaced by an account of his life by Dedekind, were published at Leipzig, second edition, 1892; an important supplement, edited by M. Nöther and W. Wirtinger, was issued in 1902. His lectures on elliptic functions, edited by H. B. L. Stahl, were published separately, Leipzig, 1899. Another short biography of Riemann has been written by E. J. Schering, Göttingen, 1867.

ifest—his investigations on functions and on geometry, in particular, initiating developments of great importance.

His earliest paper, written in 1850, was on the general theory of functions of a complex variable. This gave rise to a new method of treating the theory of functions. The development of this method is specially due to the Göttingen school. In 1854 Riemann wrote his celebrated memoir on the hypotheses on which geometry is founded: to this subject I allude below. This was succeeded by memoirs on elliptic functions and on the distribution of primes: these have been already mentioned. He also investigated the conformal representation of areas, one on the other: a problem subsequently treated by H. A. Schwarz and F. H. Schottky, both of Berlin. Lastly, in multiple periodic functions, it is hardly too much to say that in his memoir in *Borchardt's Journal* for 1857, he did for the Abelian functions what Abel had done for the elliptic functions. A posthumous fragment on linear differential equations with algebraic coefficients has served as the foundation of important work by L. Schlesinger.

I have already alluded to the researches of *Legendre, Gauss, Abel, Jacobi,* and *Riemann* on elliptic and Abelian functions. The subject has been also discussed by (among other writers) *J. G. Rosenhain* (1816–1887) of Königsberg, who wrote (in 1844) on the hyperelliptic, and double theta functions; *A. Göpel* (1812–1847) of Berlin, who discussed[1] hyperelliptic functions; *L. Kronecker*[2] of Berlin, who wrote on elliptic functions; *L. Königsberger*[3] of Heidelberg and *F. Brioschi*[4] (1824–1897) of Milan, both of whom wrote on elliptic and hyperelliptic functions; *Henry Smith* of Oxford, who discussed the transformation theory, the theta and omega functions, and certain functions of the modulus; *A. Cayley* of Cambridge, who was the first to work out (in 1845) the theory of doubly infinite products and determine their periodicity, and who has written at length on the connection between the researches of Legendre and Jacobi; and *C. Hermite* of Paris, whose researches are mostly concerned with the transformation theory and the higher development of the theta functions.

---

[1]See *Crelle's Journal*, vol. xxxv, 1847, pp. 277–312; an obituary notice, by Jacobi, is given on pp. 313–317.

[2]Kronecker's collected works in four volumes, edited by K. Hensel, are now in course of publication at Leipzig, 1895, &c.

[3]See Königsberger's lectures, published at Leipzig in 1874.

[4]His collected works were published in two volumes, Milan, 1901, 1902.

Weierstrass.[1]   The subject of higher trigonometry was put on a somewhat different footing by the researches of Weierstrass. *Karl Weierstrass*, born in Westphalia on October 31, 1815, and died at Berlin on February 19, 1897, was one of the greatest mathematicians of the nineteenth century. He took no part in public affairs; his life was uneventful; and he spent the last forty years of it at Berlin, where he was professor.

With two branches of pure mathematics—elliptic and Abelian functions, and the theory of functions—his name is inseparably connected. His earlier researches on elliptic functions related to the theta functions, which he treated under a modified form in which they are expressible in powers of the modulus. At a later period he developed a method for treating all elliptic functions in a symmetrical manner. Jacobi had shown that a function of n variables might have $2n$ periods. Accordingly Weierstrass sought the most general expressions for such functions, and showed that they enjoyed properties analogous to those of the hyperelliptic functions. Hence the properties of the latter functions could be reduced as particular cases of general results.

He was naturally led to this method of treating hyperelliptic functions by his researches on the general theory of functions; these co-ordinated and comprised various lines of investigation previously treated independently. In particular he constructed a theory of uniform analytic functions. The representation of functions by infinite products and series also claimed his especial attention. Besides functions he also wrote or lectured on the nature of the assumptions made in analysis, on the calculus of variations, and on the theory of minima surfaces. His methods are noticeable for their wide-reaching and general character. Recent investigations on elliptic functions have been largely based on Weierstrass's method.

Among other prominent mathematicians who have recently written on elliptic and hyperelliptic functions, I may mention the names of *G. H. Halphen*[2] (1844–1889), an officer in the French army, whose investigations were largely founded on Weierstrass's work; *F. C. Klein* of Göttingen, who has written on Abelian functions, elliptic modular

---

[1] Weierstrass's collected works are now in course of issue, Berlin, 1894, &c. Sketches of his career by G. Mittag-Leffler and H. Poincaré are given in *Acta Mathematica*, 1897, vol. xxi, pp. 79–82, and 1899, vol. xxii, pp. 1–18.

[2] See Halphen's collected works, 3 vols., Paris, 1916, 1918, 1921. A sketch of his life and work is given in *Liouville's Journal* for 1889, pp. 345–359, and in the *Comptes Rendus*, 1890, vol. cx, pp. 489–497.

functions, and hyperelliptic functions; *H. A. Schwarz* of Berlin; *H. Weber* of Strassburg; *M. Nöther* of Erlangen; *H. B. L. Stahl* of Tübingen; *F. G. Frobenius* of Berlin; *J. W. L. Glaisher* of Cambridge, who has in particular developed the theory of the zeta function; and *H. F. Baker* of Cambridge.

The usual text-books of to-day on elliptic functions are those by J. Tannery and J. Molk, 4 volumes, Paris, 1893–1901; by P. E. Appell and E. Lacour, Paris, 1896; by H. Weber, Brunswick, 1891; and by G. H. Halphen, 3 volumes, Paris, 1886–1891. To these I may add one by A. G. Greenhill on the *Applications of Elliptic Functions*, London, 1892.

The Theory of Functions.   I have already mentioned that the modern theory of functions is largely due to Weierstrass and H. C. R. Méray. It is a singularly attractive subject, and has proved an important and far-reaching branch of mathematics. Historically its modern presentation may be said to have been initiated by *A. Cauchy*, who laid the foundations of the theory of synectic functions of a complex variable. Work on these lines was continued by *J. Liouville*, who wrote chiefly on doubly periodic functions. These investigations were extended and connected in the work by *A. Briot* (1817–1882), and *J. C. Bouquet* (1819–1885), and subsequently were further developed by *C. Hermite*.

Next I may refer to the researches on the theory of algebraic functions which have their origin in *V. A. Puiseux's* memoir of 1851, and *G. F. B. Riemann's* papers of 1850 and 1857; in continuation of which *H. A. Schwarz* of Berlin established accurately certain theorems of which the proofs given by Riemann were open to objection. To Riemann also we are indebted for valuable work on modular functions which has been recently published in his *Nachträge*. Subsequently *F. C. Klein* of Göttingen connected Riemann's theory of functions with the theory of groups, and wrote on automorphic and modular functions; *H. Poincaré* of Paris also wrote on automorphic functions, and on the general theory with special applications to differential equations. Quite recently *K. Hensel* of Marburg has written on algebraic functions; and *W. Wirtinger* of Vienna on Abelian functions.

I have already said that the work of Weierstrass shed a new light on the whole subject. His theory of analytical functions has been developed by *G. Mittag-Leffler* of Stockholm; and *C. Hermite, P. E. Appell, C. E. Picard, E. Goursat, E. N. Laguerre,* and *J. S. Hadamard,* all of Paris, have also written on special branches of the general theory;

while *E. Borel, R. L. Baire, H. L. Lebesgue,* and *E. L. Lindelöf* have produced a series of tracts on uniform functions which have had a wide circulation and influence.

As text-books I may mention the *Theory of Functions of a Complex Variable,* by A. R. Forsyth, second edition, Cambridge, 1900; *Abel's Theorem* by H. F. Baker, Cambridge, 1897, and *Multiple Periodic Functions* by the same writer, Cambridge, 1907; the *Théorie des fonctions algébriques* by P. E. Appell and E. Goursat, Paris, 1895; parts of C. E. Picard's *Traité d'Analyse,* in 3 volumes, Paris, 1891 to 1896; the *Theory of Functions* by J. Harkness and F. Morley, London, 1893; the *Theory of Functions of a Real Variable and of Fourier's Series* by E. W. Hobson, Cambridge, 1907; and *Die Theorie des Abel'schen Functionen* by H. B. L. Stahl, Leipzig, 1896.

Higher Algebra.   The theory of numbers may be considered as a higher arithmetic, and the theory of elliptic and Abelian functions as a higher trigonometry. The theory of higher algebra (including the theory of equations) has also attracted considerable attention, and was a favourite subject of study of the mathematicians whom I propose to mention next, though the interests of these writers were by no means limited to this subject.

Cauchy.[1]   *Augustin Louis Cauchy,* the leading representative of the French school of analysis in the nineteenth century, was born at Paris on Aug. 21, 1789, and died at Sceaux on May 25, 1857. He was educated at the Polytechnic school, the nursery of so many French mathematicians of that time, and adopted the profession of a civil engineer. His earliest mathematical paper was one on polyhedra in 1811. Legendre thought so highly of it that he asked Cauchy to attempt the solution of an analogous problem which had baffled previous investigators, and his choice was justified by the success of Cauchy in 1812. Memoirs on analysis and the theory of numbers, presented in 1813, 1814, and 1815, showed that his ability was not confined to geometry alone. In one of these papers he generalised some results which had been established by Gauss and Legendre; in another of them he gave a theorem on the number of values which an algebraical function can assume when the literal constants it contains are interchanged. It was the latter theorem that enabled Abel to show that in general an alge-

---

[1] See *La Vie et les travaux de Cauchy* by L. Valson, two volumes, Paris, 1868. A complete edition of his works is now being issued by the French Government.

braic equation of a degree higher than the fourth cannot be solved by the use of a finite number of purely algebraical expressions.

To Abel, Cauchy, and Gauss we owe the scientific treatment of series which have an infinite number of terms. In particular, Cauchy established general rules for investigating the convergency and divergency of such series, rules which were extended by J. L. F. Bertrand (1822–1900) of Paris, Secretary of the French Académie des Sciences, A. Pringsheim of Munich, and considerably amplified later by E. Borel, by M. G. Servant, both of Paris, and by other writers of the modern French school. In only a few works of an earlier date is there any discussion as to the limitations of the series employed. It is said that Laplace, who was present when Cauchy read his first paper on the subject, was so impressed by the illustrations of the danger of employing such series without a rigorous investigation of their convergency, that he put on one side the work on which he was then engaged and denied himself to all visitors, in order to see if any of the demonstrations given in the earlier volumes of the *Mécanique céleste* were invalid; and he was fortunate enough to find that no material errors had been thus introduced. The treatment of series and of the fundamental conceptions of the calculus in most of the text-books then current was based on Euler's works, and was not free from objection. It is one of the chief merits of Cauchy that he placed these subjects on a stricter foundation.

On the restoration in 1816 the French Academy was purged, and, incredible though it may seem, Cauchy accepted a seat procured for him by the expulsion of Monge. He was also at the same time made professor at the Polytechnic; and his lectures there on algebraic analysis, the calculus, and the theory of curves, were published as text-books. On the revolution in 1830 he went into exile, and was first appointed professor at Turin, whence he soon moved to Prague to undertake the education of the Comte de Chambord. He returned to France in 1837; and in 1848, and again in 1851, by special dispensation of the Emperor was allowed to occupy a chair of mathematics without taking the oath of allegiance.

His activity was prodigious, and from 1830 to 1859 he published in the *Transactions* of the Academy, or the *Comptes Rendus*, over 600 original memoirs and about 150 reports. They cover an extraordinarily wide range of subjects, but are of very unequal merit.

Among the more important of his other researches are those on the legitimate use of imaginary quantities; the determination of the number of real and imaginary roots of any algebraic equation within a given contour; his method of calculating these roots approximately; his theory of the symmetric functions of the coefficients of equations of any degree; his *a priori* valuation of a quantity less than the least difference between the roots of an equation; his papers on determinants in 1841, which assisted in bringing them into general use; and his investigations on the theory of numbers. Cauchy also did something to reduce the art of determining definite integrals to a science; the rule for finding the principal values of integrals was enunciated by him. The calculus of residues was his invention. His proof of Taylor's theorem seems to have originated from a discussion of the double periodicity of elliptic functions. The means of showing a connection between different branches of a subject by giving complex values to independent variables is largely due to him.

He also gave a direct analytical method for determining planetary inequalities of long period. To physics he contributed memoirs on waves and on the quantity of light reflected from the surfaces of metals, as well as other papers on optics.

Argand.   I may mention here the name of *Jean Robert Argand*, who was born at Geneva on July 18, 1768, and died at Paris on August 13, 1822. In his *Essai*, issued in 1806, he gave a geometrical representation of a complex number, and applied it to show that every algebraic equation has a root. This was prior to the memoirs of Gauss and Cauchy on the same subject, but the essay did not attract much attention when it was first published. An even earlier demonstration that $\sqrt{(-1)}$ may be interpreted to indicate perpendicularity in two-dimensional space, and even the extension of the idea to three-dimensional space by a method foreshadowing the use of quaternions, had been given in a memoir by C. Wessel, presented to the Copenhagen Academy of Sciences in March 1797; other memoirs on the same subject had been published in the *Philosophical Transactions* for 1806, and by H. Kühn in the *Transactions* for 1750 of the Petrograd Academy.[1]

I have already said that the idea of a simple complex number like $a + bi$ where $i^2 = -1$ was extended by Kummer. The general theory has been discussed by K. Weierstrass, H. A. Schwarz of Berlin, J. W. R. Dedekind, H. Poincaré, and other writers.

Hamilton.[2]   In the opinion of some writers the theory of quater-

---

[1] See W. W. Beman in the *Proceedings of the American Association for the Advancement of Science*, vol. xlvi, 1897.

[2] See the life of Hamilton (with a bibliography of his writings) by R. P. Graves,

nions will be ultimately esteemed one of the great discoveries of the nineteenth century in pure mathematics. That discovery is due to *Sir William Rowan Hamilton*, who was born in Dublin on August 4, 1805, and died there on September 2, 1865. His education, which was carried on at home, seems to have been singularly discursive. Under the influence of an uncle who was a good linguist, he first devoted himself to linguistic studies; by the time he was seven he could read Latin, Greek, French, and German with facility; and when thirteen he was able to boast that he was familiar with as many languages as he had lived years. It was about this time that he came across a copy of Newton's *Universal Arithmetic*. This was his introduction to modern analysis, and he soon mastered the elements of analytical geometry and the calculus. He next read the *Principia* and the four volumes then published of Laplace's *Mécanique céleste*. In the latter he detected a mistake, and his paper on the subject, written in 1823, attracted considerable attention. In the following year he entered at Trinity College, Dublin. His university career is unique, for the chair of Astronomy becoming vacant in 1827, while he was yet an undergraduate, he was asked by the electors to stand for it, and was elected unanimously, it being understood that he should be left free to pursue his own line of study.

His earliest paper on optics, begun in 1823, was published in 1828 under the title of a *Theory of Systems of Rays*, to which two supplements were afterwards added; in the latter of these the phenomenon of conical refraction is predicted. This was followed by a paper in 1827 on the principle of *Varying Action*, and in 1834 and 1835 by memoirs on a *General Method in Dynamics*—the subject of theoretical dynamics being properly treated as a branch of pure mathematics. His lectures on *Quaternions* were published in 1852. Some of his results on this subject would seem to have been previously discovered by Gauss, but these were unknown and unpublished until long after Hamilton's death. Amongst his other papers, I may specially mention one on the form of the solution of the general algebraic equation of the fifth degree, which confirmed Abel's conclusion that it cannot be expressed by a finite number of purely algebraical expressions; one on fluctuating functions; one on the hodograph; and, lastly, one on the numerical solution of differential equations. His *Elements of Quaternions* was issued in 1866: of this a competent authority says that the methods of analysis there

three volumes, Dublin, 1882–89; the leading facts are given in an article in the *North British Review* for 1886.

given show as great an advance over those of analytical geometry, as the latter showed over those of Euclidean geometry. In more recent times the subject has been further developed by P. G. Tait (1831–1901) of Edinburgh, by A. Macfarlane of America, and by C. J. Joly in his *Manual of Quaternions*, London, 1905.

Hamilton was painfully fastidious on what he published, and he left a large collection of manuscripts which are now in the library of Trinity College, Dublin, some of which it is to be hoped will be ultimately printed.

Grassmann.[1]   The idea of non-commutative algebras and of quaternions seems to have occurred to Grassmann and Boole at about the same time as to Hamilton. *Hermann Günther Grassmann* was born in Stettin on April 15, 1809, and died there in 1877. He was professor at the gymnasium at Stettin. His researches on non-commutative algebras are contained in his *Ausdehnungslehre*, first published in 1844 and enlarged in 1862. This work has had great influence, especially on the continent, where Grassmann's methods have generally been followed in preference to Hamilton's. Grassmann's researches have been continued and extended, notably by S. F. V. Schlegel and G. Peano.

The scientific treatment of the fundamental principles of algebra initiated by Hamilton and Grassmann was continued by De Morgan and Boole in England, and was further developed by H. Hankel (1839–1873) in Germany in his work on complexes, 1867, and, on somewhat different lines, by G. Cantor in his memoirs on the theory of irrationals, 1871; the discussion is, however, so technical that I am unable to do more than allude to it. Of Boole and De Morgan I say a word or two in passing.

Boole.   *George Boole*, born at Lincoln on November 2, 1815, and died at Cork on December 8, 1864, independently invented a system of non-commutative algebra, and was one of the creators of symbolic or mathematical logic.[2] From his memoirs on linear transformations part of the theory of invariants has developed. His *Finite Differences* remains a standard work on that subject.

De Morgan.[3]   *Augustus de Morgan*, born in Madura (Madras) in June 1806, and died in London on March 18, 1871, was educated at

---

[1] Grassmann's collected works in three volumes, edited by F. Engel, are now in course of issue at Leipzig, 1894, &c.

[2] On the history of mathematical logic, see P. E. B. Jourdain, *Quarterly Journal of Mathematics*, vol. xliii, 1912, pp. 219–314.

[3] De Morgan's life was written by his widow, S. E. de Morgan, London, 1882.

Trinity College, Cambridge. In 1828 he became professor at the then newly-established University of London (University College). There, through his works and pupils, he exercised a wide influence on English mathematicians. He was deeply read in the philosophy and history of mathematics, but the results are given in scattered articles; of these I have made considerable use in this book. His memoirs on the foundation of algebra; his treatise on the differential calculus published in 1842, a work of great ability, and noticeable for his treatment of infinite series; and his articles on the calculus of functions and on the theory of probabilities, are worthy of special note. The article on the calculus of functions contains an investigation of the principles of symbolic reasoning, but the applications deal with the solution of functional equations rather than with the general theory of functions.

Galois.[1]   A new development of algebra—the theory of groups of substitutions—was suggested by *Evariste Galois*, who promised to be one of the most original mathematicians of the nineteenth century, born at Paris on October 26, 1811, and killed in a duel on May 30, 1832, at the early age of 20.

The theory of groups, and of invariant subgroups, has profoundly modified the treatment of the theory of equations. An immense literature has grown up on the subject. The modern theory of groups originated with the treatment by Galois, Cauchy, and J. A. Serret (1819–1885), professor at Paris; their work is mainly concerned with finite discontinuous substitution groups. This line of investigation has been pursued by M. E. C. Jordan (1838–1922) of Paris and E. Netto of Strassburg. The problem of operations with discontinuous groups, with applications to the theory of functions, has been further taken up by (among others) F. G. Frobenius of Berlin, F. C. Klein of Göttingen, and W. Burnside formerly of Cambridge and now of Greenwich.

Cayley.[2]   Another Englishman whom we may reckon among the great mathematicians of this prolific century was *Arthur Cayley*. Cayley was born in Surrey, on Aug. 16, 1821, and after education at Trinity College, Cambridge, was called to the bar. But his interests centred on mathematics; in 1863 he was elected Sadlerian Professor at Cambridge, and he spent there the rest of his life. He died on Jan. 26, 1895.

---

[1]On Galois's investigations, see the edition of his works with an introduction by E. Picard, Paris, 1897.

[2]Cayley's collected works in thirteen volumes were issued at Cambridge, 1889–1898.

Cayley's writings deal with considerable parts of modern pure mathematics. I have already mentioned his writings on the partition of numbers and on elliptic functions treated from Jacobi's point of view; his later writings on elliptic functions dealt mainly with the theory of transformation and the modular equation. It is, however, by his investigations in analytical geometry and on higher algebra that he will be best remembered.

In analytical geometry the conception of what is called (perhaps, not very happily) the *absolute* is due to Cayley. As stated by himself, the "theory, in effect, is that the metrical properties of a figure are not the properties of the figure considered *per se* ... but its properties when considered in connection with another figure, namely, the conic termed the absolute"; hence metric properties can be subjected to descriptive treatment. He contributed largely to the general theory of curves and surfaces, his work resting on the assumption of the necessarily close connection between algebraical and geometrical operations.

In higher algebra the theory of invariants is due to Cayley; his ten classical memoirs on binary and ternary forms, and his researches on matrices and non-commutative algebras, mark an epoch in the development of the subject.

Sylvester.[1]   Another teacher of the same time was *James Joseph Sylvester*, born in London on Sept. 3, 1814, and died on March 15, 1897. He too was educated at Cambridge. Like Cayley, with whom he was on intimate terms of friendship, he was called to the bar, and yet preserved all his interests in mathematics. He held professorships successively at Woolwich, Baltimore, and Oxford. He had a strong personality and was a stimulating teacher, but it is difficult to describe his writings, for they are numerous, disconnected, and discursive.

On the theory of numbers Sylvester wrote valuable papers on the distribution of primes and on the partition of numbers. On analysis he wrote on the calculus and on differential equations. But perhaps his favourite study was higher algebra, and from his numerous memoirs on this subject I may in particular single out those on canonical forms, on the theory of contravariants, on reciprocants or differential invariants, and on the theory of equations, notably on Newton's rule. I may also add that he created the language and notation of considerable parts of those subjects on which he wrote.

---

[1]Sylvester's collected works, edited by H. F. Baker, are in course of publication at Cambridge; 2 volumes are already issued.

The writings of Cayley and Sylvester stand in marked contrast: Cayley's are methodical, precise, formal, and complete; Sylvester's are impetuous, unfinished, but none the less vigorous and stimulating. Both mathematicians found the greatest attraction in higher algebra, and to both that subject in its modern form is deeply indebted.

Lie.[1]   Among the great analysts of the nineteenth century to whom I must allude here, is *Marius Sophus Lie*, born on Dec. 12, 1842, and died on Feb. 18, 1899. Lie was educated at Christiania, whence he obtained a travelling scholarship, and in the course of his journeys made the acquaintance of Klein, Darboux, and Jordan, to whose influence his subsequent career is largely due.

In 1870 he discovered the transformation by which a sphere can be made to correspond to a straight line, and, by the use of which theorems on aggregates of lines can be translated into theorems on aggregates of spheres. This was followed by a thesis on the theory of tangential transformations for space.

In 1872 he became professor at Christiania. His earliest researches here were on the relations between differential equations and infinitesimal transformations. This naturally led him to the general theory of finite continuous groups of substitutions; the results of his investigations on this subject are embodied in his *Theorie der Transformationsgruppen*, Leipzig, three volumes, 1888–1893. He proceeded next to consider the theory of infinite continuous groups, and his conclusions, edited by G. Scheffers, were published in 1893. About 1879 Lie turned his attention to differential geometry; a systematic exposition of this is in course of issue in his *Geometrie der Berührungstransformationen.*

Lie seems to have been disappointed and soured by the absence of any general recognition of the value of his results. Reputation came, but it came slowly. In 1886 he moved to Leipzig, and in 1898 back to Christiania, where a post had been created for him. He brooded, however, over what he deemed was the undue neglect of the past, and the happiness of the last decade of his life was much affected by it.

Hermite.[2]   Another great algebraist of the century was *Charles Hermite*, born in Lorraine on December 24, 1822, and died at Paris, January 14, 1901. From 1869 he was professor at the Sorbonne, and

[1]See the obituary notice by A. R. Forsyth in the *Year-Book of the Royal Society*, London, 1901.

[2]Hermite's collected works, edited by E. Picard, were issued at Paris in four volumes, 1905 to 1917.

through his pupils exercised a profound influence on the mathematicians of to-day.

While yet a student he wrote to Jacobi on Abelian functions, and the latter embodied the results in his works. Hermite's earlier papers were largely on the transformation of these functions, a problem which he finally effected by the use of modular functions. He applied elliptic functions to find solutions of the quintic equation and of Lamé's differential equation.

Later he took up the subject of algebraic continued fractions, and this led to his celebrated proof, given in 1873, that e cannot be the root of an algebraic equation, from which it follows that e is a transcendental number. F. Lindemann showed in a similar way in 1882 that π is transcendental. The proofs have been subsequently improved and simplified by K. Weierstrass, D. Hilbert, and F. C. Klein.[1]

To the end of his life Hermite maintained his creative interest in the subjects of the integral calculus and the theory of functions. He also discussed the theory of associated covariants in binary quantics and the theory of ternary quantics.

So many other writers have treated the subject of *Higher Algebra* (including therein the theory of forms and the theory of equations) that it is difficult to summarise their conclusions.

The convergency of series has been discussed by *J. L. Raabe* (1801–1859) of Zürich, *J. L. F. Bertrand*, the secretary of the French Academy; *E. E. Kummer* of Berlin; *U. Dini* of Pisa; *A. Pringsheim* of Munich;[2] and *Sir George Gabriel Stokes* (1819–1903) of Cambridge,[3] to whom the well-known theorem on the critical values of the sums of periodic series is due. The last-named writer introduced the important conception of non-uniform convergence; a subject subsequently treated by P. L. Seidel.

---

[1] The value of π was calculated to 707 places of decimals by W. Shanks in 1873; see *Proceedings of the Royal Society*, vol. xxi, p. 318, vol. xxii, p. 45. The value of e was calculated to 225 places of decimals by F. Tichanek; see F. J. Studnicka, *Vorträge über monoperiodische Functionen*, Prague, 1892, and *L'Intermédiaire des Mathématiciens*, Paris, 1912, vol. xix, p. 247.

[2] On the researches of Raabe, Bertrand, Kummer, Dini, and Pringsheim, see the *Bulletin* of the New York (American) Mathematical Society, vol. ii, 1892–3, pp. 1–10.

[3] Stokes's collected mathematical and physical papers in five volumes, and his memoir and scientific correspondence in two volumes, were issued at Cambridge, 1880 to 1907.

Perhaps here, too, I may allude in passing to the work of *G. F. B. Riemann, G. G. Stokes, H. Hankel,* and *G. Darboux* on asymptotic expansions; of *H. Poincaré* on the application of such expansions to differential equations; and of *E. Borel* and *E. Cesàro* on divergent series.

On the theory of groups of substitutions I have already mentioned the work, on the one hand, of Galois, Cauchy, Serret, Jordan, and Netto, and, on the other hand, of Frobenius, Klein, and Burnside in connection with discontinuous groups, and that of Lie in connection with continuous groups.

I may also mention the following writers: *C. W. Borchardt*[1] (1817–1880) of Berlin, who in particular discussed generating functions in the theory of equations, and arithmetic-geometric means. *C. Hermite*, to whose work I have alluded above. *Enrico Betti* of Pisa and *F. Brioschi* of Milan, both of whom discussed binary quantics; the latter applied hyperelliptic functions to give a general solution of a sextic equation. *S. H. Aronhold* (1819–1884) of Berlin, who developed symbolic methods in connection with the invariant theory of quantics. *P. A. Gordan*[2] of Erlangen, who has written on the theory of equations, the theories of groups and forms, and shown that there are only a finite number of concomitants of quantics. *R. F. A. Clebsch*[3] (1833–1872) of Göttingen, who independently investigated the theory of binary forms in some papers collected and published in 1871; he also wrote on Abelian functions. *P. A. MacMahon*, formerly an officer in the British army, who has written on the connection of symmetric functions, invariants and covariants, the concomitants of binary forms, and combinatory analysis. *F. C. Klein* of Göttingen, who, in addition to his researches, already mentioned, on functions and on finite discontinuous groups, has written on differential equations. *A. R. Forsyth* of Cambridge, who has developed the theory of invariants and the general theory of differential equations, ternariants, and quaternariants. *P. Painlevé* of Paris, who has written on the theory of differential equations. And, lastly, *D. Hilbert* of Göttingen, who has treated the theory of homogeneous

[1] A collected edition of Borchardt's works, edited by G. Hettner, was issued at Berlin in 1888.

[2] An edition of Gordan's work on invariants (determinants and binary forms), edited by G. Kerschensteiner, was issued at Leipzig in three volumes, 1885, 1887, 1908.

[3] An account of Clebsch's life and works is printed in the *Mathematische Annalen*, 1873, vol. vi, pp. 197–202, and 1874, vol. vii, pp. 1–55.

forms.

No account of contemporary writings on higher algebra would be complete without a reference to the admirable *Higher Algebra* by G. Salmon (1819–1904), provost of Trinity College, Dublin, and the *Cours d'algèbre supérieure* by J. A. Serret, in which the chief discoveries of their respective authors are embodied. An admirable historical summary of the theory of the complex variable is given in the *Vorlesungen über die complexen Zahlen*, Leipzig, 1867, by H. Hankel, of Tübingen.

Analytical Geometry. It will be convenient next to call attention to another division of pure mathematics—analytical geometry—which has been greatly developed in recent years. It has been studied by a host of modern writers, but I do not propose to describe their investigations, and I shall content myself by merely mentioning the names of the following mathematicians.

James Booth[1] (1806–1878) and *James MacCullagh*[2] (1809–1846), both of Dublin, were two of the earliest British writers in this century to take up the subject of analytical geometry, but they worked mainly on lines already studied by others. Fresh developments were introduced by *Julius Plücker*[3] (1801–1868) of Bonn, who devoted himself especially to the study of algebraic curves, of a geometry in which the line is the element in space, and to the theory of congruences and complexes; his equations connecting the singularities of curves are well known; in 1847 he exchanged his chair for one of physics, and subsequently gave up most of his time to researches on spectra and magnetism.

The majority of the memoirs on analytical geometry by *A. Cayley* and by *Henry Smith* deal with the theory of curves and surfaces; the most remarkable of those of *L. O. Hesse* (1811–1874) of Munich are on the plane geometry of curves; of those of *J. G. Darboux* of Paris are on the geometry of surfaces; of those of *G. H. Halphen* (1844–1889) of Paris are on the singularities of surfaces and on tortuous curves; and of those of *P. O. Bonnet* are on ruled surfaces, curvature, and torsion. The singularities of curves and surfaces have also been considered by *H. G. Zeuthen* of Copenhagen, and by *H. C. H. Schubert*[4] of Hamburg. The theory of tortuous curves has been discussed by *M. Nöther* of Er-

---

[1] See Booth's *Treatise on some new Geometrical Methods*, London, 1873.

[2] See MacCullagh's collected works edited by Jellett and Haughton, Dublin, 1880.

[3] Plücker's collected works in two volumes, edited by A. Schoenflies and F. Pockels, were published at Leipzig, 1875, 1896.

[4] Schubert's lectures were published at Leipzig, 1879.

langen; and *R. F. A. Clebsch*[1] of Göttingen has applied Abel's theorem to geometry.

Among more recent text-books on analytical geometry are J. G. Darboux's *Théorie générale des surfaces*, and *Les Systèmes orthogonaux et les coordonnées curvilignes*; R. F. A. Clebsch's *Vorlesungen über Geometrie*, edited by F. Lindemann; and G. Salmon's *Conic Sections, Geometry of Three Dimensions*, and *Higher Plane Curves*; in which the chief discoveries of these writers are embodied.

Plücker suggested in 1846 that the straight line should be taken as the element of space. This formed the subject of investigations by *G. Battaglini* (1826–1892) of Rome, *F. C. Klein*, and *S. Lie*.[2] Recent works on it are R. Sturm's *Die Gebilde ersten und zweiten Grades der Liniengeometrie*, 3 volumes, Leipzig, 1892, 1893, 1896, and C. M. Jessop's *Treatise on the Line Complex*, Cambridge, 1903.

Finally, I may allude to the extension of the subject-matter of analytical geometry in the writings of *A. Cayley* in 1844, *H. G. Grassmann* in 1844 and 1862, *G. F. B. Riemann* in 1854, whose work was continued by *G. Veronese* of Padua, *H. C. H. Schubert* of Hamburg, *C. Segre* of Turin, *G. Castelnuovo* of Rome, and others, by the introduction of the idea of space of n dimensions.

Analysis.   Among those who have extended the range of analysis (including the calculus and differential equations) or whom it is difficult to place in any of the preceding categories are the following, whom I mention in alphabetical order. *P. E. Appell*[3] of Paris; *J. L. F. Bertrand* of Paris; *G. Boole* of Cork; *A. L. Cauchy* of Paris; *J. G. Darboux*[3] of Paris; *A. R. Forsyth* of Cambridge; *F. G. Frobenius* of Berlin; *J. Lazarus Fuchs* (1833–1902) of Berlin; *G. H. Halphen* of Paris; *C. G. J. Jacobi* of Berlin; *M. E. C. Jordan* (1838–1922) of Paris; *L. Königsberger* of Heidelberg; *Sophie Kowalevski*[4] (1850–1891) of Stockholm; *M. S. Lie* of Leipzig; *E. Picard*[3] of Paris; *H. Poincaré*[3] of Paris; *G. F. B. Riemann* of Göttingen; *H. A. Schwarz* of Berlin;

---

[1]Clebsch's lectures have been published by F. Lindemann, two volumes, Leipzig, 1875, 1891.

[2]On the history of this subject see G. Loria, *Il passato ed il presente delle principali teorie geometriche*, Turin, 1st ed. 1887; 2nd ed. 1896.

[3]Biographies of Appell, Darboux, Picard, and Poincaré, with bibliographies, by E. Lebon, were issued in Paris in 1909, 1910.

[4]See the *Bulletin des sciences mathématiques*, vol. xv, pp. 212–220.

J. J. Sylvester; and *K. Weierstrass* of Berlin, who developed the calculus of variations.

The subject of *differential equations* should perhaps have been separated and treated by itself. But it is so vast that it is difficult—indeed impossible—to describe recent researches in a single paragraph. It will perhaps suffice to refer to the admirable series of treatises, seven volumes, on the subject by A. R. Forsyth, which give a full presentation of the subjects treated.

A recent development on integral equations, or the inversion of a definite integral, has attracted considerable attention. It originated in a single instance given by Abel, and has been treated by V. Volterra of Rome, J. Fredholm of Stockholm, D. Hilbert of Göttingen, and numerous other recent writers.

Synthetic Geometry.   The writers I have mentioned above mostly concerned themselves with analysis. I will next describe some of the more important works produced in this century on synthetic geometry.[1]

Modern synthetic geometry may be said to have had its origin in the works of Monge in 1800, Carnot in 1803, and Poncelet in 1822, but these only foreshadowed the great extension it was to receive in Germany, of which Steiner and von Staudt are perhaps the best known exponents.

Steiner.[2]   *Jacob Steiner,* "the greatest geometrician since the time of Apollonius," was born at Utzensdorf on March 18, 1796, and died at Bern on April 1, 1863. His father was a peasant, and the boy had no opportunity to learn reading and writing till the age of fourteen. He subsequently went to Heidelberg and thence to Berlin, supporting himself by giving lessons. His *Systematische Entwickelungen* was published in 1832, and at once made his reputation: it contains a full discussion of the principle of duality, and of the projective and homographic relations of rows, pencils, &c., based on metrical properties. By the influence of Crelle, Jacobi, and the von Humboldts, who were impressed by the

---

[1]The *Aperçu historique sur l'origine et le développement des méthodes en géométrie,* by M. Chasles, Paris, second edition, 1875; and *Die synthetische Geometrie im Alterthum und in der Neuzeit,* by Th. Reye, Strassburg, 1886, contain interesting summaries of the history of geometry, but Chasles's work is written from an exclusively French point of view.

[2]Steiner's collected works, edited by Weierstrass, were issued in two volumes, Berlin, 1881–82. A sketch of his life is contained in the *Erinnerung an Steiner,* by C. F. Geiser, Schaffhausen, 1874.

power of this work, a chair of geometry was created for Steiner at Berlin, and he continued to occupy it till his death. The most important of his other researches are contained in papers which appeared in *Crelle's Journal*: these relate chiefly to properties of algebraic curves and surfaces, pedals and roulettes, and maxima and minima: the discussion is purely geometrical. Steiner's works may be considered as the classical authority on recent synthetic geometry.

Von Staudt.   A system of pure geometry, quite distinct from that expounded by Steiner, was proposed by *Karl Georg Christian von Staudt*, born at Rothenburg on Jan. 24, 1798, and died in 1867, who held the chair of mathematics at Erlangen. In his *Geometrie der Lage*, published in 1847, he constructed a system of geometry built up without any reference to number or magnitude, but, in spite of its abstract form, he succeeded by means of it alone in establishing the non-metrical projective properties of figures, discussed imaginary points, lines, and planes, and even obtained a geometrical definition of a number: these views were further elaborated in his *Beiträge zur Geometrie der Lage*, 1856–1860. This geometry is curious and brilliant, and has been used by Culmann as the basis of his graphical statics.

As usual text-books on synthetic geometry I may mention M. Chasles's *Traité de géométrie supérieure*, 1852; J. Steiner's *Vorlesungen über synthetische Geometrie*, 1867; L. Cremona's *Éléments de géométrie projective*, English translation by C. Leudesdorf, Oxford, second edition, 1893; and Th. Reye's *Geometrie der Lage*, Hanover, 1866–1868, English translation by T. F. Holgate, New York, part i, 1898. A good presentation of the modern treatment of pure geometry is contained in the *Introduzione ad una teoria geometrica delle curve piane*, 1862, and its continuation *Preliminari di una teoria geometrica delle superficie* by Luigi Cremona (1830–1903): his collected works, in three volumes, may be also consulted.

The differences in ideas and methods formerly observed in analytic and synthetic geometries tend to disappear with their further development.

Non-Euclidean Geometry.   Here I may fitly add a few words on recent investigations on the foundations of geometry.

The question of the truth of the assumptions usually made in our geometry had been considered by J. Saccheri as long ago as 1733; and in more recent times had been discussed by N. I. Lobatschewsky (1793–1856) of Kasan, in 1826 and again in 1840; by Gauss, perhaps as early as 1792, certainly in 1831 and in 1846; and by J. Bolyai (1802–1860) in 1832 in the appendix to the first volume of his father's *Tentamen*, but Riemann's memoir of 1854 attracted general attention to the subject of non-Euclidean geometry, and the theory has been since extended and simplified by various writers, notably by A. Cayley of Cambridge, E. Beltrami[1] (1835–1900) of Pavia, by H. L. F. von Helmholtz (1821–1894) of Berlin, by S. P. Tannery (1843–1904) of Paris, by F. C. Klein of Göttingen, and by A. N. Whitehead of Cambridge in his *Universal Algebra*. The subject is so technical that I confine myself to a bare sketch of the argument[2] from which the idea is derived.

The Euclidean system of geometry, with which alone most people are acquainted, rests on a number of independent axioms and postulates. Those which are necessary for Euclid's geometry have, within recent years, been investigated and scheduled. They include not only those explicitly given by him, but some others which he unconsciously used. If these are varied, or other axioms are assumed, we get a different series of propositions, and any consistent body of such propositions constitutes a system of geometry. Hence there is no limit to the number of possible Non-Euclidean geometries that can be constructed.

Among Euclid's axioms and postulates is one on parallel lines, which is usually stated in the form that if a straight line meets two straight lines, so as to make the sum of the two interior angles on the same side of it taken together less than two right angles, then these straight lines being continually produced will at length meet upon that side on which are the angles which are less than two right angles. Expressed in this form the axiom is far from obvious, and from early times numerous attempts have been made to prove it.[3] All such attempts failed, and it is now known that the axiom cannot be deduced from the other axioms assumed by Euclid.

The earliest conception of a body of Non-Euclidean geometry was

---

[1] Beltrami's collected works are (1908) in course of publication at Milan. A list of his writings is given in the *Annali di matematica*, March 1900.

[2] For references see my *Mathematical Recreations and Essays*, London, ninth edition, 1920, chaps, xv, xxi. A historical summary of the treatment of non-Euclidean geometry is given in *Die Theorie der Parallellinien* by F. Engel and P. Stäckel, Leipzig, 1895, 1899; see also J. Frischauf's *Elemente der absoluten Geometrie*, Leipzig, 1876; and a report by G. B. Halsted on progress in the subject is printed in *Science*, N.S., vol. x, New York, 1899, pp. 545–557.

[3] Some of the more interesting and plausible attempts have been collected by T. P. Thompson in his *Geometry without Axioms*, London, 1833, and later by J. Richard in his *Philosophie de mathématique*, Paris, 1903.

due to the discovery, made independently by Saccheri, Lobatschewsky, and John Bolyai, that a consistent system of geometry of two dimensions can be produced on the assumption that the axiom on parallels is not true, and that through a point a number of straight (that is, geodetic) lines can be drawn parallel to a given straight line. The resulting geometry is called *hyperbolic*.

Riemann later distinguished between boundlessness of space and its infinity, and showed that another consistent system of geometry of two dimensions can be constructed in which all straight lines are of a finite length, so that a particle moving along a straight line will return to its original position. This leads to a geometry of two dimensions, called *elliptic geometry*, analogous to the hyperbolic geometry, but characterised by the fact that through a point no straight line can be drawn which, if produced far enough, will not meet any other given straight line. This can be compared with the geometry of figures drawn on the surface of a sphere.

Thus according as no straight line, or only one straight line, or a pencil of straight lines can be drawn through a point parallel to a given straight line, we have three systems of geometry of two dimensions known respectively as elliptic, parabolic or homaloidal or Euclidean, and hyperbolic.

In the parabolic and hyperbolic systems straight lines are infinitely long. In the elliptic they are finite. In the hyperbolic system there are no similar figures of unequal size; the area of a triangle can be deduced from the sum of its angles, which is always less than two right angles; and there is a finite maximum to the area of a triangle. In the elliptic system all straight lines are of the same finite length; any two lines intersect; and the sum of the angles of a triangle is greater than two right angles.

In spite of these and other peculiarities of hyperbolic and elliptical geometries, it is impossible to prove by observation that one of them is not true of the space in which we live. For in measurements in each of these geometries we must have a unit of distance; and if we live in a space whose properties are those of either of these geometries, and such that the greatest distances with which we are acquainted (*ex. gr.* the distances of the fixed stars) are immensely smaller than any unit, natural to the system, then it may be impossible for us by our observations to detect the discrepancies between the three geometries. It might indeed be possible by observations of the parallaxes of stars to prove that the parabolic system and either the hyperbolic or elliptic system were false, but never can it be proved by measurements that Euclidean geometry is true. Similar difficulties might arise in connection with excessively minute quantities. In short, though the results of Euclidean geometry are more exact than present experiments can verify for finite things, such as those with which we have to deal, yet for much larger things or much smaller things or for parts of space at present inaccessible to us they may not be true.

Other systems of Non-Euclidean geometry might be constructed by changing other axioms and assumptions made by Euclid. Some of these are interesting, but those mentioned above have a special importance from the somewhat sensational fact that they lead to no results inconsistent with the properties of the space in which we live.

We might also approach the subject by remarking that in order that a space of two dimensions should have the geometrical properties with which we are familiar, it is necessary that it should be possible at any place to construct a figure congruent to a given figure; and this is so only if the product of the principal radii of curvature at every point of the space or surface be constant. This product is constant in the case (i) of spherical surfaces, where it is positive; (ii) of plane surfaces (which lead to Euclidean geometry), where it is zero; and (iii) of pseudo-spherical surfaces, where it is negative. A tractroid is an instance of a pseudo-spherical surface; it is saddle-shaped at every point. Hence on spheres, planes, and tractroids we can construct normal systems of geometry. These systems are respectively examples of elliptic, Euclidean, and hyperbolic geometries. Moreover, if any surface be bent without dilation or contraction, the measure of curvature remains unaltered. Thus these three species of surfaces are types of three kinds on which congruent figures can be constructed. For instance a plane can be rolled into a cone, and the system of geometry on a conical surface is similar to that on a plane.

In the preceding sketch of the foundations of Non-Euclidean geometry I have assumed tacitly that the measure of a distance remains the same everywhere.

The above refers only to hyper-space of two dimensions. Naturally there arises the question whether there are different kinds of hyper-space of three or more dimensions. Riemann showed that there are three kinds of hyper-space of three dimensions having properties analogous to the three kinds of hyper-space of two dimensions already discussed. These are differentiated by the test whether at every point no geodetical surfaces, or one geodetical surface, or a fasciculus of geodetical surfaces can be drawn parallel to a given surface; a geodetical surface being defined as such that every geodetic line joining two points on it lies wholly on the surface.

Foundations of Mathematics. Assumptions made in the Subject. The discussion on the Non-Euclidean geometry brought into prominence the logical foundations of the subject. The questions of the principles of and underlying assumptions made in mathematics have been discussed of late by J. W. R. Dedekind of Brunswick, G. Cantor of Halle, G. Frege of Jena, G. Peano of Turin, the Hon. B. A. W. Russell and A. N. Whitehead, both of Cambridge.

Kinematics. The theory of kinematics, that is, the investigation of the properties of motion, displacement, and deformation, considered independently of force, mass, and other physical conceptions, has been treated by various writers. It is a branch of pure mathematics, and forms a fitting introduction to the study of natural philosophy. Here I do no more than allude to it.

I shall conclude the chapter with a few notes—more or less discursive—on branches of mathematics of a less abstract character and concerned with problems that occur in nature. I commence by mentioning the subject of *Mechanics.* The subject may be treated graphically or analytically.

Graphics. In the science of graphics rules are laid down for solving various problems by the aid of the drawing-board: the modes of calculation which are permissible are considered in modern projective geometry, and the subject is closely connected with that of modern geometry. This method of attacking questions has been hitherto applied chiefly to problems in mechanics, elasticity, and electricity; it is especially useful in engineering, and in that subject an average draughtsman ought to be able to obtain approximate solutions of most of the equations, differential or otherwise, with which he is likely to be concerned, which will not involve errors greater than would have to be allowed for in any case in consequence of our imperfect knowledge of the structure of the materials employed.

The theory may be said to have originated with Poncelet's work, but I believe that it is only within the last twenty years that systematic expositions of it have been published. Among the best known of such works I may mention the *Graphische Statik*, by *C. Culmann*, Zürich, 1875, recently edited by W. Ritter; the *Lezioni di statica grafica*, by

A. Favaro, Padua, 1877 (French translation annotated by P. Terrier in 2 volumes, 1879–85); the *Calcolo grafico*, by *L. Cremona*, Milan, 1879 (English translation by T. H. Beare, Oxford, 1889), which is largely founded on Möbius's work; *La statique graphique*, by *M. Levy*, Paris, 4 volumes, 1886–88; and *La statica grafica*, by *C. Sairotti*, Milan, 1888.

The general character of these books will be sufficiently illustrated by the following note on the contents of Culmann's work. Culmann commences with a description of the geometrical representation of the four fundamental processes of addition, subtraction, multiplication, and division; and proceeds to evolution and involution, the latter being effected by the use of the equiangular spiral. He next shows how the quantities considered—such as volumes, moments, and moments of inertia—may be represented by straight lines; thence deduces the laws for combining forces, couples, &c.; and then explains the construction and use of the ellipse and ellipsoid of inertia, the neutral axis, and the kern; the remaining and larger part of the book is devoted to showing how geometrical drawings, made on these principles, give the solutions of many practical problems connected with arches, bridges, frameworks, earth pressure on walls and tunnels, &c.

The subject has been treated during the last twenty years by numerous writers, especially in Italy and Germany, and applied to a large number of problems. But as I stated at the beginning of this chapter that I should as far as possible avoid discussion of the works of living authors I content myself with a bare mention of the subject.[1]

Analytical Mechanics.   I next turn to the question of mechan-

---

[1] In an English work, I may add here a brief note on Clifford, who was one of the earliest British mathematicians of later times to advocate the use of graphical and geometrical methods in preference to analysis. *William Kingdon Clifford*, born at Exeter on May 4, 1845, and died at Madeira on March 3, 1879, was educated at Trinity College, Cambridge, of which society he was a fellow. In 1871 he was appointed professor of applied mathematics at University College, London, a post which he retained till his death. His remarkable felicity of illustration and power of seizing analogies made him one of the most brilliant expounders of mathematical principles. His health failed in 1876, when the writer of this book undertook his work for a few months; Clifford then went to Algeria and returned at the end of the year, but only to break down again in 1878. His most important works are his *Theory of Biquaternions, On the Classification of Loci* (unfinished), and *The Theory of Graphs* (unfinished). His *Canonical Dissection of a Riemann's Surface* and the *Elements of Dynamic* also contain much interesting matter. For further details of Clifford's life and work see the authorities quoted in the article on him in the *Dictionary of National Biography*, vol. xi.

ics treated analytically. The knowledge of mathematical mechanics of solids attained by the great mathematicians of the last century may be said to be summed up in the admirable *Mécanique analytique* by Lagrange and *Traité de mécanique* by Poisson, and the application of the results to astronomy forms the subject of Laplace's *Mécanique céleste*. These works have been already described. The mechanics of fluids is more difficult than that of solids and the theory is less advanced.

Theoretical Statics, especially the theory of *the potential* and *attractions*, has received considerable attention from the mathematicians of this century.

I have previously mentioned that the introduction of the idea of the potential is due to Lagrange, and it occurs in a memoir of a date as early as 1773. The idea was at once grasped by Laplace, who, in his memoir of 1784, used it freely and to whom the credit of the invention was formerly, somewhat unjustly, attributed. In the same memoir Laplace also extended the idea of zonal harmonic analysis which had been introduced by Legendre in 1783. Of *Gauss's* work on attractions I have already spoken. The theory of level surfaces and lines of force is largely due to *Chasles*, who also determined the attraction of an ellipsoid at any external point. I may also here mention the *Barycentrisches Calcul*, published in 1826 by *A. F. Möbius*[1] (1790–1868), who was one of the best known of Gauss's pupils. Attention must also be called to the important memoir, published in 1828, on the potential and its properties, by G. Green[2] (1793–1841) of Cambridge. Similar results were independently established, in 1839, by Gauss, to whom their general dissemination was due.

Theoretical Dynamics, which was cast into its modern form by Jacobi, has been studied by most of the writers above mentioned. I may also here repeat that the principle of "Varying Action" was elaborated by Sir William Hamilton in 1827, and the "Hamiltonian equations"

---

[1] Möbius's collected works were published at Leipzig in four volumes, 1885–87.

[2] A collected edition of Green's works was published at Cambridge in 1871. Other papers of Green which deserve mention here are those in 1832 and 1833 on the equilibrium of fluids, on attractions in space of n dimensions, and on the motion of a fluid agitated by the vibrations of a solid ellipsoid; and those in 1837 on the motion of waves in a canal, and on the reflexion and refraction of sound and light. In the last of these, the geometrical laws of sound and light are deduced by the principle of energy from the undulatory theory, the phenomenon of total reflexion is explained physically, and certain properties of the vibrating medium are deduced. Green also discussed the propagation of light in any crystalline medium.

were given in 1835; and I may further call attention to the dynamical investigations of J. E. E. Bour (1832–1866), of Liouville, and of J. L. F. Bertrand, all of Paris. The use of generalised co-ordinates, introduced by Lagrange, has now become the customary means of attacking dynamical (as well as many physical) problems.

As usual text-books I may mention those on particle and rigid dynamics by E. J. Routh, Cambridge; *Leçons sur l'intégration des équations différentielles de la mécanique* by P. Painlevé, Paris, 1895; *Intégration des équations de la mécanique* by J. Graindorge, Brussels, 1889; and C. E. Appell's *Traité de mécanique rationnelle*, Paris, 2 vols., 1892, 1896. Allusion to the treatise on Natural Philosophy by Sir William Thomson (later known as Lord Kelvin) of Glasgow, and P. G. Tait of Edinburgh, may be also here made.

On the mechanics of fluids, liquids, and gases, apart from the physical theories on which they rest, I propose to say nothing, except to refer to the memoirs of Green, Sir George Stokes, Lord Kelvin, and von Helmholtz. The fascinating but difficult theory of vortex rings is due to the two writers last mentioned. One problem in it has been also considered by Sir J. J. Thomson, of Cambridge, but it is a subject which is as yet beyond our powers of analysis. The subject of sound may be treated in connection with hydrodynamics, but on this I would refer the reader who wishes for further information to the work first published at Cambridge in 1877 by Lord Rayleigh.

Theoretical Astronomy is included in, or at any rate closely connected with, theoretical dynamics. Among those who in this century have devoted themselves to the study of theoretical astronomy the name of *Gauss* is one of the most prominent; to his work I have already alluded.

Bessel.[1]   The best known of Gauss's contemporaries was *Friedrich Wilhelm Bessel*, who was born at Minden on July 22, 1784, and died at Königsberg on March 17, 1846. Bessel commenced his life as a clerk on board ship, but in 1806 he became an assistant in the observatory at Lilienthal, and was thence in 1810 promoted to be director of the new Prussian Observatory at Königsberg, where he continued to live during the remainder of his life. Bessel introduced into pure mathematics those functions which are now called by his name (this was in

---

[1] See pp. 35–53 of A. M. Clerke's *History of Astronomy*, Edinburgh, 1887. Bessel's collected works and correspondence have been edited by R. Engelmann and published in four volumes at Leipzig, 1875–82.

1824, though their use is indicated in a memoir seven years earlier); but his most notable achievements were the reduction (given in his *Fundamenta Astronomiae*, Königsberg, 1818) of the Greenwich observations by Bradley of 3222 stars, and his determination of the annual parallax of 61 Cygni. Bradley's observations have been recently reduced again by A. Auwers of Berlin.

Leverrier.[1]   Among the astronomical events of this century the discovery of the planet Neptune by Leverrier and Adams is one of the most striking. *Urbain Jean Joseph Leverrier*, the son of a petty Government employé in Normandy, was born at St. Lô on March 11, 1811, and died at Paris on September 23, 1877. He was educated at the Polytechnic school, and in 1837 was appointed as lecturer on astronomy there. His earliest researches in astronomy were communicated to the Academy in 1839: in these he calculated, within much narrower limits than Laplace had done, the extent within which the inclinations and eccentricities of the planetary orbits vary. The independent discovery in 1846 by Leverrier and Adams of the planet Neptune by means of the disturbance it produced on the orbit of Uranus attracted general attention to physical astronomy, and strengthened the opinion as to the universality of gravity. In 1855 Leverrier succeeded Arago as director of the Paris observatory, and reorganised it in accordance with the requirements of modern astronomy. Leverrier now set himself the task of discussing the theoretical investigations of the planetary motions and of revising all tables which involved them. He lived just long enough to sign the last proof-sheet of this work.

Adams.[2]   The co-discoverer of Neptune was *John Couch Adams*, who was born in Cornwall on June 5, 1819, educated at St. John's College, Cambridge, subsequently appointed Lowndean professor in the University, and director of the Observatory, and who died at Cambridge on January 21, 1892.

There are three important problems which are specially associated with the name of Adams. The first of these is his discovery of the planet Neptune from the perturbations it produced on the orbit of Uranus: in point of time this was slightly earlier than Leverrier's investigation.

---

[1]For further details of his life see Bertrand's *éloge* in vol. xli of the *Mémoires de l'académie*; and for an account of his work see Adams's address in vol. xxxvi of the *Monthly Notices* of the Royal Astronomical Society.

[2]Adams's collected papers, with a biography, were issued in two volumes, Cambridge, 1896, 1900.

The second is his memoir of 1855 on the secular acceleration of the moon's mean motion. Laplace had calculated this on the hypothesis that it was caused by the eccentricity of the earth's orbit, and had obtained a result which agreed substantially with the value deduced from a comparison of the records of ancient and modern eclipses. Adams shewed that certain terms in an expression had been neglected, and that if they were taken into account the result was only about one-half that found by Laplace. The results agreed with those obtained later by Delaunay in France and Cayley in England, but their correctness has been questioned by Plana, Pontécoulant, and other continental astronomers. The point is not yet definitely settled.

The third investigation connected with the name of Adams, is his determination in 1867 of the orbit of the Leonids or shooting stars which were especially conspicuous in November, 1866, and whose period is about thirty-three years. H. A. Newton (1830–1896) of Yale, had shewn that there were only five possible orbits. Adams calculated the disturbance which would be produced by the planets on the motion of the node of the orbit of a swarm of meteors in each of these cases, and found that this disturbance agreed with observation for one of the possible orbits, but for none of the others. Hence the orbit was known.

Other well-known astronomers of this century are *G. A. A. Plana* (1781–1864), whose work on the motion of the moon was published in 1832; *Count P. G. D. Pontécoulant* (1795–1871); *C. E. Delaunay* (1816–1872), whose work on the lunar theory indicates the best method yet suggested for the analytical investigations of the whole problem, and whose (incomplete) lunar tables are among the astronomical achievements of this century; *P. A. Hansen*[1] (1795–1874), head of the observatory at Gotha, who compiled the lunar tables published in London in 1857 which are still used in the preparation of the Nautical Almanack, and elaborated the methods employed for the determination of lunar and planetary perturbations; *F. F. Tisserand* (1845–1896) of Paris, whose *Mécanique céleste* is now a standard authority on dynamical astronomy; and *Simon Newcomb* (1835–1909), superintendent of the *American Ephemeris*, who re-examined the Greenwich observations from the earliest times, applied the results to the lunar theory, and revised Hansen's tables.

Other notable work is associated with the names of Hill, Darwin,

---

[1] For an account of Hansen's numerous memoirs see the *Transactions of the Royal Society of London for* 1876–77.

and Poincaré. *G. W. Hill*,[1] until recently on the staff of the *American Ephemeris*, determined the inequalities of the moon's motion due to the non-spherical figure of the earth—an investigation which completed Delaunay's lunar theory.[2] Hill also dealt with the secular motion of the moon's perigee and the motion of a planet's perigee under certain conditions; and wrote on the analytical theory of the motion of Jupiter and Saturn, with a view to the preparation of tables of their positions at any given time. *Sir G. H. Darwin* (1845–1912), of Cambridge, wrote on the effect of tides on viscous spheroids, the development of planetary systems by means of tidal friction, the mechanics of meteoric swarms, and the possibility of pear-shaped planetary figures. *H. Poincaré* (1854–1912), of Paris, discussed the difficult problem of three bodies, and the form assumed by a mass of fluid under its own attraction, and is the author of an admirable treatise, the *Mécanique céleste*, three volumes. The treatise on the lunar theory by *E. W. Brown*, Cambridge, 1896; his memoir on *Inequalities in the Motion of the Moon* due to Planetary Action, Cambridge, 1908; and a report (printed in the *Report of the British Association*, London, 1899, vol. LXIX, pp. 121–159) by *E. T. Whittaker* on researches connected with the solution of the problem of three bodies, contain valuable accounts of recent progress in the lunar and planetary theories.

Within the last half century the results of spectrum analysis have been applied to determine the constitution of the heavenly bodies, and their directions of motions to and from the earth. The early history of spectrum analysis will be always associated with the names of *G. R. Kirchhoff* (1824–1887) of Berlin, of *A. J. Ångström* (1814–1874) of Upsala, and of *George G. Stokes* of Cambridge, but it pertains to optics rather than to astronomy. How unexpected was the application to astronomy is illustrated by the fact that A. Comte in 1842, when discussing the study of nature, regretted the waste of time due to some astronomers paying attention to the fixed stars, since, he said, nothing could possibly be learnt about them; and indeed a century ago it would have seemed incredible that we could investigate the chemical constitution of worlds in distant space.

During the last few years the range of astronomy has been still

---

[1] G. W. Hill's collected works have been issued in four volumes, Washington, 1905.

[2] On recent development of the lunar theory, see the *Transactions of the British Association*, vol. lxv, London, 1895, p. 614.

further extended by the art of photography. To what new results this may lead it is as yet impossible to say. In particular we have been thus enabled to trace the forms of gigantic spiral nebulae which seem to be the early stages of vast systems now in process of development.

The constitution of the universe, in which the solar system is but an insignificant atom, has long attracted the attention of thoughtful astronomers, and noticeably was studied by William Herschel. Recently J. C. Kapteyn of Groningen has been able to shew that all the stars whose proper motions can be detected belong to one or other of two streams moving in different directions, one with a velocity about three times as great as the other. The solar system is in the slower stream. These results have been confirmed by A. S. Eddington and F. W. Dyson. It would appear likely that we are on the threshold of wide-reaching discoveries about the constitution of the visible universe.

Mathematical Physics.   An account of the history of mathematics and allied sciences in the last century would be misleading if there were no reference to the application of mathematics to numerous problems in heat, elasticity, light, electricity, and other physical subjects. The history of mathematical physics is, however, so extensive that I could not pretend to do it justice, even were its consideration properly included in a history of mathematics. At any rate I consider it outside the limits I have laid down for myself in this chapter. I abandon its discussion with regret because the Cambridge school has played a prominent part in its development, as witness (to mention only three or four of those concerned) the names of Sir George G. Stokes, professor from 1849 to 1903, Lord Kelvin, J. Clerk Maxwell (1831–1879), professor from 1871 to 1879, Lord Rayleigh, professor from 1879 to 1884, Sir J. J. Thomson, professor from 1884, Sir Joseph Larmor, professor from 1903, and Sir Ernest Rutherford, professor from 1919.

INDEX.

THE END